AF356651

Yeast, Biofuels, and Value-Added Products

Yeast, Biofuels, and Value-Added Products

Timothy Tse

Basel • Beijing • Wuhan • Barcelona • Belgrade • Novi Sad • Cluj • Manchester

Editor
Timothy Tse
Department of Food and
Bioproduct Sciences
University of Saskatchewan
Saskatoon
Canada

Editorial Office
MDPI AG
Grosspeteranlage 5
4052 Basel, Switzerland

This is a reprint of articles from the Special Issue published online in the open access journal *Fermentation* (ISSN 2311-5637) (available at: www.mdpi.com/journal/fermentation/special_issues/yeast_value).

For citation purposes, cite each article independently as indicated on the article page online and as indicated below:

Lastname, A.A.; Lastname, B.B. Article Title. *Journal Name* **Year**, *Volume Number*, Page Range.

ISBN 978-3-7258-2264-5 (Hbk)
ISBN 978-3-7258-2263-8 (PDF)
doi.org/10.3390/books978-3-7258-2263-8

Contents

About the Editor vii

Viviani Tadioto, Anderson Giehl, Rafael Dorighello Cadamuro, Iara Zanella Guterres,
Angela Alves dos Santos and Stefany Kell Bressan et al.
Bioactive Compounds from and against Yeasts in the One Health Context: A Comprehensive
Review
Reprinted from: *Fermentation* **2023**, *9*, 363, doi:10.3390/fermentation9040363 1

Ane Catarine Tosi Costa, Mariano Russo, A. Alberto R. Fernandes, James R. Broach and
Patricia M. B. Fernandes
Transcriptional Response of Multi-Stress-Tolerant *Saccharomyces cerevisiae* to Sequential Stresses
Reprinted from: *Fermentation* **2023**, *9*, 195, doi:10.3390/fermentation9020195 22

Zora S. Rerop, Nikolaus I. Stellner, Petra Graban, Martina Haack, Norbert Mehlmer and
Mahmoud Masri et al.
Bioconversion of a Lignocellulosic Hydrolysate to Single Cell Oil for Biofuel Production in a
Cost-Efficient Fermentation Process
Reprinted from: *Fermentation* **2023**, *9*, 189, doi:10.3390/fermentation9020189 38

Javier Ricardo Gómez Cardozo, Jean-Baptiste Beigbeder, Julia Maria de Madeiros Dantas
and Jean-Michel Lavoie
High-Gravity Fermentation for Bioethanol Production from Industrial Spent Black Cherry Brine
Supplemented with Whey
Reprinted from: *Fermentation* **2023**, *9*, 170, doi:10.3390/fermentation9020170 57

Touijer Hanane, Benchemsi Najoua, Hamdi Salsabil, Janati Idrissi Abdellatif, Bousta Dalila
and Irfan Ahmad et al.
Qualitative Screening of Yeast Biodiversity for Hydrolytic Enzymes Isolated from the
Gastrointestinal Tract of a Coprophage *"Gymnopleurus sturmi"* and Dung of Ruminants
Reprinted from: *Fermentation* **2022**, *8*, 692, doi:10.3390/fermentation8120692 68

Efri Mardawati, Agus T. Hartono, Bambang Nurhadi, Hana Nur Fitriana, Euis Hermiati and
Riksfardini Annisa Ermawar
Xylitol Production from Pineapple Cores (*Ananas comosus* (L.) *Merr*) by Enzymatic and Acid
Hydrolysis Using Microorganisms *Debaryomyces hansenii* and *Candida tropicalis*
Reprinted from: *Fermentation* **2022**, *8*, 694, doi:10.3390/fermentation8120694 86

Yue Zhang, Rasool Kamal, Qing Li, Xue Yu, Qian Wang and Zongbao Kent Zhao
Comparative Fatty Acid Compositional Profiles of *Rhodotorula toruloides* Haploid and Diploid
Strains under Various Storage Conditions
Reprinted from: *Fermentation* **2022**, *8*, 467, doi:10.3390/fermentation8090467 98

Sreyden Hor, Mallika Boonmee Kongkeitkajorn and Alissara Reungsang
Sugarcane Bagasse-Based Ethanol Production and Utilization of Its Vinasse for Xylitol
Production as an Approach in Integrated Biorefinery
Reprinted from: *Fermentation* **2022**, *8*, 340, doi:10.3390/fermentation8070340 109

Getari Kasmiarti, Dwita Oktiarni, Poedji Loekitowati Hariani, Novia Novia and
Hermansyah Hermansyah
Isolation of Novel Yeast from Coconut (*Cocos nucifera* L.) Water and Phenotypic Examination as
the Potential Parameters in Bioethanol Production
Reprinted from: *Fermentation* **2022**, *8*, 283, doi:10.3390/fermentation8060283 127

Kalliopi Basa, Seraphim Papanikolaou, Maria Dimopoulou, Antonia Terpou, Stamatina Kallithraka and George-John E. Nychas
Trials of Commercial- and Wild-Type *Saccharomyces cerevisiae* Strains under Aerobic and Microaerophilic/Anaerobic Conditions: Ethanol Production and Must Fermentation from Grapes of Santorini (Greece) Native Varieties
Reprinted from: *Fermentation* **2022**, *8*, 249, doi:10.3390/fermentation8060249 . . . . **138**

Wendy Al Sahyouni, Sally El Kantar, Anissa Khelfa, Young-Kyoung Park, Jean-Marc Nicaud and Nicolas Louka et al.
Optimization of *cis*-9-Heptadecenoic Acid Production from the Oleaginous Yeast *Yarrowia lipolytica*
Reprinted from: *Fermentation* **2022**, *8*, 245, doi:10.3390/fermentation8060245 . . . . **161**

About the Editor

Timothy Tse

Dr. Timothy Tse is an accomplished researcher specializing in environmental and agri-food research, with a particular emphasis on sustainability, fermentation technologies, and value-added products. His research contributions have led to over 30 scientific publications and 7 book chapters. With extensive experience in analytical methodologies, and next-generation sequencing, Dr. Tse has helped to develop advancements in ethanol purification technologies and investigated enhancements of agri-waste by-products through fermentation to create high-value products (e.g., therapeutic compounds). His research is driven by a commitment to sustainability and innovation, using cutting-edge fermentation processes to foster impactful collaborations and contribute to the advancement of bio-based research.

fermentation

MDPI

Review

Bioactive Compounds from and against Yeasts in the One Health Context: A Comprehensive Review

Viviani Tadioto [1,2,3], Anderson Giehl [1,2], Rafael Dorighello Cadamuro [1,3], Iara Zanella Guterres [3,4], Angela Alves dos Santos [5], Stefany Kell Bressan [2], Larissa Werlang [2], Boris U. Stambuk [1,5], Gislaine Fongaro [1,3], Izabella Thaís Silva [1,3,4] and Sérgio Luiz Alves, Jr. [1,2,*]

[1] Graduate Program in Biotechnology and Biosciences, Federal University of Santa Catarina, Florianópolis 88040-900, SC, Brazil
[2] Laboratory of Yeast Biochemistry, Federal University of Fronteira Sul, Campus Chapecó 89815-899, SC, Brazil
[3] Laboratory of Applied Virology, Department of Microbiology, Immunology, and Parasitology, Federal University of Santa Catarina, Florianópolis 88040-900, SC, Brazil
[4] Graduate Program in Pharmacy, Federal University of Santa Catarina, Florianópolis 88040-900, SC, Brazil
[5] Laboratory of Yeast Molecular Biology and Biotechnology, Department of Biochemistry, Federal University of Santa Catarina, Florianópolis 88040-900, SC, Brazil
* Correspondence: slalvesjr@uffs.edu.br

Abstract: Yeasts are the most used microorganisms for biotechnological purposes. Although they have been mainly recognized for their application in the beverage and bioethanol industries, these microorganisms can be efficiently employed in pharmaceutical and food production companies. In these industrial sectors, yeasts are highly desirable for their capacity to produce bioactive compounds from simple substrates, including wastes. In this review, we present the state of the art of bioactive compound production in microbial cell factories and analyze the avenues to increase the productivity of these molecules, which benefit human and environmental health. The article addresses their vast biological activities, from preventing to treating human diseases and from pre to postharvest control on agroindustrial streams. Furthermore, different yeast species, genetically engineered or not, are herein presented not only as biofactories of the referred to compounds but also as their targets. This comprehensive analysis of the literature points out the significant roles of biodiversity, bioprospection, and genome editing tools on the microbial production of bioactive compounds and reveals the value of these approaches from the one health perspective.

Keywords: anticancer; antifungal; anti-inflammatory; antimicrobial; antioxidant; antiviral; biocontrol; ecological food production; killer toxins; volatile compounds

check for **updates**

Citation: Tadioto, V.; Giehl, A.; Cadamuro, R.D.; Guterres, I.Z.; dos Santos, A.A.; Bressan, S.K.; Werlang, L.; Stambuk, B.U.; Fongaro, G.; Silva, I.T.; et al. Bioactive Compounds from and against Yeasts in the One Health Context: A Comprehensive Review. *Fermentation* **2023**, *9*, 363. https://doi.org/10.3390/fermentation9040363

Academic Editor: Timothy Tse

Received: 18 February 2023
Revised: 1 April 2023
Accepted: 5 April 2023
Published: 7 April 2023

1. Introduction

The concept of one health recognizes that human health relies on the health of other living beings and the environment [1]. In this context, considering the biotechnological potential of yeasts, alternatives to combating bacterial antibiotic resistance, improving the nutritional content of foods, increasing sustainable management of crops, and replacing conventional raw materials with residual biomass are envisaged, thus ensuring a more sustainable production chain. As stated by Elnaiem et al. [2], there is a need for urgent investment in strategies that reduce the risk of the appearance of new diseases, which will improve animal, human and environmental health and act in the fight against climate change.

Plant-associated yeasts can control their host plant's microbiota, thus presenting as an alternative to synthetic pesticides and as inducers of vegetal growth. They are, therefore, potential substitutes for conventional fertilizers or growth hormones. Additionally, second-generation biorefinery strategies have shown proven efficiency in converting residual substrates (e.g., agroindustrial wastes) into a myriad of bioproducts through the metabolism of yeasts, whether genetically engineered or not [3–13]. Therefore, the bioactive compounds produced by yeasts allow the achievement of the one health goals, as they bring benefits not only to human health but also to animal and plant health, helping to maintain environmental balance. In the following sections, we present a comprehensive review of the most valuable bioactive compounds produced by (and against) yeasts and how they can contribute to a more sustainable future.

2. Bioactive Compounds Naturally Produced by Non-Engineered Yeasts

Polysaccharides, polyphenols, carotenoids, flavonoids, and terpenoids are part of the wide range of compounds that have biological activities, including antimicrobial, immuno-suppressive, anticancer, and anti-inflammatory activity [14,15]. Although some of them are complex molecules, these bioactive compounds can be produced by microorganisms through biosynthetic processes from simple substrates [16]. Some yeasts have stood out in this context, namely *Saccharomyces cerevisiae* [17,18], *Komagataella phaffii* (formerly *Pichia pastoris*) [19], *Hansenula polymorpha* [20], *Rhodosporidium diobovatum* [21], *Aureobasidium pullulans* [22], *Rhodotorula glutinis* [23,24], *Saccharomyces bayanus*, *Candida stellata*, *Kluyveromyces thermotolerants*, and *Hanseniaspora uvarum* [25,26]. Mostly, yeast fermentation products are employed in the nutritional improvement of food products. However, based on the versatility of their metabolic routes, these microorganisms synthesize bioactive compounds of different molecular classes, such as peptides, polyphenols, and terpenoids [27,28]. Table 1 summarizes the main bioactive compounds produced by nongenetically modified yeasts of different species and points out their commercial interests and market values.

Table 1. Bioactive compounds produced by different nongenetically modified yeasts, their commercial interest, and market value.

Compound	Formula	Commercial Interest	Market Value in Millions of USD (Year)	Reference
Glutathione	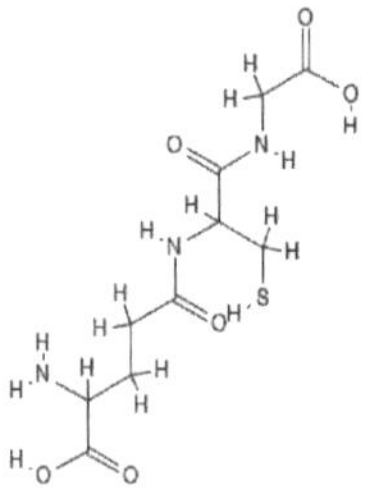	Antioxidant	316.56 (2020)	[17,18]
Melatonin	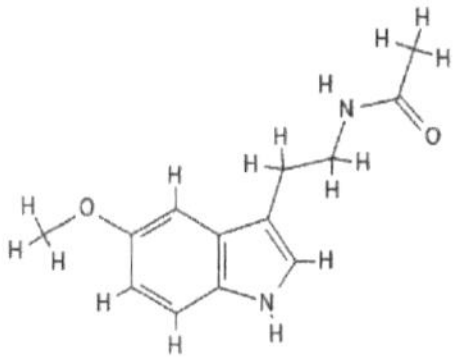	Sleep cycle receptor mediator and antioxidant	437.9 (2021)	[26,29]

Table 1. *Cont.*

Compound	Formula	Commercial Interest	Market Value in Millions of USD (Year)	Reference
Riboflavin	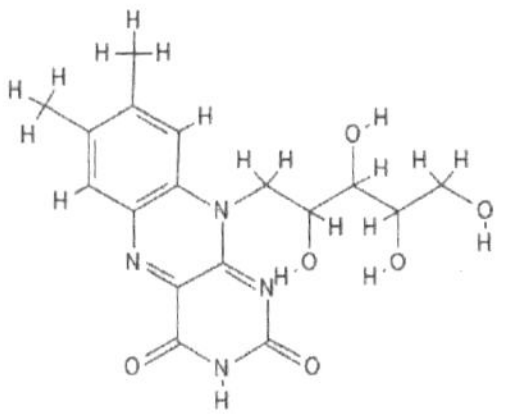	Food coloring and supplement	397 (2023)	[30,31]
Toxin K1	*KEX1* e *KEX2* proteases encoded	Antifungal	-	[32]
Toxin K28		Antifungal	-	[33]
Toxin K2		Antifungal	-	[34]
Toxin KP	Decapeptide KP (AKVTMTCSAS)	Toxoplasmosis and antifungal treatment	-	[35]
Rhodotorulic Acid	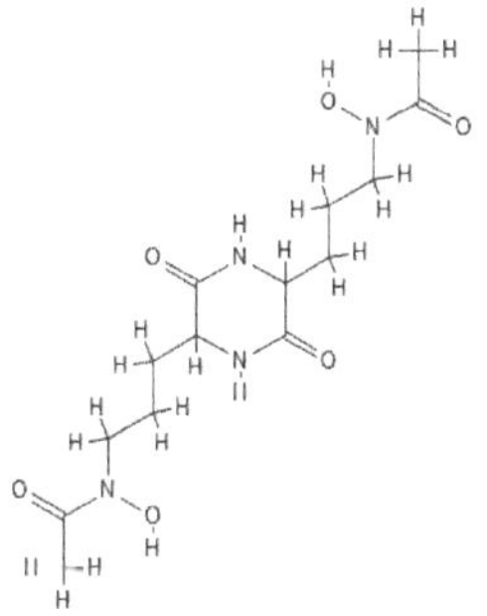	Iron bioavailability and regulator of iron-mediated membrane transporters	-	[36]
Ferrichrome	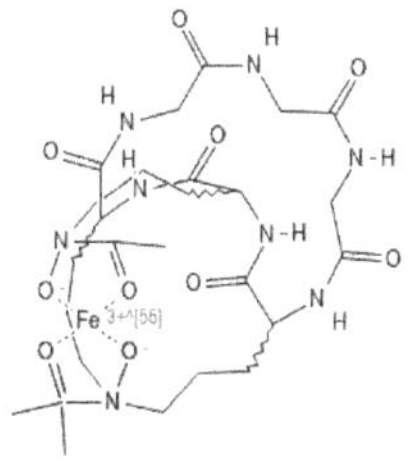	Bioavailability and regulator of iron and other metals	-	[36,37]
Acetic Acid	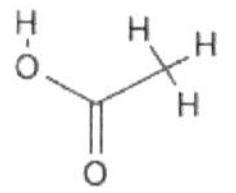	Preservative and food additive; drug production, industrial and laboratory solvent	2900 (2021)	[38,39]
Citronellol	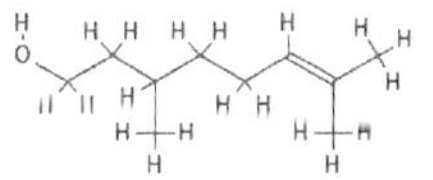	Antifungal	~146.8 (2030)	[40,41]

Table 1. *Cont.*

Compound	Formula	Commercial Interest	Market Value in Millions of USD (Year)	Reference
Phenylethyl Alcohol		Antimicrobial, antiseptic, disinfectant, aromatic essence, and preservative of pharmaceuticals and cosmetics	255.3 (2021)	[40,42]
Indol		Fragrance component and flavor aggregator	35.7 (2021)	[25,43]
Quercetin-3-glucoside		Antiviral, antioxidant	-	[44]
Caffeic Acid		Antioxidant, anti-inflammatory		[44]
Naringenin		Antioxidant, anti-inflammatory, antifungal		[44]

2.1. Pharmacological Outlooks

Notably, the main species used in the production of bioactive compounds is *Saccharomyces cerevisiae* [45,46]. The glutathione tripeptide, formed by a thiol of cysteine, glycine, and glutamine, is a major example of these products [47,48]. Although humans naturally produce it, the significant antioxidant effect of this molecule makes it a product of commercial interest, being sold mainly in retail chains and pharmacies as a supplement [49,50]. The interest in glutathione, however, is not restricted to combating free radicals. In humans, this tripeptide acts in the metabolism of peroxides [51] and xenobiotics [52], the inflammatory/immunological response [53], and the activation of some enzymes [54]. Currently, like most bioactive compounds, glutathione is also mainly produced by strains of *S. cerevisiae* due to its safety in food production [17,18]. In addition to *S. cerevisiae*, though, other yeasts can be used for the same purpose, such as the methylotrophic yeasts *K. phaffii* [19] and *H. polymorpha* [20], as well as the marine yeast *Rhodosporidium diobovatum* [21].

Melatonin is another major antioxidant that can be produced by yeast species, such as *S. cerevisiae*, *S. bayanus*, and *Saccharomyces uvarum*. This compound, an indole amine, is part of the composition of wines and beers, being produced naturally during the fermentation process [26,55–59]. Another important role of melatonin in the human and animal body is the maintenance of sleep rhythms [60,61]. Cardinali [62] reports the prevention of neurodegenerative diseases such as Alzheimer's and Parkinson's in experimental models when melatonin is administered. Reiter et al. [63] report the reduction of stroke-provoked neural damage in experimental models. Commercially, melatonin is sold as a sleep hormone, being found in drugstores and retail. Despite being naturally produced by the body, it can often be out of balance due to biological factors or habits that inhibit its production—that is why it is commercially sought after as a replacement.

Biologically active molecules can also be peptides that act as toxins depending on the environmental conditions. Bioactive peptides are protein segments abundantly found in yeast extracts, especially those generated by *S. cerevisiae*, *Kluyveromyces marxianus*, and *Candida utilis*, since the cells of these microorganisms harbor 35 to 60% of protein in their dry mass [64]. These yeasts are able to produce antimicrobial peptides that fight other yeasts, bacteria, and molds; in addition, they produce mycocins, extracellular proteins that hinder the growth of Gram-negative and Gram-positive bacteria [65].

The so-called killer yeasts are also extensively studied for the production of bioactive compounds. In addition to acting in the biological control of phytopathogens (as discussed in the following section), these yeasts can serve as producers of new human and animal health drugs. In this context, a killer decapeptide produced by *Pichia anomala* stands out, showing in vitro and in vivo inhibitory activities against influenza A and human immunodeficiency type-1 (HIV-1) viruses [66,67]. Decapeptides such as this one have already shown activity against infections caused by *Candida albicans* (also a yeast), thanks to the destruction of its biofilm in invaded tissues [68], and *Toxoplasma gondii* (the etiological agent of toxoplasmosis) [69].

Tested for pharmacological purposes, the yeast *Pichia kudriavzevii* showed the production of a toxin with activity against human-infective bacteria, such as *Escherichia coli*, *Enterococcus faecalis*, *Klebsiella* spp., *Staphylococcus aureus*, *Pseudomonas aeruginosa*, and *Pseudomonas alcaligenes*. The toxin produced by *P. kudriavzevii* has a high specificity of action and high toxicity against the above pathogens, which are extremely desirable characteristics for drug development [70]. Alternatively, the toxin can also be used as a prophylactic measure, preventing these pathogens from reaching humans through the ingestion of contaminated food [71].

Yeasts are also able to accumulate lipids for the pharmaceutical industry [72,73]. The so-called single-cell oils (SCO) include long-chain fatty acids of nutritional interest, such as the polyunsaturated acids Eicosapentaenoic (produced by *Meyerozyma guilliermondii*), Linoleic (produced by *Apiotrichum brassicae*, *Candida tropicalis*, *Metschnikowia pulcherrima*, *P. kudriavzevii*, *M. guilliermondii*, *Lodderomyces elongisporus*, and *Rhodotorula mucilaginosa*), and γ-Linolenic (produced by *A. brassicae*, *M. pulcherrima*, *M. guilliermondii*, *L. elongisporus*, and *R. mucilaginosa*) [72,74–76].

The tropical yeast *Saccharomyces boulardii*, isolated from lychee and mangosteen peels, has interesting properties for supplementation in the human microbiota, given its ability to adapt to the inhospitable conditions of the gastrointestinal system. This yeast can produce high levels of acetic acid, inhibiting virulent strains of *E. coli* [38], which are eventually responsible for human food poisoning. *Saccharomyces boulardii* is also considered efficient in treating inflammatory bowel diseases [77,78] and improving inflammation caused by *Citrobacter rodentium* [79]. This yeast also demonstrated protection against infection generated by the yeast *C. albicans* [80].

Riboflavin is also an important bioactive compound produced by yeasts. Capable of preventing deficiency symptoms such as dermatitis, this vitamin also plays a key role in the food industry. Its chemical synthesis requires organic solvents, generates waste that is harmful to the environment, and requires a lot of energy compared to the yeast fermentation process [30]. Currently, the yeast *M. guilliermondii* stands out in the large-scale production of riboflavin [81]. In addition, the marine yeast *Candida membranifaciens* subsp. *flavinogenie* W14-3, isolated from sea water in China, also showed a high potential for producing this bioactive compound [30].

2.2. Microbial Competition and Ecological Food Production

Due to multiple adverse situations, microorganisms are not always inserted in a natural habitat with abundant nutrients from which they can benefit. In these cases, microbial competition is inevitable, and the competition tools of each microorganism can be exploited as agroecological strategies. In this context, yeasts seem to have a prominent place [82,83]. Due to interactions with other microorganisms in natural environments, yeasts can use competitive approaches that guarantee advantages against other opposing organisms, either through mechanisms of enzyme secretion, toxin production, or the release of volatile organic compounds (VOCs) [84].

Volatiles produced by yeasts can display antifungal and bacteriostatic action. Two of these compounds are ethanol and 2,3-butanediol, both produced by the yeasts *S. cerevisiae* and *P. kudriavzevii*, which have already been proven capable of reducing the growth of the filamentous fungi *Aspergillus flavus* and *Aspergillus parasiticus* [85]. The volatile compound 2-phenylethanol, produced by *P. anomala*, has also demonstrated a fungus-control effect; in one case, this effect was against *Aspergillus ochraceus* [86], a well-known generator of toxins in food [87].

The species *Sporidiobolus pararoseus* has the ability to control the plant illness caused by the gray fungus *Botrytis cinerea*. Huang et al. [88] attributed this activity to 2-ethyl-1-hexanol. The yeast *Candida intermedia* was also able to control this phytopathogen by producing the VOCs 1,3,5,7-cyclooctatetraene, 3-methyl-1-butanol, 2-nonanone, and phenylethyl alcohol [89]. *Cyberlindnera jadinii*, *Candida fryrichii*, *C. intermedia*, and *Lachancea thermotolerans* have also been identified as producers of 2-phenyl ethanol, inhibiting the molds *Aspergillus carbonarius* and *A. ochraceus* [90–92].

Against postharvest bacterial diseases, the polymorphic yeast *Aureobasidium pullulans* has been shown to produce secondary metabolites such as aureobasidins, liamocins, 2-propyl acrylic acid, and 2-methylene succinic acid, which act against bacteria such as *Salmonella typhi*, *Proteus vulgaris*, *E. coli*, *P. aeruginosa*, *Klebsiella pneumoniae*, *Bacillus subtilis*, *S. aureus*, and *Sarcina ventriculi*. Liamocins also showed selective activity against bacteria of the genus *Streptococcus* [93–96]. The biocontrol strategy using VOCs produced by yeasts is interesting to use to maintain the shelf life of fruits such as grapes, especially with compounds produced by *Wickerhamomyces anomalus*, *M. pulcherrima*, and *A. pullulans* [97]. Furthermore, VOCs produced by *Candida saque* were also efficient in controlling the proliferation of fungi in apples [40].

Marine yeasts are often found in hostile conditions to maintain their lives. Marine species, such as the aforementioned *A. pullulans*, can produce siderophores—secondary metabolites used by microorganisms as chelators of essential trace elements such as iron [98–103]. The so-referred yeast produces a high concentration of the siderophore hydroxamate, a bioactive compound capable of inhibiting the cell growth of the bacteria *Vibrio anguillarum* and *Vibrio parahaemolyticus*, which are found in sick marine animals. In fact, siderophores are effective in controlling pathogenic bacteria in marine fish, presenting as an important alternative for pisciculture and the remediation of sites with radioactive waste [104].

In addition to marine yeasts, yeasts associated with plants and fruits can also produce siderophores, such as *R. glutinis*, *Rhodotorula rubra*, and species of the genus *Meyerozyma*, such as *M. guilliermondii* [23,105]. Although some siderophores have shown pharmacological potential, allowing the control of pathogenic microorganisms [98,101], it is in agriculture that they primarily receive attention. These compounds are quite efficient in protecting plants of agronomic interest against fungal and bacterial infections [99]. The chelating action of these compounds also contributes to the availability of essential micronutrients for plant development [106].

Competition for micronutrients, especially iron, limits the development of microorganisms [83]. The species *R. glutinis*, isolated from a rhizosphere environment, uses the siderophore rhodotorulic acid to inhibit the growth of its competitor *B. cinerea* [107]. Along the same line, *M. pulcherrima*, also found in rhizospheres [108], inhibits its competitors through pulcherriminic acid [109]. Still in the context of the rhizosphere, it should be noted that the growth of plant roots can benefit from the presence of yeasts in the rhizosphere [110,111].

In terms of microbial competition, the production of killer toxins is another yeast strategy. This trait has been widely identified in different yeast genera [112]. In nature, these killer microorganisms are found in different environments, whether in fruits, water, or soil [113]. Toxins can be active against filamentous fungi [114] and bacteria [115]. In shrimp culture, for example, wild yeasts can also confer advantages to the production system. The psychrotolerant yeast *Mrakia frigida*, found in marine sediments in Antarctica, produces a killer toxin against the shrimp-pathogenic yeast *Metschnikowia bicuspidata*, and can be inoculated in shrimp breeding tanks to improve this animal's health [116]. Still in the context of animal production, the yeast *Williopsis mrakii* deserves attention in silage conservation. While lactic fermentation (by lactic acid bacteria) of silage is desirable, subsequent aerobic consumption of lactic acid (by other microorganisms) causes its deterioration. In this scenario, Lowes et al. [117] demonstrated that the HMK mycocin produced by *W. mrakii* was able to prevent the development of spoilage microorganisms.

Enzymes secreted by yeasts are also part of the list of mechanisms used to control phytopathogenic fungi. Their antifungal action is often due to their hydrolytic capacity; this is the case, for example, of the chitinases produced by the genera *Pichia*, *Debaryomyces*, *Metschnikowia*, *Meyerozyma*, and *Saccharomycopsis*, which act on fungal cell walls composed of chitin. Among the hydrolases, proteases from *Aureobasidium*, *Pichia*, and *Saccharomycopsis*, and glucanases from *Wickerhamomyces* and *Pichia* also have a fungicidal or fungistatic effect [118–122].

Strains of *W. anomalus* and *M. guilliermondii* have significant effects in reducing the postharvest disease caused by the fungus *Colletotrichum gloeosporioides* in papaya [123]. The yeast *W. anomalus* can also control the cherry tomato gray mold caused by the fungus *Lycopersicon esculentum* [124]. Likewise, *R. glutinis* strains ensure the significant control of blue rot caused by *Penicillium expansum* in postharvested apples [125].

Therefore, alternatives for improving the ecological practices of food production arise from the ecological relationships of yeasts. Bioactive compounds produced by yeasts can act in pre and postharvest control, replacing synthetic agricultural defensives, which have a high environmental impact on and pose high environmental risk to human and animal health [113,124].

3. Other Microorganisms' Bioactive Compounds Acting against Yeast

In natural environments, the production of organic acids such as acetic, lactic, and propionic acids affects the interspecific relationships among microbial cells. The elucidation of their mechanisms of inhibition, though, remains not fully understood. Acetic acid, for example, exhibits a synergistic effect with lactic acid with the capability of preventing fungal/yeast growth, being probably more potent due to its higher pKa value (4.76), which facilitates its uptake by cells and leads to a higher level of acid dissociation in the cytoplasm, damaging the cell [126]. The same holds true for other organic acids e.g., propionic acid displays a pKa value of 4.87. Mixtures of acids usually appear in the natural ecosystem and have been applied as fungal/yeast inhibitors. A mix of formic, acetic, propionic, butyric, caproic, and n-valeric acid, for example, is naturally produced by lactic acid bacteria (LAB) [127]. On the other hand, in case these acids are thought to be used as biocontrol agents of yeasts, it should be noted that their exacerbated use may lead to a contamination process in soil, plants, water bodies, and air, thus also affecting animal and human health. Thus, there is a demand for new biocompounds in aiming to control yeasts without environmental contamination [128].

In this sense, LAB may also offer alternatives. Reuterin, for example, produced by *Lactobacillus reuteri*, is a low-molecular-weight compound that exhibits antimicrobial activity against yeasts such as *C. albicans* [129]. The cyclic dipeptides produced by LAB are also good examples of bioactive compounds that act against yeasts. *Lactobacillus plantarum* has been cultivated, seeking metabolites to biocontrol mold and yeasts, and two compounds were isolated with this aim: cyclo(L-Phe-L-Pro) and cyclo(L-Phe-trans-4-OH-L-Pro). The former's minimal inhibitory concentration (MIC) was 20 mg/mL in assays with *Penicillium roqueforti* and *Aspergillus fumigatus*. Additionally, a synergistic effect was observed when this compound was associated with phenyllactic acid, which reduced the MIC to 10 mg/mL [130]. Several lactobacilli can produce these compounds, but the mechanism of action and biochemical pathways of cyclic dipeptide acting as inhibitors of fungi are still unknown [131].

Fatty acids (FAs) are also good examples of molecules with antifungal activity. The literature demonstrates that the chain length of the fatty acid appears to be relevant in this role. The FA identified as lauric (C12) and capric (C10) seems to be the most toxic against *C. albicans* [132]. Four hydroxylated fatty acids produced by *Lactobacillus plantaram* (3-hydroxydecanoic, 3-hydroxy-5-cis-dodecenoic, 3-(R)-hydroxydodecanoic, and 3-(R)-hydroxytetradecanoic acid) also proved to be relevant in acting against yeasts, with MICs ranging from 10 to 100 μg/mL [133]. However, the mechanisms of how fatty acids inhibit the growth of yeasts are still unclear. It is believed that fatty acids generate the partition of lipid bilayers present on the membranes of fungi, resulting in instability and a loss of permeability control. In these scenarios, fluidity results in the liberation of intracellular proteins, enzymes, and electrolytes, disintegrating the fungal cell [134].

Endophytic bacteria and filamentous fungi are also important sources of bioactive compounds. These microorganisms live within plant tissues, taking their nutrients from this habitat and producing metabolites with some capability of providing protection to the host [135]. In this regard, an endophytic fungus isolated from the leaves of *Psidium guajava* L., identified as *Alternaria tenuissima*, demonstrated activity against the human pathogenic yeast *C. albicans* [136]. The study focused on exploiting the metabolites extracted with ethyl acetate (EA) from the endophytic mold. Different concentrations of the EA extract were analyzed against *C. albicans*, and a maximum inhibition zone of 14 ± 1.0 mm at a concentration of 10 mg/mL was observed. Moreover, a decrease in the number of colony-forming units (CFUs) of a *C. albicans* culture treated with this EA extract was observed. The number of CFUs counted suggested that the MIC value of the so-referred extract against this pathogen was 1400 μg/mL. The EA extract from the culture of the endophytic fungus *Xylaria* sp. PSU-D14, isolated from the leaves of *Garcinia dulcis*, also exhibited antifungal activity against *C. albicans*, with a MIC value of 128 μg/mL [137]. From the broth extract of this endophytic fungus, a glucoside derivative, xylarosides A, along with another known compound, sordaricin were isolated. This metabolite demonstrated antifungal activity, against *C. albicans*, with a MIC value of 32 μg/mL [138].

The alkaloid aspernigrin A, which has been isolated from *Cladosporium herbarum* (an endophyte of the *Cynodondactylon*), is also effective in acting against *C. albicans*. Zhang et al. [139] showed that this compound is able to inhibit this yeast's growth with a MIC value of 75.0 μg/mL. Another compound, cryptocandin A, isolated from the endophytic fungus *Cryptosporiopsis quercina*, was characterized as a lipopeptide with antifungal properties. Cryptocandin A demonstrated activity against some important fungal pathogens of humans and showed MIC values of 0.03–0.07 µg/mL against isolates of *C. albicans*, *Trichophyton mentagrophytes*, and *Trichophyton rubrum*. The most significant result was against *C. albicans*, with a MIC value of 0.03 µg/mL. This value was very similar to the values obtained with amphotericin B, a well-known antifungal agent used in most cases—clinically expressing the potential of this new substance. Cryptocandin A contains some hydroxylated amino acids, for example, the 3-hydroxy-4-hydroxymethyl proline, which can contribute to its activity [140]. The main bioactive compounds produced by bacteria and molds against yeasts are summarized in Table 2.

Table 2. Relation between microorganisms, biocompounds produced, targets, and MIC.

Producing Microorganism	Biocompound	Target Yeast Tested	MIC	Reference
Alternaria tenuissima	-	*Candida albicans*	1400 µg/mL	[136]
Cladosporium herbarum	Alkaloid-aspernigrin A	*Candida albicans*	75.0 µg/mL	[139]
Cryptosporiopsis quercina	Lipopeptide cryptocandin A	*Candida albicans*	0.03 µg/mL	[140]
Cryptosporiopsis quercina	Lipopeptide cryptocandin A	*Cryptococcus neoformans*	-	[140]
Pestalotiopsis foedan	Isobenzofuranone, Pestaphtalides A	*Candida albicans*	-	[141]
Xylaria sp.	-	*Candida albicans*	128 µg/mL	[137]
Xylaria sp.	Sordaricin	*Candida albicans*	32 µg/mL	[138]
Nannocystis	-	*Candida albicans* and *Pichia anomala*	-	[142]
Streptomyces sp.	Cyclo(l-leucyl–l-prolyl)	*Candida albicans*; *Candida glabrata*; *Candida tropicalis*;	32 mg/mL 16 mg/mL 8 mg/mL	[143]
Streptomyces sp.	Cyclo(l-phenylalanyl–l-prolyl)	*Candida albicans*; *Candida glabrata*; *Candida tropicalis*;	64 mg/mL 256 mg/mL 32 mg/mL	[143]
Lactibacillus plantaram	3-hydroxy tetradecanoic acid	*Kluyveromyces marxianus*	10 µg/mL	[133]
Lactibacillus plantaram	3-hydroxy dodecanoic acid	*Kluyveromyces marxianus*	25 µg/mL	[133]
Lactibacillus plantaram	3-hydroxy undecanoic acid	*Kluyveromyces marxianus*	25 µg/mL	[133]
Mycosphaerella sp.	2-amino-3,4-dihydroxy-2-25 (hydroxymethyl)-14-oxo-6,12-eicosenoic acid	*Cryptococcus neoformans*	1.95–7.82 µM	[144]
Mycosphaerella sp.	Myriocin	*Cryptococcus neoformans*	0.48–1.95 µM	[144]

MIC: minimum inhibitory concentration; -: concentration not mentioned.

Candida auris and Cryptococcus neoformans: Two Emerging Pathogenic Yeasts to Be Tackled with Bioactive Compounds

Among the yeast pathogens that infect humans and cause them diseases, *Candida auris* and *Cryptococcus neoformans* are definitely worth noting. Recently, the World Health Organization (WHO) considered both species critically harmful to human health, showing resistance against antifungals and being responsible for increased death rates [145].

Cryptococcus neoformans infect the lungs, causing pulmonary infections such as pneumonia that can happen before dissemination to the central nervous system [146]. Cryptococcosis and meningitis, caused by *C. neoformans*, are responsible for a considerable number of deaths in the world [147]. This yeast usually infects immunocompromised people, and it has recently gained attention during the COVID-19 pandemic in patients with SARS-CoV-2 co-infection, especially in those over sixty years old [148].

Two compounds, eicosanoic acid and myriocin, isolated from the endophytic fungus *Mycosphaerella* sp. (found to be associated with the plant *Eugenia bimarginata*) were tested against *Cr. neoformans*, and their MIC values were obtained by microdilution assays. For eicosanoic acid, the MIC values ranged from 1.95 to 7.82 µM, and myriocin's MIC values ranged from 0.48 to 1.95 µM [144]. Other compounds that demonstrated activity against this yeast were isolated from endophytic fungi associated with *Acanthus ilicifolius* var. *xiamenensis* [149] and *Tripterigeum wilfordii* [140], demonstrating the potential of endophytic molds as a source of novel compounds with bioactivity against yeasts.

Candida auris was described as a pathogenic yeast during the last decade. Over 25% of its strains naturally present multidrug resistance against antifungal compounds, which may be related to their surface stability. Also, the widespread presence of *C. auris* worldwide may facilitate its contact with different compounds and thus increase its resistance [150]. The first report was made in 2009, receiving attention from medical journals and also mass media [151–153].

Despite the search for biocompounds that could inhibit the propagation of *C. auris*, it remains a challenge to obtain them from sources such as filamentous fungi. However, studies report possible synthetic or semisynthetic options, from natural sources, that can inhibit *C. auris* growth [154,155].

4. Using Yeasts as "Microbial Cell Factories" of Bioactive Compounds

Yeasts are widely known for producing natural compounds and have been applied in several industrial sectors, such as the production of proteins, pigments, vitamins, fuels, beverages, and foods [156]. *Saccharomyces cerevisiae*, the main microorganism used in alcoholic fermentation environments, in addition to producing ethanol, has been associated with the release of secondary metabolites responsible for flavors and aromas; for example, in the production of wine, sake, and cachaça, different *S. cerevisiae* strains have already been correlated with the formation of acetate esters, ethyl esters, acids and higher alcohols [157]. Nevertheless, although *S. cerevisiae* plays a central role in winemaking, yeasts such as *Torulaspora delbrueckii* have been gaining prominence due to their greater release of aromatic compounds and the greater intensity of the color of the final product [158,159]. In table olive production, various yeasts, through the generation of metabolites such as glycerol, ethanol, higher alcohols, esters, and other volatile compounds, shape the texture and flavor characteristics of the final product [160]. In addition to food and beverage production, platform molecules have also been explored in yeast-based bioprocesses; for example, isolates of *S. cerevisiae* and *Yarrowia lipolytica* have already been tested for the production of succinic acid—an important industrial chemical that is currently produced by petrochemical processing [161,162].

As stated before, bioactive compounds are produced by organisms such as plants, fungi, and bacteria [14,15]. These metabolites can be isolated directly from their natural producers, from fungal and bacterial cultures, or from their accumulation in plant biomass. However, extracting these compounds through direct isolation is difficult, especially in plants with low concentrations, which makes the natural extraction of these products environmentally destructive and economically unsustainable [163]. For example, the natural concentration of taxol (a diterpene alkaloid used against several types of cancer) is, on average, 0.01–0.06% of the dry weight of the plant *Taxus* spp. [164]; however, the estimated amount of purified taxol needed to treat 500 cancer patients is 1 kg, which is equivalent to about 10 tons of bark or the felling of 700 trees [165]. In addition to this low concentration, seasonal variability and government policies can also influence the productivity of plant crops and, consequently, the extraction of the compounds of interest [166]. Some of these natural products can also be obtained through chemical synthesis, but their structural complexity generally makes this synthetic route difficult, and there are often toxic agents present as byproducts [167,168]—among them, especially, are solvents such as benzene, dichloromethane, and chloroform [169]. In this context, producing bioactive compounds using genetically modified microorganisms represents an advantageous technological alternative in environmental, economic, and health terms.

With the recent advances in genetic engineering (or genome editing, to use today's more popular term), bioprocesses based on recombinant microorganisms have become increasingly sought after as an alternative to traditional techniques of direct isolation of compounds or their chemical synthesis [170,171]. Fast- and controlled-growth microorganisms, such as the bacterium *E. coli* and the yeast *S. cerevisiae*, are already widely used as platform cells to produce high levels of chemicals of human interest, such as biofuels and proteins [172,173]. The yeast *S. cerevisiae*, in particular, has been shown to be suitable to host several heterologous biosynthetic pathways, given the following traits: (i) being a model organism for molecular biology research, (ii) being the first eukaryotic organism to have its genome completely sequenced, (iii) having the properties of robust growth and high productivity, (iv) being the best-studied organism among lower eukaryotes, and (v) having a wide variety of commercially available and ready-made mutant strains for laboratory use [174,175]. In addition, yeasts have a eukaryotic subcellular organization capable of post-translationally processing many proteins from more complex eukaryotic organisms [176,177]. These factors have led to the use of *S. cerevisiae* strains as veritable versatile biofactories.

Arguably, *S. cerevisiae* has a long history of producing compounds of commercial interest. This yeast is widely used to produce fermented foods, beverages, and biofuels, as well as pharmaceutical products such as recombinant human insulin—which since the early 1980s has been massively produced using this microorganism [178]—and the industrial-scale production of the antimalarial precursor artemisinic acid [179]. With the advent of the CRISPR-Cas9 genome editing system [180], rapid and precise genetic manipulation has favored the insertion of a myriad of new biosynthetic pathways in *S. cerevisiae*, many of which provide products originally found and produced by plants [181]. With CRISPR-Cas9, strains of *S. cerevisiae* have already been developed to produce the opioids thebaine and hydrocodone from sugars. To achieve this purpose, 21 enzymes were expressed for producing the former and 23 were expressed for producing the latter, with genes from plants, mammals, and bacteria [182]. This yeast has also been modified to produce resveratrol directly from glucose or ethanol [183]. The CRISPR-Cas9 tool was also used to express the last five steps of the carotenoid biosynthetic pathway in *S. cerevisiae* [184]. More recently, *S. cerevisiae* has been designed as a platform for the production of daidzein, a key chemical for isoflavonoid biosynthesis [185].

During the last decade, the genomes of several nonconventional yeast species have also been sequenced and made available in databases [186]. In many cases, the interest in new yeasts exists due to the difficulties in using *S. cerevisiae* as a biofactory, especially when exploring alternative substrates. For example, this yeast mainly uses hexoses as carbon sources, which is an obstacle in the fermentation of sustainable substrates, such as lignocellulosic hydrolysates, which contain, in addition to hexoses, a high concentration of pentoses [187]. Because of this, nonconventional yeasts have been developed as alternative platforms for the production of several recombinant proteins. In fact, the species *Y. lipolytica*, *Kluyveromyces lactis*, *K. marxianus*, *H. polymorpha*, *C. tropicalis* and *K. phaffii* have successfully fulfilled this role (Table 3). This success is mainly due to their (i) capacity to grow under inhibitory conditions, (ii) tolerance to high temperatures, (iii) ability to use different carbon sources, and (iv) high potential for extracellular protein secretion [188]. Among these yeasts, the oleaginous *Y. lipolytica* has been genetically modified to produce some bioactive compounds, such as linalool [189], β-carotene [190], and lycopene [191]. Recently, the oleaginous *C. tropicalis* was modified to produce α-Humulene, a terpenoid with multiple pharmacological activities [192].

Table 3. Examples of biocompounds produced by engineered non-*Saccharomyces* yeasts.

Yeast	Compound	Application	Reference
Yarrowia lipolytica	β-carotene	Antioxidant	[190]
	Linalool	Vitamin precursor, antifungal, antimicrobial and fragrance fixative	[189]
	Lycopene	Antioxidant	[191]
Kluyveromyces lactis	Brazzein	Sweetener compound	[193]
Kluyveromyces marxianus	Astaxanthin	Antioxidant	[194]
Hansenula polymorpha	Gamma-linolenic acid	Anti-inflammatory and imunorregulatory	[195]
Candida tropicalis	α-Humulene	Anti-inflammatory	[192]
	Lycopene	Antioxidant	[196]
Komagataella phaffii (formerly *Pichia pastoris*)	Lovastatin	Antihypercolesterolemia	[197]
	Δ9-Tetrahydrocannabinol	Compound with analgesic properties	[198]
	Astaxanthin	Antioxidant	[199]
	Citrinin	Antibiotic and antifungal	[200]

Concomitantly with genome editing, evolutionary engineering has also proven to be a resource for directing and selecting the best microbial cell factories. During stages of adaptive evolution, the microorganism of choice is grown for an extended period of time (which can be weeks or even years) under specific conditions that push it towards a desired trait. Afterward, whether or not a more fit and enriched strain with new advantageous mutations has been developed is checked [201,202]. Adaptive laboratory evolution has already been performed, for example, to increase carotenoid production in a recombinant *S. cerevisiae* strain through its prolonged exposure to hydrogen peroxide stress, since carotenoids have antioxidant properties [203]. In another study, buoyancy was used to select mutants of a recombinant *Y. lipolytica* strain with a higher ability to accumulate lipids [204]—once lipids reduced cell density.

In short, the ability to heterologously produce bioactive compounds in yeast demonstrates advantages over plant extraction and chemical synthesis. In addition to reducing environmental damage (caused by the need for large areas of cultivation for plant extraction), the biosynthesis of complex molecules can be achieved through new molecular biology techniques and increased knowledge about the genetic characteristics of yeasts. Furthermore, compounds that normally exist at low natural levels can be re-engineered in microbial biofactories for increased productivity [205]. Finally, the vast and unexplored biodiversity of nonconventional yeasts can further optimize the production of bioactive compounds, especially considering the efficient use of renewable and low-cost raw materials as substrates for these biotransformations.

5. Final Considerations

Although higher eukaryotes may produce bioactive compounds, the great demand for these molecules makes it unfeasible and unsustainable to rely on plants and animals for this aim on industrial scale. On the other hand, microbial cell factories perfectly fulfill this role; in this regard, yeasts stand out as the most successful microorganism employed in bioprocesses [11].

Despite *Saccharomyces* yeasts being seen as the first option for biotechnological purposes until the late 20th century, many studies have recently highlighted the potential applications of unconventional yeasts in the pharmaceutical and food production sectors, as we have discussed above. In this sense, the recent discoveries of new valuable microorganisms should call attention to the social, economic, and environmental importance of biodiversity and bioprospection.

Thus, in the light of circular economy approaches and the one health concept, new public policies and international laws are imperative to induce (i) the development of alternatives to the current environment-impacting production chains, (ii) optimization strategies to increase the production of bioactive compounds from residual feedstocks using yeasts as microbial cell factories, and (iii) the substitution of synthetic pesticides and plant growth hormones in food production systems.

Author Contributions: Conceptualization and supervision, S.L.A.J.; writing, V.T., A.G., R.D.C., I.Z.G., A.A.d.S., S.K.B. and L.W.; review and editing, B.U.S., G.F., I.T.S. and S.L.A.J. All authors have read and agreed to the published version of the manuscript.

Funding: This work is part of the project "INCT Yeasts: Biodiversity, preservation and biotechnological innovation", supported by grants and fellowships from the Brazilian National Council for Scientific and Technological Development (CNPq, grant #406564/2022-1). It is also funded by the Brazilian Coordination for the Improvement of Higher Education Personnel (CAPES), and the Research Promotion Program and the Support Program for Scientific and Technological Initiation from the Federal University of Fronteira Sul (UFFS, grant #PES-2022-0221 and #PES-2022-0224).

Institutional Review Board Statement: Not applicable.

Informed Consent Statement: Not applicable.

Data Availability Statement: Not applicable.

Conflicts of Interest: The authors declare no conflict of interest.

References

1. World Health Organization (WHO). One Health. Available online: https://www.who.int/news-room/questions-and-answers/item/one-health (accessed on 22 January 2023).
2. Elnaiem, A.; Mohamed-Ahmed, O.; Zumla, A.; Mecaskey, J.; Charron, N.; Abakar, M.F.; Raji, T.; Bahalim, A.; Manikam, L.; Risk, O.; et al. Global and Regional Governance of One Health and Implications for Global Health Security. *Lancet* **2023**, *401*, 688–704. [CrossRef]
3. Vargas, A.C.G.; Dresch, A.P.; Schmidt, A.R.; Tadioto, V.; Giehl, A.; Fogolari, O.; Mibielli, G.M.; Alves, S.L.; Bender, J.P. Batch Fermentation of Lignocellulosic Elephant Grass Biomass for 2G Ethanol and Xylitol Production. *Bioenergy Res.* **2023**. [CrossRef]
4. Scapini, T.; Bonatto, C.; Dalastra, C.; Bazoti, S.F.; Camargo, A.F.; Alves, S.L., Jr.; Venturin, B.; Steinmetz, R.L.R.; Kunz, A.; Fongaro, G.; et al. Bioethanol and Biomethane Production from Watermelon Waste: A Circular Economy Strategy. *Biomass Bioenergy* **2023**, *170*, 106719. [CrossRef]
5. Fenner, E.D.; Scapini, T.; da Costa Diniz, M.; Giehl, A.; Treichel, H.; Álvarez-Pérez, S.; Alves, S.L. Nature's Most Fruitful Threesome: The Relationship between Yeasts, Insects, and Angiosperms. *J. Fungi* **2022**, *8*, 984. [CrossRef]
6. Dalastra, C.; Scapini, T.; Kubeneck, S.; Camargo, A.F.; Klanovicz, N.; Alves, S.L., Jr.; Shah, M.P.; Treichel, H. Wastewater as a Feasible Feedstock for Biorefineries. In *Biorefinery for Water and Wastewater Treatment*; Shah, M.P., Ed.; Springer International Publishing: Cham, Switzerland, 2023; pp. 1–25. [CrossRef]
7. Colet, R.; Hassemer, G.; Alves, S.L., Jr.; Paroul, N.; Zeni, J.; Backes, G.T.; Valduga, E.; Cansian, R.L. *Pichia*: From Supporting Actors to the Leading Roles. In *Yeasts: From Nature to Bioprocesses*; Alves, S.L., Jr., Treichel, H., Basso, T.O., Stambuk, B.U., Eds.; Bentham Science: Singapore, 2022; pp. 148–191. [CrossRef]
8. Achilles, K.A.; Camargo, A.F.; Reichert Júnior, F.W.; Lerin, L.; Scapini, T.; Stefanski, F.S.; Dalastra, C.; Treichel, H.; Mossi, A.J. Improvement of Organic Agriculture with Growth-Promoting and Biocontrol Yeasts. In *Yeasts: From Nature to Bioprocesses*; Alves, S.L., Jr., Treichel, H., Basso, T.O., Stambul, B.U., Eds.; Bentham Science: Singapore, 2022; pp. 378–395. [CrossRef]
9. Louhasakul, Y.; Cheirsilp, B. Biotechnological Applications of Oleaginous Yeasts. In *Yeasts: From Nature to Bioprocesses*; Alves, S.L., Jr., Treichel, H., Basso, T., Stambuk, B.U., Eds.; Bentham Science: Singapore, 2022; pp. 357–377. [CrossRef]
10. Giehl, A.; Scapini, T.; Treichel, H.; Alves, S.L., Jr. Production of volatile organic compounds by yeasts in biorefineries: Ecological, environmental, and biotechnological outlooks. In *Ciências Ambientais e da Saúde na Atualidade: Insights Para Alcançar os Objetivos para o Desenvolvimento Sustentável*; Michelon, W., Viancelli, A., Eds.; Instituto de Inteligência em Pesquisa e Consultoria Cientifica Ltda: Concórdia/SC, Brazil, 2022; pp. 64–78. [CrossRef]
11. Alves, S.L., Jr.; Treichel, H.; Basso, T.O.; Stambuk, B.U. Are Yeasts "Humanity's Best Friends"? In *Yeasts: From Nature to Bioprocesses*; Alves, S.L., Jr., Treichel, H., Basso, T., Stambuk, B.U., Eds.; Bentham Science: Singapore, 2022; pp. 431–458. [CrossRef]
12. Alves, S.L., Jr.; Scapini, T.; Warken, A.; Klanovicz, N.; Procópio, D.P.; Tadioto, V.; Stambuk, B.U.; Basso, T.O.; Treichel, H. Engineered *Saccharomyces* or Prospected Non-*Saccharomyces*: Is There Only One Good Choice for Biorefineries? In *Yeasts: From Nature to Bioprocesses*; Alves, S.L., Jr., Treichel, H., Basso, T., Stambuk, B.U., Eds.; Bentham Science: Singapore, 2022; pp. 243–283. [CrossRef]
13. Pais, C.; Franco-Duarte, R.; Sampaio, P.; Wildner, J.; Carolas, A.; Figueira, D.; Ferreira, B.S. Production of Dicarboxylic Acid Platform Chemicals Using Yeasts. In *Biotransformation of Agricultural Waste and By-Products*; Poltronieri, P., D'Urso, O.F., Eds.; Elsevier: Amsterdam, The Netherlands, 2016; pp. 237–269. [CrossRef]

14. Veeresham, C. Natural Products Derived from Plants as a Source of Drugs. *J. Adv. Pharm Technol. Res.* **2012**, *3*, 200–201. [CrossRef]
15. Cragg, G.M.; Newman, D.J. Natural Products: A Continuing Source of Novel Drug Leads. *Biochim. Biophys. Acta Gen. Subj.* **2013**, *1830*, 3670–3695. [CrossRef] [PubMed]
16. Hubbard, B.K.; Walsh, C.T. Vancomycin Assembly: Nature's Way. *Angew. Chem. Int. Ed.* **2003**, *42*, 730–765. [CrossRef] [PubMed]
17. Hong, J.-Y.; Son, S.-H.; Hong, S.-P.; Yi, S.-H.; Kang, S.H.; Lee, N.-K.; Paik, H.-D. Production of β-Glucan, Glutathione, and Glutathione Derivatives by Probiotic *Saccharomyces cerevisiae* Isolated from *Cucumber jangajji*. *LWT* **2019**, *100*, 114–118. [CrossRef]
18. Santos, L.O.; Silva, P.G.P.; Lemos Junior, W.J.F.; de Oliveira, V.S.; Anschau, A. Glutathione Production by *Saccharomyces cerevisiae*: Current State and Perspectives. *Appl. Microbiol. Biotechnol.* **2022**, *106*, 1879–1894. [CrossRef] [PubMed]
19. Rollini, M.; Pagani, H.; Riboldi, S.; Manzoni, M. Influence of Carbon Source on Glutathione Accumulation in Methylotrophic Yeasts. *Ann. Microbiol.* **2005**, *55*, 199–203.
20. Ubiyvovk, V.M.; Ananin, V.M.; Malyshev, A.Y.; Kang, H.A.; Sibirny, A.A. Optimization of Glutathione Production in Batch and Fed-Batch Cultures by the Wild-Type and Recombinant Strains of the Methylotrophic Yeast *Hansenula polymorpha* DL-1. *BMC Biotechnol.* **2011**, *11*, 8. [CrossRef] [PubMed]
21. Kong, M.; Wang, F.; Tian, L.; Tang, H.; Zhang, L. Functional Identification of Glutamate Cysteine Ligase and Glutathione Synthetase in the Marine Yeast *Rhodosporidium diobovatum*. *Sci. Nat.* **2018**, *105*, 1–9. [CrossRef] [PubMed]
22. Chi, Z.; Liu, G.L.; Lu, Y.; Jiang, H.; Chi, Z.M. Bio-Products Produced by Marine Yeasts and Their Potential Applications. *Bioresour. Technol.* **2016**, *202*, 244–252. [CrossRef] [PubMed]
23. Calvente, V.; De Orellano, M.E.; Sansone, G.; Benuzzi, D.; Sanz de Tosetti, M.I. A Simple Agar Plate Assay for Screening Siderophore Producer Yeasts. *J. Microbiol. Methods* **2001**, *47*, 273–279. [CrossRef] [PubMed]
24. Sineli, P.E.; Maza, D.D.; Aybar, M.J.; Figueroa, L.I.C.; Viñarta, S.C. Bioconversion of Sugarcane Molasses and Waste Glycerol on Single Cell Oils for Biodiesel by the Red Yeast *Rhodotorula glutinis* R4 from Antarctica. *Energy Convers. Manag. X* **2022**, *16*, 100331. [CrossRef]
25. Arevalo-Villena, M.; Bartowsky, E.J.; Capone, D.; Sefton, M.A. Production of Indole by Wine-Associated Microorganisms under Oenological Conditions. *Food Microbiol.* **2010**, *27*, 685–690. [CrossRef]
26. Rodriguez-Naranjo, M.I.; Torija, M.J.; Mas, A.; Cantos-Villar, E.; Garcia-Parrilla, M.D.C. Production of Melatonin by *Saccharomyces* Strains under Growth and Fermentation Conditions. *J. Pineal Res.* **2012**, *53*, 219–224. [CrossRef]
27. Ganeva, V.; Angelova, B.; Galutzov, B.; Goltsev, V.; Zhiponova, M. Extraction of Proteins and Other Intracellular Bioactive Compounds From Baker's Yeasts by Pulsed Electric Field Treatment. *Front. Bioeng. Biotechnol.* **2020**, *8*, 1433. [CrossRef]
28. Gómez-Mejía, E.; Rosales-Conrado, N.; León-González, M.E.; Madrid, Y. Determination of Phenolic Compounds in Residual Brewing Yeast Using Matrix Solid-Phase Dispersion Extraction Assisted by Titanium Dioxide Nanoparticles. *J. Chromatogr. A* **2019**, *1601*, 255–265. [CrossRef]
29. Melatonin Market | Size, Trends, Forecast | 2022–2027. Available online: https://www.marketdataforecast.com/market-reports/melatonin-market (accessed on 7 February 2023).
30. Wang, L.; Chi, Z.; Wang, X.; Ju, L.; Chi, Z.; Guo, N. Isolation and Characterization of *Candida Membranifaciens* subsp. *flavinogenie* W14-3, a Novel Riboflavin-Producing Marine Yeast. *Microbiol. Res.* **2008**, *163*, 255–266. [CrossRef]
31. Riboflavin Market Analysis—Industry Report—Trends, Size & Share. Available online: https://www.mordorintelligence.com/industry-reports/riboflavin-market (accessed on 7 February 2023).
32. Schmitt, M.J.; Breinig, F. Yeast Viral Killer Toxins: Lethality and Self-Protection. *Nat. Rev. Microbiol.* **2006**, *4*, 212–221. [CrossRef]
33. Becker, B.; Schmitt, M.J. Yeast Killer Toxin K28: Biology and Unique Strategy of Host Cell Intoxication and Killing. *Toxins* **2017**, *9*, 333. [CrossRef]
34. Orentaite, I.; Poranen, M.M.; Oksanen, H.M.; Daugelavicius, R.; Bamford, D.H. K2 Killer Toxin-Induced Physiological Changes in the Yeast *Saccharomyces cerevisiae*. *FEMS Yeast Res.* **2016**, *16*, fow003. [CrossRef] [PubMed]
35. Ciociola, T.; Pertinhez, T.A.; De Simone, T.; Magliani, W.; Ferrari, E.; Belletti, S.; D'Adda, T.; Conti, S.; Giovati, L. In Vitro and In Vivo Anti-*Candida* Activity and Structural Analysis of Killer Peptide (KP)-Derivatives. *J. Fungi* **2021**, *7*, 129. [CrossRef] [PubMed]
36. Yun, C.W.; Tiedeman, J.S.; Moore, R.E.; Philpott, C.C. Siderophore-Iron Uptake in *Saccharomyces cerevisiae*: Identification of ferrichrome and fusarinine transporters. *J. Biol. Chem.* **2000**, *275*, 16354–16359. [CrossRef] [PubMed]
37. Chiu, P.C.; Nakamura, Y.; Nishimura, S.; Tabuchi, T.; Yashiroda, Y.; Hirai, G.; Matsuyama, A.; Yoshida, M. Ferrichrome, a Fungal-Type Siderophore, Confers High Ammonium Tolerance to Fission Yeast. *Sci. Rep.* **2022**, *12*, 1–10. [CrossRef]
38. Offei, B.; Vandecruys, P.; De Graeve, S.; Foulquié-Moreno, M.R.; Thevelein, J.M. Unique Genetic Basis of the Distinct Antibiotic Potency of High Acetic Acid Production in the Probiotic Yeast *Saccharomyces cerevisiae* var. *boulardii*. *Genome Res.* **2019**, *29*, 1478–1494. [CrossRef]
39. Acetic Acid Market Size, Share & Trends Report, 2022–2030. Available online: https://www.grandviewresearch.com/industry-analysis/acetic-acid-market (accessed on 7 February 2023).
40. Arrarte, E.; Garmendia, G.; Rossini, C.; Wisniewski, M.; Vero, S. Volatile Organic Compounds Produced by Antarctic Strains of *Candida sake* Play a Role in the Control of Postharvest Pathogens of Apples. *Biol. Control* **2017**, *109*, 14–20. [CrossRef]
41. Citronellol Market | Global Industry Report, 2030. Available online: https://www.transparencymarketresearch.com/citronellol-market.html (accessed on 7 February 2023).

42. 2-Phenylethanol Market Share Statistics 2022–2028. Available online: https://www.gminsights.com/industry-analysis/2-phenylethanol-market (accessed on 7 February 2023).

43. Indole Market Size, Share, Trends, Growth Analysis. Available online: https://precisionbusinessinsights.com/market-reports/indole-market/ (accessed on 7 February 2023).

44. de Lourdes Chaves Macêdo, E.; Colombo Pimentel, T.; de Sousa Melo, D.; Cristina de Souza, A.; Santos de Morais, J.; dos Santos Lima, M.; Ribeiro Dias, D.; Freitas Schwan, R.; Magnani, M. Yeasts from Fermented Brazilian Fruits as Biotechnological Tools for Increasing Phenolics Bioaccessibility and Improving the Volatile Profile in Derived Pulps. *Food Chem.* **2023**, *401*, 134200. [CrossRef]

45. Ho, C.H.; Piotrowski, J.; Dixon, S.J.; Baryshnikova, A.; Costanzo, M.; Boone, C. Combining Functional Genomics and Chemical Biology to Identify Targets of Bioactive Compounds. *Curr. Opin. Chem. Biol.* **2011**, *15*, 66–78. [CrossRef]

46. Moglia, A.; Goitre, L.; Gianoglio, S.; Baldini, E.; Trapani, E.; Genre, A.; Scattina, A.; Dondo, G.; Trabalzini, L.; Beekwilder, J.; et al. Evaluation of the Bioactive Properties of Avenanthramide Analogs Produced in Recombinant Yeast. *Biofactors* **2015**, *41*, 15–27. [CrossRef] [PubMed]

47. Dienes-Nagy, Á.; Vuichard, F.; Belcher, S.; Blackford, M.; Rösti, J.; Lorenzini, F. Simultaneous Quantification of Glutathione, Glutathione Disulfide and Glutathione-S-Sulfonate in Grape and Wine Using LC-MS/MS. *Food Chem.* **2022**, *386*, 132756. [CrossRef] [PubMed]

48. Meister, A.; Anderson, M.E. Glutathione. *Annu. Rev. Biochem.* **1983**, *52*, 711–760. [CrossRef] [PubMed]

49. Thompson, J.A.; Franklin, C.C. Enhanced Glutathione Biosynthetic Capacity Promotes Resistance to As3+-Induced Apoptosis. *Toxicol. Lett.* **2010**, *193*, 33–40. [CrossRef]

50. Penninckx, M. A Short Review on the Role of Glutathione in the Response of Yeasts to Nutritional, Environmental, and Oxidative Stresses. *Enzym. Microb. Technol.* **2000**, *26*, 737–742. [CrossRef]

51. Bonnefoy, M.; Drai, J.; Kostka, T. Antioxidants to Slow Aging, Facts and Perspectives. *Presse Med.* **2002**, *31*, 1174–1184.

52. Huber, P.C.; Almeida, W.P.; Fátima, Â. de Glutationa e Enzimas Relacionadas: Papel Biológico e Importância Em Processos Patológicos. *Quim. Nova* **2008**, *31*, 1170–1179. [CrossRef]

53. Perricone, C.; De Carolis, C.; Perricone, R. Glutathione: A Key Player in Autoimmunity. *Autoimmun Rev.* **2009**, *8*, 697–701. [CrossRef]

54. You, B.R.; Park, W.H. Gallic Acid-Induced Lung Cancer Cell Death Is Related to Glutathione Depletion as Well as Reactive Oxygen Species Increase. *Toxicol. Vitr.* **2010**, *24*, 1356–1362. [CrossRef]

55. Slominski, A.; Fischer, T.W.; Zmijewski, M.A.; Wortsman, J.; Semak, I.; Zbytek, B.; Slominski, R.M.; Tobin, D.J. On the Role of Melatonin in Skin Physiology and Pathology. *Endocrine* **2005**, *27*, 137–148. [CrossRef]

56. Arnao, M.B.; Hernández-Ruiz, J. The Physiological Function of Melatonin in Plants. *Plant Signal Behav.* **2006**, *1*, 89. [CrossRef]

57. Sharafati-Chaleshtori, R.; Shirzad, H.; Rafieian-Kopaei, M.; Soltani, A. Melatonin and Human Mitochondrial Diseases. *J. Res. Med. Sci.* **2017**, *22*, 2. [CrossRef] [PubMed]

58. Jockers, R.; Delagrange, P.; Dubocovich, M.L.; Markus, R.P.; Renault, N.; Tosini, G.; Cecon, E.; Zlotos, D.P. Update on Melatonin Receptors: IUPHAR Review 20. *Br. J. Pharm.* **2016**, *173*, 2702–2725. [CrossRef] [PubMed]

59. Reiter, R.J.; Rosales-Corral, S.; Tan, D.X.; Jou, M.J.; Galano, A.; Xu, B. Melatonin as a Mitochondria-Targeted Antioxidant: One of Evolution's Best Ideas. *Cell. Mol. Life Sci.* **2017**, *74*, 3863–3881. [CrossRef]

60. Emet, M.; Ozcan, H.; Ozel, L.; Yayla, M.; Halici, Z.; Hacimuftuoglu, A. A Review of Melatonin, Its Receptors and Drugs. *Eurasian J. Med.* **2016**, *48*, 135. [CrossRef]

61. Hagenauer, M.H.; Perryman, J.I.; Lee, T.M.; Carskadon, M.A. Adolescent Changes in the Homeostatic and Circadian Regulation of Sleep. *Dev. Neurosci.* **2009**, *31*, 276. [CrossRef]

62. Cardinali, D.P. Melatonin: Clinical Perspectives in Neurodegeneration. *Front. Endocrinol.* **2019**, *10*, 480. [CrossRef] [PubMed]

63. Reiter, R.J.; Tan, D.; Leon, J.; Kilic, Ü.; Kilic, E. When Melatonin Gets on Your Nerves: Its Beneficial Actions in Experimental Models of Stroke. *Exp. Biol. Med.* **2005**, *230*, 104–117. [CrossRef]

64. Amorim, M.; Marques, C.; Pereira, J.O.; Guardão, L.; Martins, M.J.; Osório, H.; Moura, D.; Calhau, C.; Pinheiro, H.; Pintado, M. Antihypertensive Effect of Spent Brewer Yeast Peptide. *Process Biochem.* **2019**, *76*, 213–218. [CrossRef]

65. Mirzaei, M.; Shavandi, A.; Mirdamadi, S.; Soleymanzadeh, N.; Motahari, P.; Mirdamadi, N.; Moser, M.; Subra, G.; Alimoradi, H.; Goriely, S. Bioactive Peptides from Yeast: A Comparative Review on Production Methods, Bioactivity, Structure-Function Relationship, and Stability. *Trends Food Sci. Technol.* **2021**, *118*, 297–315. [CrossRef]

66. Conti, G.; Magliani, W.; Conti, S.; Nencioni, L.; Sgarbanti, R.; Palamara, A.T.; Polonelli, L. Therapeutic Activity of an Anti-Idiotypic Antibody-Derived Killer Peptide against Influenza a Virus Experimental Infection. *Antimicrob. Agents Chemother.* **2008**, *52*, 4331–4337. [CrossRef]

67. Magliani, W.; Conti, S.; Ciociola, T.; Giovati, L.; Zanello, P.P.; Pertinhez, T.; Spisni, A.; Polonelli, L. Killer Peptide: A Novel Paradigm of Antimicrobial, Antiviral and Immunomodulatory Auto-Delivering Drugs. *Future Med. Chem.* **2011**, *3*, 1209–1231. [CrossRef]

68. Paulone, S.; Ardizzoni, A.; Tavanti, A.; Piccinelli, S.; Rizzato, C.; Lupetti, A.; Colombari, B.; Pericolini, E.; Polonelli, L.; Magliani, W.; et al. The Synthetic Killer Peptide KP Impairs *Candida albicans* Biofilm in vitro. *PLoS ONE* **2017**, *12*, e0181278. [CrossRef]

69. Giovati, L.; Santinoli, C.; Mangia, C.; Vismarra, A.; Belletti, S.; D'Adda, T.; Fumarola, C.; Ciociola, T.; Bacci, C.; Magliani, W.; et al. Novel Activity of a Synthetic Decapeptide Against *Toxoplasma gondii* Tachyzoites. *Front. Microbiol.* **2018**, *9*, 753. [CrossRef]
70. Bajaj, B.K.; Raina, S.; Singh, S. Killer Toxin from a Novel Killer Yeast *Pichia kudriavzevii* RY55 with Idiosyncratic Antibacterial Activity. *J. Basic Microbiol.* **2013**, *53*, 645–656. [CrossRef]
71. Pretscher, J.; Fischkal, T.; Branscheidt, S.; Jäger, L.; Kahl, S.; Schlander, M.; Thines, E.; Claus, H. Yeasts from Different Habitats and Their Potential as Biocontrol Agents. *Fermentation* **2018**, *4*, 31. [CrossRef]
72. Huang, C.; Chen, X.; Xiong, L.; Chen, X.; Ma, L.; Chen, Y. Single Cell Oil Production from Low-Cost Substrates: The Possibility and Potential of Its Industrialization. *Biotechnol. Adv.* **2013**, *31*, 129–139. [CrossRef] [PubMed]
73. Miranda, C.; Bettencourt, S.; Pozdniakova, T.; Pereira, J.; Sampaio, P.; Franco-Duarte, R.; Pais, C. Modified High-Throughput Nile Red Fluorescence Assay for the Rapid Screening of Oleaginous Yeasts Using Acetic Acid as Carbon Source. *BMC Microbiol.* **2020**, *20*, 60. [CrossRef] [PubMed]
74. Bettencourt, S.; Miranda, C.; Pozdniakova, T.A.; Sampaio, P.; Franco-Duarte, R.; Pais, C. Single Cell Oil Production by Oleaginous Yeasts Grown in Synthetic and Waste-Derived Volatile Fatty Acids. *Microorganisms* **2020**, *8*, 1809. [CrossRef] [PubMed]
75. Fabricio, M.F.; Valente, P.; Záchia Ayub, M.A. Oleaginous Yeast *Meyerozyma guilliermondii* Shows Fermentative Metabolism of Sugars in the Biosynthesis of Ethanol and Converts Raw Glycerol and Cheese Whey Permeate into Polyunsaturated Fatty Acids. *Biotechnol. Prog.* **2019**, *35*, e2895. [CrossRef] [PubMed]
76. Adel, A.; El-Baz, A.; Shetaia, Y.; Sorour, N.M. Biosynthesis of Polyunsaturated Fatty Acids by Two Newly Cold-Adapted Egyptian Marine Yeast. *3 Biotech* **2021**, *11*, 461. [CrossRef] [PubMed]
77. Kelesidis, T.; Pothoulakis, C. Efficacy and Safety of the Probiotic *Saccharomyces boulardii* for the Prevention and Therapy of Gastrointestinal Disorders. *Ther. Adv. Gastroenterol.* **2012**, *5*, 111. [CrossRef] [PubMed]
78. Sen, S.; Mansell, T.J. Yeasts as Probiotics: Mechanisms, Outcomes, and Future Potential. *Fungal Genet. Biol.* **2020**, *137*, 103333. [CrossRef] [PubMed]
79. Wu, X.; Vallance, B.A.; Boyer, L.; Bergstrom, K.S.B.; Walker, J.; Madsen, K.; O'Kusky, J.R.; Buchan, A.M.; Jacobson, K. Saccharomyces Boulardii Ameliorates Citrobacter Rodentium-Induced Colitis through Actions on Bacterial Virulence Factors. *Am. J. Physiol. Gastrointest Liver Physiol.* **2008**, *294*, G295–G306. [CrossRef]
80. Ducluzeau, R.; Bensaada, M. Comparative Effect of a Single or Continuous Administration of *Saccharomyces Boulardii* on the Establishment of Various Strains of *Candida* in the Digestive Tract of Gnotobiotic Mice. *Ann. Microbiol.* **1982**, *133*, 491–501.
81. Fayura, L.R.; Fedorovych, D.V.; Prokopiv, T.M.; Boretsky, Y.R.; Sibirny, A.A. The Pleiotropic Nature of Rib80, Hit1, and Red6 Mutations Affecting Riboflavin Biosynthesis in the Yeast *Pichia guilliermondii*. *Microbiology* **2007**, *76*, 55–59. [CrossRef]
82. Schaible, U.E.; Kaufmann, S.H.E. A Nutritive View on the Host–Pathogen Interplay. *Trends Microbiol.* **2005**, *13*, 373–380. [CrossRef]
83. Spadaro, D.; Droby, S. Development of Biocontrol Products for Postharvest Diseases of Fruit: The Importance of Elucidating the Mechanisms of Action of Yeast Antagonists. *Trends Food Sci. Technol.* **2016**, *47*, 39–49. [CrossRef]
84. Freimoser, F.M.; Rueda-Mejia, M.P.; Tilocca, B.; Migheli, Q. Biocontrol Yeasts: Mechanisms and Applications. *World J. Microbiol. Biotechnol.* **2019**, *35*, 1–19. [CrossRef]
85. Sampaolesi, S.; Briand, L.E.; De Antoni, G.; León Peláez, A. The Synthesis of Soluble and Volatile Bioactive Compounds by Selected Brewer's Yeasts: Antagonistic Effect against Enteropathogenic Bacteria and Food Spoiler–Toxigenic *Aspergillus* sp. *Food Chem. X* **2022**, *13*, 100193. [CrossRef]
86. Hua, S.S.T.; Beck, J.J.; Sarreal, S.B.L.; Gee, W. The Major Volatile Compound 2-Phenylethanol from the Biocontrol Yeast, *Pichia anomala*, Inhibits Growth and Expression of Aflatoxin Biosynthetic Genes of *Aspergillus flavus*. *Mycotoxin Res.* **2014**, *30*, 71–78. [CrossRef] [PubMed]
87. Masoud, W.; Poll, L.; Jakobsen, M. Influence of Volatile Compounds Produced by Yeasts Predominant during Processing of Coffea Arabica in East Africa on Growth and Ochratoxin A (OTA) Production by *Aspergillus ochraceus*. *Yeast* **2005**, *22*, 1133–1142. [CrossRef]
88. Huang, R.; Che, H.J.; Zhang, J.; Yang, L.; Jiang, D.H.; Li, G.Q. Evaluation of *Sporidiobolus pararoseus* Strain YCXT3 as Biocontrol Agent of *Botrytis cinerea* on Post-Harvest Strawberry Fruits. *Biol. Control* **2012**, *62*, 53–63. [CrossRef]
89. Huang, R.; Li, G.Q.; Zhang, J.; Yang, L.; Che, H.J.; Jiang, D.H.; Huang, H.C. Control of Postharvest Botrytis Fruit Rot of Strawberry by Volatile Organic Compounds of *Candida intermedia*. *Dis. Control Pest Manag.* **2011**, *101*, 859–869. [CrossRef] [PubMed]
90. Farbo, M.G.; Urgeghe, P.P.; Fiori, S.; Marcello, A.; Oggiano, S.; Balmas, V.; Hassan, Z.U.; Jaoua, S.; Migheli, Q. Effect of Yeast Volatile Organic Compounds on Ochratoxin A-Producing *Aspergillus carbonarius* and *A. ochraceus*. *Int. J. Food Microbiol.* **2018**, *284*, 1–10. [CrossRef]
91. Fiori, S.; Urgeghe, P.P.; Hammami, W.; Razzu, S.; Jaoua, S.; Migheli, Q. Biocontrol Activity of Four Non- and Low-Fermenting Yeast Strains against *Aspergillus carbonarius* and Their Ability to Remove Ochratoxin A from Grape Juice. *Int. J. Food Microbiol.* **2014**, *189*, 45–50. [CrossRef] [PubMed]
92. Tilocca, B.; Balmas, V.; Hassan, Z.U.; Jaoua, S.; Migheli, Q. A Proteomic Investigation of *Aspergillus carbonarius* Exposed to Yeast Volatilome or to Its Major Component 2-Phenylethanol Reveals Major Shifts in Fungal Metabolism. *Int. J. Food Microbiol.* **2019**, *306*, 108265. [CrossRef] [PubMed]

93. Holb, I.J.; Kunz, S. Integrated Control of Brown Rot Blossom Blight by Combining Approved Chemical Control Options with *Aureobasidium pullulans* in Organic Cherry Production. *Crop Prot.* **2013**, *54*, 114–120. [CrossRef]
94. Weiss, A.; Weißhaupt, S.; Krawiec, P.; Kunz, S. Use of *Aureobasidium pullulans* for Resistance Management in Chemical Control of *Botrytis cinerea* in Berries. *Acta Hortic.* **2014**, *1017*, 237–242. [CrossRef]
95. Prasongsuk, S.; Lotrakul, P.; Ali, I.; Bankeeree, W.; Punnapayak, H. The Current Status of *Aureobasidium pullulans* in Biotechnology. *Folia Microbiol.* **2017**, *63*, 129–140. [CrossRef]
96. Price, N.P.; Bischoff, K.M.; Leathers, T.D.; Cossé, A.A.; Manitchotpisit, P. Polyols, not sugars, determine the structural diversity of anti-streptococcal liamocins produced by *Aureobasidium pullulans* strain NRRL 50380. *J. Antibiot.* **2016**, *70*, 136–141. [CrossRef]
97. Parafati, L.; Vitale, A.; Restuccia, C.; Cirvilleri, G. Biocontrol Ability and Action Mechanism of Food-Isolated Yeast Strains against *Botrytis cinerea* Causing Post-Harvest Bunch Rot of Table Grape. *Food Microbiol.* **2015**, *47*, 85–92. [CrossRef] [PubMed]
98. Miethke, M.; Marahiel, M.A. Siderophore-Based Iron Acquisition and Pathogen Control. *Microbiol. Mol. Biol. Rev.* **2007**, *71*, 413–451. [CrossRef] [PubMed]
99. Cassat, J.E.; Skaar, E.P. Iron in Infection and Immunity. *Cell Host Microbe* **2013**, *13*, 509. [CrossRef]
100. Barber, M.F.; Elde, N.C. Buried Treasure: Evolutionary Perspectives on Microbial Iron Piracy. *Trends Genet.* **2015**, *31*, 627–636. [CrossRef]
101. Sheldon, J.R.; Heinrichs, D.E. Recent Developments in Understanding the Iron Acquisition Strategies of Gram Positive Pathogens. *FEMS Microbiol. Rev.* **2015**, *39*, 592–630. [CrossRef]
102. Boiteau, R.M.; Mende, D.R.; Hawco, N.J.; McIlvin, M.R.; Fitzsimmons, J.N.; Saito, M.A.; Sedwick, P.N.; DeLong, E.F.; Repeta, D.J. Siderophore-Based Microbial Adaptations to Iron Scarcity across the Eastern Pacific Ocean. *Proc. Natl. Acad. Sci. USA* **2016**, *113*, 14237–14242. [CrossRef] [PubMed]
103. Kramer, J.; Özkaya, Ö.; Kümmerli, R. Bacterial Siderophores in Community and Host Interactions. *Nat. Rev. Microbiol.* **2019**, *18*, 152–163. [CrossRef]
104. Wang, W.; Chi, Z.; Liu, G.; Buzdar, M.A.; Chi, Z.; Gu, Q. Chemical and Biological Characterization of Siderophore Produced by the Marine-Derived *Aureobasidium pullulans* HN6.2 and Its Antibacterial Activity. *BioMetals* **2009**, *22*, 965–972. [CrossRef]
105. de Lima Targino, H.M.; Silva, V.S.L.; Escobar, I.E.C.; de Almeida Ribeiro, P.R.; Gava, C.A.T.; Fernandes-Júnior, P.I. Maize-Associated *Meyerozyma* from the Brazilian Semiarid Region Are Effective Plant Growth-Promoting Yeasts. *Rhizosphere* **2022**, *22*, 100538. [CrossRef]
106. Puig, S.; Ramos-Alonso, L.; Romero, A.M.; Martínez-Pastor, M.T. The Elemental Role of Iron in DNA Synthesis and Repair. *Metallomics* **2017**, *9*, 1483–1500. [CrossRef]
107. Sansone, G.; Rezza, I.; Calvente, V.; Benuzzi, D.; Tosetti, M.I.S. de Control of *Botrytis cinerea* Strains Resistant to Iprodione in Apple with Rhodotorulic Acid and Yeasts. *Postharvest Biol. Technol.* **2005**, *35*, 245–251. [CrossRef]
108. Vadkertiová, R.; Sláviková, E. Killer Activity of Yeasts Isolated from Natural Environments against Some Medically Important *Candida* Species. *Pol. J. Microbiol.* **2007**, *56*, 39–43. [PubMed]
109. Sipiczki, M. Metschnikowia Strains Isolated from Botrytized Grapes Antagonize Fungal and Bacterial Growth by Iron Depletion. *Appl. Environ. Microbiol.* **2006**, *72*, 6716–6724. [CrossRef]
110. El-Tarabily, K.A.; Sivasithamparam, K. Potential of Yeasts as Biocontrol Agents of Soil-Borne Fungal Plant Pathogens and as Plant Growth Promoters. *Mycoscience* **2006**, *47*, 25–35. [CrossRef]
111. Cloete, K.J.; Valentine, A.J.; Stander, M.A.; Blomerus, L.M.; Botha, A. Evidence of Symbiosis between the Soil Yeast *Cryptococcus laurentii* and a Sclerophyllous Medicinal Shrub, Agathosma Betulina (Berg.) Pillans. *Microb. Ecol.* **2009**, *57*, 624–632. [CrossRef] [PubMed]
112. Klassen, R.; Schaffrath, R.; Buzzini, P.; Ganter, P.F. Antagonistic Interactions and Killer Yeasts. In *Yeasts in Natural Ecosystems: Ecology*; Springer: Cham, Switzerland, 2017; pp. 229–275. [CrossRef]
113. Mannazzu, I.; Domizio, P.; Carboni, G.; Zara, S.; Zara, G.; Comitini, F.; Budroni, M.; Ciani, M. Yeast Killer Toxins: From Ecological Significance to Application. *Crit. Rev. Biotechnol.* **2019**, *39*, 603–617. [CrossRef] [PubMed]
114. Izgu, D.A.; Kepekci, R.A.; Izgu, F. Inhibition of *Penicillium digitatum* and *Penicillium italicum* in Vitro and in Planta with Panomycocin, a Novel Exo-β-1,3-Glucanase Isolated from *Pichia anomala* NCYC 434. *Antonie Van Leeuwenhoek* **2011**, *99*, 85–91. [CrossRef]
115. Perez, M.F.; Contreras, L.; Garnica, N.M.; Fernández-Zenoff, M.V.; Farías, M.E.; Sepulveda, M.; Ramallo, J.; Dib, J.R. Native Killer Yeasts as Biocontrol Agents of Postharvest Fungal Diseases in Lemons. *PLoS ONE* **2016**, *11*, e0165590. [CrossRef]
116. Hua, M.X.; Chi, Z.; Liu, G.L.; Buzdar, M.A.; Chi, Z.M. Production of a Novel and Cold-Active Killer Toxin by *Mrakia frigida* 2E00797 Isolated from Sea Sediment in Antarctica. *Extremophiles* **2010**, *14*, 515–521. [CrossRef]
117. Lowes, K.F.; Shearman, C.A.; Payne, J.; MacKenzie, D.; Archer, D.B.; Merry, R.J.; Gasson, M.J. Prevention of Yeast Spoilage in Feed and Food by the Yeast Mycocin HMK. *Appl. Environ. Microbiol.* **2000**, *66*, 1066. [CrossRef] [PubMed]
118. Zhang, D.; Spadaro, D.; Garibaldi, A.; Gullino, M.L. Potential Biocontrol Activity of a Strain of *Pichia guilliermondii* against Grey Mold of Apples and Its Possible Modes of Action. *Biol. Control* **2011**, *57*, 193–201. [CrossRef]
119. Zhang, D.; Spadaro, D.; Valente, S.; Garibaldi, A.; Gullino, M.L. Cloning, Characterization, Expression and Antifungal Activity of an Alkaline Serine Protease of *Aureobasidium pullulans* PL5 Involved in the Biological Control of Postharvest Pathogens. *Int. J. Food Microbiol.* **2012**, *153*, 453–464. [CrossRef] [PubMed]

120. Parafati, L.; Cirvilleri, G.; Restuccia, C.; Wisniewski, M. Potential Role of Exoglucanase Genes (WaEXG1 and WaEXG2) in the Biocontrol Activity of *Wickerhamomyces anomalus*. *Microb. Ecol.* **2017**, *73*, 876–884. [CrossRef]
121. Junker, K.; Chailyan, A.; Hesselbart, A.; Forster, J.; Wendland, J. Multi-Omics Characterization of the Necrotrophic Mycoparasite *Saccharomycopsis schoenii*. *PLoS Pathog.* **2019**, *15*, e1007692. [CrossRef]
122. Zajc, J.; Gostinčar, C.; Černoša, A.; Gunde-Cimerman, N. Stress-Tolerant Yeasts: Opportunistic Pathogenicity Versus Biocontrol Potential. *Genes* **2019**, *10*, 42. [CrossRef] [PubMed]
123. Lima, J.R.; Gondim, D.M.F.; Oliveira, J.T.A.; Oliveira, F.S.A.; Gonçalves, L.R.B.; Viana, F.M.P. Use of Killer Yeast in the Management of Postharvest Papaya Anthracnose. *Postharvest Biol. Technol.* **2013**, *83*, 58–64. [CrossRef]
124. Raynaldo, F.A.; Dhanasekaran, S.; Ngea, G.L.N.; Yang, Q.; Zhang, X.; Zhang, H. Investigating the Biocontrol Potentiality of *Wickerhamomyces anomalus* against Postharvest Gray Mold Decay in Cherry Tomatoes. *Sci. Hortic.* **2021**, *285*, 110137. [CrossRef]
125. Calvente, V.; Benuzzi, D.; De Tosetti, M.I.S. Antagonistic Action of Siderophores from *Rhodotorula glutinis* upon the Postharvest Pathogen *Penicillium expansum*. *Int. Biodeterior Biodegrad.* **1999**, *43*, 167–172. [CrossRef]
126. Dang, T.D.T.; Vermeulen, A.; Ragaert, P.; Devlieghere, F. A Peculiar Stimulatory Effect of Acetic and Lactic Acid on Growth and Fermentative Metabolism of *Zygosaccharomyces bailii*. *Food Microbiol.* **2009**, *26*, 320–327. [CrossRef]
127. Corsetti, A.; De Angelis, M.; Dellaglio, F.; Paparella, A.; Fox, P.F.; Settanni, L. Characterization of Sourdough Lactic Acid Bacteria Based on Genotypic and Cell-Wall Protein Analyses. *J. Appl. Microbiol.* **2003**, *94*, 641–654. [CrossRef] [PubMed]
128. Hollomon, D. Does Agricultural Use of Azole Fungicides Contribute to Resistance in the Human Pathogen *Aspergillus fumigatus*? *Pest Manag. Sci.* **2017**, *73*, 1987–1993. [CrossRef] [PubMed]
129. Axelsson, L.T.; Chung, T.C.; Dobrogosz, W.J.; Lindgren, S.E. Production of a Broad Spectrum Antimicrobial Substance by *Lactobacillus reuteri*. *Microb. Ecol. Health Dis.* **2009**, *2*, 131–136. [CrossRef]
130. Dal Bello, F.; Clarke, C.I.; Ryan, L.A.M.; Ulmer, H.; Schober, T.J.; Ström, K.; Sjögren, J.; van Sinderen, D.; Schnürer, J.; Arendt, E.K. Improvement of the Quality and Shelf Life of Wheat Bread by Fermentation with the Antifungal Strain *Lactobacillus plantarum* FST 1.7. *J. Cereal Sci.* **2007**, *45*, 309–318. [CrossRef]
131. Crowley, S.; Mahony, J.; Van Sinderen, D. Current Perspectives on Antifungal Lactic Acid Bacteria as Natural Bio-Preservatives. *Trends Food Sci. Technol.* **2013**, *33*, 93–109. [CrossRef]
132. Bergsson, G.; Arnfinnsson, J.; Steingrímsson, O.; Thormar, H. In Vitro Killing of *Candida albicans* by Fatty Acids and Monoglycerides. *Antimicrob. Agents Chemother.* **2001**, *45*, 3209–3212. [CrossRef]
133. Magnusson, J.; Ström, K.; Roos, S.; Sjögren, J.; Schnürer, J. Broad and Complex Antifungal Activity among Environmental Isolates of Lactic Acid Bacteria. *FEMS Microbiol. Lett* **2003**, *219*, 129–135. [CrossRef]
134. And, T.J.A.; Bé Langer, R.R. Specificity and Mode of Action of the Antifungal Fatty Acid Cis-9-Heptadecenoic Acid Produced by *Pseudozyma flocculosa*. *Appl. Environ. Microbiol.* **2001**, *67*, 956–960. [CrossRef]
135. Strobel, G.; Daisy, B. Bioprospecting for Microbial Endophytes and Their Natural Products. *Microbiol. Mol. Biol. Rev.* **2003**, *67*, 491–502. [CrossRef] [PubMed]
136. Chatterjee, S.; Ghosh, S.; Mandal, N.C. Potential of an Endophytic Fungus *Alternaria tenuissima* PE2 Isolated from *Psidium guajava* L. for the Production of Bioactive Compounds. *S. Afr. J. Bot.* **2022**, *150*, 658–670. [CrossRef]
137. Phongpaichit, S.; Rungjindamai, N.; Rukachaisirikul, V.; Sakayaroj, J. Antimicrobial Activity in Cultures of Endophytic Fungi Isolated from *Garcinia* Species. *FEMS Immunol. Med. Microbiol.* **2006**, *48*, 367–372. [CrossRef]
138. Pongcharoen, W.; Rukachaisirikul, V.; Phongpaichit, S.; Kühn, T.; Pelzing, M.; Sakayaroj, J.; Taylor, W.C. Metabolites from the Endophytic Fungus *Xylaria* sp. PSU-D14. *Phytochemistry* **2008**, *69*, 1900–1902. [CrossRef]
139. Zhang, H.W.; Song, Y.C.; Tan, R.X. Biology and Chemistry of Endophytes. *Nat. Prod. Rep.* **2006**, *23*, 753–771. [CrossRef]
140. Strobel, G.A.; Miller, R.V.; Martinez-Miller, C.; Condron, M.M.; Teplow, D.B.; Hess, W.M. Cryptocandin, a Potent Antimycotic from the Endophytic Fungus *Cryptosporiopsis* cf. *quercina*. *Microbiol. Read.* **1999**, *145 Pt 8*, 1919–1926. [CrossRef]
141. Ding, G.; Liu, S.; Guo, L.; Zhou, Y.; Che, Y. Antifungal Metabolites from the Plant Endophytic Fungus *Pestalotiopsis foedan*. *J. Nat. Prod.* **2008**, *71*, 615–618. [CrossRef]
142. Moradi, A.; Yaghoubi-Avini, M.; Wink, J. Isolation of *Nannocystis* Species from Iran and Exploring Their Natural Products. *Arch Microbiol.* **2022**, *204*, 123. [CrossRef]
143. Rhee, K.H. Cyclic Dipeptides Exhibit Synergistic, Broad Spectrum Antimicrobial Effects and Have Anti-Mutagenic Properties. *Int. J. Antimicrob. Agents* **2004**, *24*, 423–427. [CrossRef]
144. Pereira, C.B.; Pereira de Sá, N.; Borelli, B.M.; Rosa, C.A.; Barbeira, P.J.S.; Cota, B.B.; Johann, S. Antifungal Activity of Eicosanoic Acids Isolated from the Endophytic Fungus *Mycosphaerella* sp. against *Cryptococcus neoformans* and *C. gattii*. *Microb. Pathog.* **2016**, *100*, 205–212. [CrossRef]
145. Zafar, H.; Altamirano, D.S.; Ballou, E.R.; Nielsen, K. A Titanic Drug Resistance Threat in Cryptococcus Neoformans. *Curr. Opin. Microbiol.* **2019**, *52*, 158–164. [CrossRef]
146. Botts, M.R.; Hull, C.M. Dueling in the Lung: How *Cryptococcus* Spores Race the Host for Survival. *Curr. Opin. Microbiol.* **2010**, *13*, 437–442. [CrossRef]
147. Perfect, J.R.; Bicanic, T. Cryptococcosis Diagnosis and Treatment: What Do We Know Now. *Fungal Genet. Biol.* **2015**, *78*, 49–54. [CrossRef]

148. Salehi, M.; Ahmadikia, K.; Badali, H.; Khodavaisy, S. Opportunistic Fungal Infections in the Epidemic Area of COVID-19: A Clinical and Diagnostic Perspective from Iran. *Mycopathologia* **2020**, *185*, 607–611. [CrossRef] [PubMed]
149. Chi, W.C.; Pang, K.L.; Chen, W.L.; Wang, G.J.; Lee, T.H. Antimicrobial and INOS Inhibitory Activities of the Endophytic Fungi Isolated from the Mangrove Plant *Acanthus ilicifolius* var. *xiamenensis*. *Bot. Stud.* **2019**, *60*, 1–8. [CrossRef] [PubMed]
150. Sherry, L.; Ramage, G.; Kean, R.; Borman, A.; Johnson, E.M.; Richardson, M.D.; Rautemaa-Richardson, R. Biofilm-Forming Capability of Highly Virulent, Multidrug-Resistant *Candida auris*. *Emerg. Infect. Dis.* **2017**, *23*, 328. [CrossRef] [PubMed]
151. Calvo, B.; Melo, A.S.A.; Perozo-Mena, A.; Hernandez, M.; Francisco, E.C.; Hagen, F.; Meis, J.F.; Colombo, A.L. First Report of *Candida auris* in America: Clinical and Microbiological Aspects of 18 Episodes of Candidemia. *J. Infect.* **2016**, *73*, 369–374. [CrossRef]
152. Adam, R.D.; Revathi, G.; Okinda, N.; Fontaine, M.; Shah, J.; Kagotho, E.; Castanheira, M.; Pfaller, M.A.; Maina, D. Analysis of *Candida auris* Fungemia at a Single Facility in Kenya. *Int. J. Infect. Dis.* **2019**, *85*, 182–187. [CrossRef]
153. Taori, S.K.; Khonyongwa, K.; Hayden, I.; Athukorala, G.D.A.; Letters, A.; Fife, A.; Desai, N.; Borman, A.M. *Candida auris* Outbreak: Mortality, Interventions and Cost of Sustaining Control. *J. Infect.* **2019**, *79*, 601–611. [CrossRef]
154. Wall, G.; Herrera, N.; Lopez-Ribot, J.L. Repositionable Compounds with Antifungal Activity against Multidrug Resistant *Candida auris* Identified in the Medicines for Malaria Venture's Pathogen Box. *J. Fungi* **2019**, *5*, 92. [CrossRef]
155. Treviño-Rangel, R.D.J.; González, G.M.; Montoya, A.M.; Rojas, O.C.; Elizondo-Zertuche, M.; Álvarez-Villalobos, N.A. Recent Antifungal Pipeline Developments against *Candida auris*: A Systematic Review. *J. Fungi* **2022**, *8*, 1144. [CrossRef]
156. Demain, A.L.; Martens, E. Production of Valuable Compounds by Molds and Yeasts. *J. Antibiot.* **2017**, *70*, 347–360. [CrossRef]
157. Mendes, I.; Sanchez, I.; Franco-Duarte, R.; Camarasa, C.; Schuller, D.; Dequin, S.; Sousa, M.J. Integrating Transcriptomics and Metabolomics for the Analysis of the Aroma Profiles of *Saccharomyces cerevisiae* Strains from Diverse Origins. *BMC Genom.* **2017**, *18*, 455. [CrossRef]
158. Fernandes, T.; Silva-Sousa, F.; Pereira, F.; Rito, T.; Soares, P.; Franco-Duarte, R.; Sousa, M.J. Biotechnological Importance of *Torulaspora delbrueckii*: From the Obscurity to the Spotlight. *J. Fungi* **2021**, *7*, 712. [CrossRef]
159. Silva-Sousa, F.; Fernandes, T.; Pereira, F.; Rodrigues, D.; Rito, T.; Camarasa, C.; Franco-Duarte, R.; Sousa, M.J. *Torulaspora delbrueckii* Phenotypic and Metabolic Profiling towards Its Biotechnological Exploitation. *J. Fungi* **2022**, *8*, 569. [CrossRef]
160. Arroyo-López, F.N.; Querol, A.; Bautista-Gallego, J.; Garrido-Fernández, A. Role of Yeasts in Table Olive Production. *Int. J. Food Microbiol.* **2008**, *128*, 189–196. [CrossRef]
161. Franco-Duarte, R.; Bessa, D.; Gonçalves, F.; Martins, R.; Silva-Ferreira, A.C.; Schuller, D.; Sampaio, P.; Pais, C. Genomic and Transcriptomic Analysis of *Saccharomyces cerevisiae* Isolates with Focus in Succinic Acid Production. *FEMS Yeast Res.* **2017**, *17*, fox057. [CrossRef]
162. Li, C.; Ong, K.L.; Cui, Z.; Sang, Z.; Li, X.; Patria, R.D.; Qi, Q.; Fickers, P.; Yan, J.; Lin, C.S.K. Promising Advancement in Fermentative Succinic Acid Production by Yeast Hosts. *J. Hazard. Mater.* **2021**, *401*, 123414. [CrossRef]
163. Atanasov, A.G.; Waltenberger, B.; Pferschy-Wenzig, E.M.; Linder, T.; Wawrosch, C.; Uhrin, P.; Temml, V.; Wang, L.; Schwaiger, S.; Heiss, E.H.; et al. Discovery and Resupply of Pharmacologically Active Plant-Derived Natural Products: A Review. *Biotechnol. Adv.* **2015**, *33*, 1582–1614. [CrossRef]
164. Ye, V.M.; Bhatia, S.K. Metabolic Engineering for the Production of Clinically Important Molecules: Omega-3 Fatty Acids, Artemisinin, and Taxol. *Biotechnol. J.* **2012**, *7*, 20–33. [CrossRef]
165. Barrales-Cureño, H.J.; Ramos Valdivia, A.C.; Soto Hernández, M. Increased Production of Taxoids in Suspension Cultures of Taxus Globosa after Elicitation. *Future Pharmacol.* **2022**, *2*, 45–54. [CrossRef]
166. Zaheer, K.; Humayoun Akhtar, M. An Updated Review of Dietary Isoflavones: Nutrition, Processing, Bioavailability and Impacts on Human Health. *Crit. Rev. Food Sci. Nutr.* **2017**, *57*, 1280–1293. [CrossRef]
167. Cue, B.W.; Zhang, J. Green Process Chemistry in the Pharmaceutical Industry. *Green Chem. Lett. Rev.* **2009**, *2*, 193–211. [CrossRef]
168. Sun, H.; Liu, Z.; Zhao, H.; Ang, E.L. Recent Advances in Combinatorial Biosynthesis for Drug Discovery. *Drug Des. Dev. Ther.* **2015**, *9*, 823–833. [CrossRef]
169. Byrne, F.P.; Jin, S.; Paggiola, G.; Petchey, T.H.M.; Clark, J.H.; Farmer, T.J.; Hunt, A.J.; Robert McElroy, C.; Sherwood, J. Tools and Techniques for Solvent Selection: Green Solvent Selection Guides. *Sustain. Chem. Process.* **2016**, *4*, 1–24. [CrossRef]
170. Keasling, J.D. Manufacturing Molecules through Metabolic Engineering. *Science* **2010**, *330*, 1355–1358. [CrossRef]
171. Luo, Y.; Li, B.Z.; Liu, D.; Zhang, L.; Chen, Y.; Jia, B.; Zeng, B.X.; Zhao, H.; Yuan, Y.J. Engineered Biosynthesis of Natural Products in Heterologous Hosts. *Chem. Soc. Rev.* **2015**, *44*, 5265–5290. [CrossRef]
172. Song, M.C.; Kim, E.J.; Kim, E.; Rathwell, K.; Nam, S.J.; Yoon, Y.J. Microbial Biosynthesis of Medicinally Important Plant Secondary Metabolites. *Nat. Prod. Rep.* **2014**, *31*, 1497–1509. [CrossRef] [PubMed]
173. Nielsen, J.; Keasling, J.D. Engineering Cellular Metabolism. *Cell* **2016**, *164*, 1185–1197. [CrossRef]
174. Goffeau, A.; Barrell, B.G.; Bussey, H.; Davis, R.W.; Dujon, B.; Feldmann, H.; Galibert, F.; Hoheisel, J.D.; Jacq, C.; Johnston, M.; et al. Life with 6000 Genes. *Science* **1996**, *274*, 563–567. [CrossRef] [PubMed]
175. Kim, H.; Yoo, S.J.; Kang, H.A. Yeast Synthetic Biology for the Production of Recombinant Therapeutic Proteins. *FEMS Yeast Res.* **2015**, *15*, 1–16. [CrossRef]
176. Porro, D.; Gasser, B.; Fossati, T.; Maurer, M.; Branduardi, P.; Sauer, M.; Mattanovich, D. Production of Recombinant Proteins and Metabolites in Yeasts. *Appl. Microbiol. Biotechnol.* **2011**, *89*, 939–948. [CrossRef]

177. Buckholz, R.G.; Gleeson, M.A.G. Yeast Systems for the Commercial Production of Heterologous Proteins. *Biotechnology* **1991**, *9*, 1067–1072. [CrossRef] [PubMed]
178. Meehl, M.A.; Stadheim, T.A. Biopharmaceutical Discovery and Production in Yeast. *Curr. Opin. Biotechnol.* **2014**, *30*, 120–127. [CrossRef] [PubMed]
179. Paddon, C.J.; Westfall, P.J.; Pitera, D.J.; Benjamin, K.; Fisher, K.; McPhee, D.; Leavell, M.D.; Tai, A.; Main, A.; Eng, D.; et al. High-Level Semi-Synthetic Production of the Potent Antimalarial Artemisinin. *Nature* **2013**, *496*, 528–532. [CrossRef] [PubMed]
180. Jinek, M.; Chylinski, K.; Fonfara, I.; Hauer, M.; Doudna, J.A.; Charpentier, E. A Programmable Dual-RNA-Guided DNA endonuclease in adaptive bacterial immunity. *Science* **2012**, *337*, 816–822. [CrossRef]
181. Dicarlo, J.E.; Norville, J.E.; Mali, P.; Rios, X.; Aach, J.; Church, G.M. Genome Engineering in *Saccharomyces cerevisiae* Using CRISPR-Cas Systems. *Nucleic. Acids Res.* **2013**, *41*, 4336–4343. [CrossRef]
182. Galanie, S.; Thodey, K.; Trenchard, I.J.; Interrante, M.F.; Smolke, C.D. Complete Biosynthesis of Opioids in Yeast. *Science* **2015**, *349*, 1095–1100. [CrossRef] [PubMed]
183. Li, M.; Kildegaard, K.R.; Chen, Y.; Rodriguez, A.; Borodina, I.; Nielsen, J. De Novo Production of Resveratrol from Glucose or Ethanol by Engineered *Saccharomyces cerevisiae. Metab. Eng.* **2015**, *32*, 1–11. [CrossRef]
184. EauClaire, S.F.; Zhang, J.; Rivera, C.G.; Huang, L.L. Combinatorial Metabolic Pathway Assembly in the Yeast Genome with RNA-Guided Cas9. *J. Ind. Microbiol. Biotechnol.* **2016**, *43*, 1001–1015. [CrossRef] [PubMed]
185. Liu, Q.; Liu, Y.; Li, G.; Savolainen, O.; Chen, Y.; Nielsen, J. De Novo Biosynthesis of Bioactive Isoflavonoids by Engineered Yeast Cell Factories. *Nat. Commun.* **2021**, *12*, 6085. [CrossRef]
186. Shen, X.X.; Opulente, D.A.; Kominek, J.; Zhou, X.; Steenwyk, J.L.; Buh, K.V.; Haase, M.A.B.; Wisecaver, J.H.; Wang, M.; Doering, D.T.; et al. Tempo and Mode of Genome Evolution in the Budding Yeast Subphylum. *Cell* **2018**, *175*, 1533–1545.e20. [CrossRef]
187. Gírio, F.M.; Fonseca, C.; Carvalheiro, F.; Duarte, L.C.; Marques, S.; Bogel-Łukasik, R. Hemicelluloses for Fuel Ethanol: A Review. *Bioresour. Technol.* **2010**, *101*, 4775–4800. [CrossRef]
188. Rebello, S.; Abraham, A.; Madhavan, A.; Sindhu, R.; Binod, P.; Karthika Bahuleyan, A.; Aneesh, E.M.; Pandey, A. Non-Conventional Yeast Cell Factories for Sustainable Bioprocesses. *FEMS Microbiol. Lett.* **2018**, *365*, 222. [CrossRef]
189. Cao, X.; Wei, L.J.; Lin, J.Y.; Hua, Q. Enhancing Linalool Production by Engineering Oleaginous Yeast *Yarrowia lipolytica. Bioresour. Technol.* **2017**, *245*, 1641–1644. [CrossRef] [PubMed]
190. Gao, S.; Tong, Y.; Zhu, L.; Ge, M.; Zhang, Y.; Chen, D.; Jiang, Y.; Yang, S. Iterative Integration of Multiple-Copy Pathway Genes in *Yarrowia lipolytica* for Heterologous β-Carotene Production. *Metab. Eng.* **2017**, *41*, 192–201. [CrossRef] [PubMed]
191. Luo, Z.; Liu, N.; Lazar, Z.; Chatzivasileiou, A.; Ward, V.; Chen, J.; Zhou, J.; Stephanopoulos, G. Enhancing Isoprenoid Synthesis in *Yarrowia lipolytica* by Expressing the Isopentenol Utilization Pathway and Modulating Intracellular Hydrophobicity. *Metab. Eng.* **2020**, *61*, 344–351. [CrossRef]
192. Zhang, L.; Yang, H.; Xia, Y.; Shen, W.; Liu, L.; Li, Q.; Chen, X. Engineering the Oleaginous Yeast *Candida tropicalis* for α-Humulene Overproduction. *Biotechnol. Biofuels Bioprod.* **2022**, *15*, 1–12. [CrossRef]
193. Yun, C.R.; Kong, J.N.; Chung, J.H.; Kim, M.C.; Kong, K.H. Improved Secretory Production of the Sweet-Tasting Protein, Brazzein, in *Kluyveromyces lactis. J. Agric. Food Chem.* **2016**, *64*, 6312–6316. [CrossRef]
194. Lin, Y.J.; Chang, J.J.; Lin, H.Y.; Thia, C.; Kao, Y.Y.; Huang, C.C.; Li, W.H. Metabolic Engineering a Yeast to Produce Astaxanthin. *Bioresour. Technol.* **2017**, *245*, 899–905. [CrossRef] [PubMed]
195. Khongto, B.; Laoteng, K.; Tongta, A. Fermentation Process Development of Recombinant Hansenula Polymorpha for Gamma-Linolenic Acid Production. *J. Microbiol. Biotechnol.* **2010**, *20*, 1555–1562. [CrossRef]
196. Bhataya, A.; Schmidt-Dannert, C.; Lee, P.C. Metabolic Engineering of *Pichia pastoris* X-33 for Lycopene Production. *Process Biochem.* **2009**, *44*, 1095–1102. [CrossRef]
197. Liu, Y.; Tu, X.; Xu, Q.; Bai, C.; Kong, C.; Liu, Q.; Yu, J.; Peng, Q.; Zhou, X.; Zhang, Y.; et al. Engineered Monoculture and Co-Culture of Methylotrophic Yeast for de Novo Production of Monacolin J and Lovastatin from Methanol. *Metab. Eng.* **2018**, *45*, 189–199. [CrossRef] [PubMed]
198. Zirpel, B.; Degenhardt, F.; Zammarelli, C.; Wibberg, D.; Kalinowski, J.; Stehle, F.; Kayser, O. Optimization of Δ9-Tetrahydrocannabinolic Acid Synthase Production in *Komagataella phaffii* via Post-Translational Bottleneck Identification. *J. Biotechnol.* **2018**, *272–273*, 40–47. [CrossRef]
199. Araya-Garay, J.M.; Ageitos, J.M.; Vallejo, J.A.; Veiga-Crespo, P.; Sánchez-Pérez, A.; Villa, T.G. Construction of a Novel *Pichia pastoris* Strain for Production of Xanthophylls. *AMB Express* **2012**, *2*, 24. [CrossRef] [PubMed]
200. Xue, Y.; Kong, C.; Shen, W.; Bai, C.; Ren, Y.; Zhou, X.; Zhang, Y.; Cai, M. Methylotrophic Yeast *Pichia pastoris* as a Chassis Organism for Polyketide Synthesis via the Full Citrinin Biosynthetic Pathway. *J. Biotechnol.* **2017**, *242*, 64–72. [CrossRef]
201. Mavrommati, M.; Daskalaki, A.; Papanikolaou, S.; Aggelis, G. Adaptive Laboratory Evolution Principles and Applications in Industrial Biotechnology. *Biotechnol. Adv.* **2022**, *54*, 107795. [CrossRef] [PubMed]
202. Fernandes, T.; Osório, C.; Sousa, M.J.; Franco-Duarte, R. Contributions of Adaptive Laboratory Evolution towards the Enhancement of the Biotechnological Potential of Non-Conventional Yeast Species. *J. Fungi* **2023**, *9*, 186. [CrossRef] [PubMed]
203. Reyes, L.H.; Gomez, J.M.; Kao, K.C. Improving Carotenoids Production in Yeast via Adaptive Laboratory Evolution. *Metab. Eng.* **2014**, *21*, 26–33. [CrossRef] [PubMed]

204. Liu, L.; Pan, A.; Spofford, C.; Zhou, N.; Alper, H.S. An Evolutionary Metabolic Engineering Approach for Enhancing Lipogenesis in *Yarrowia lipolytica*. *Metab. Eng.* **2015**, *29*, 36–45. [CrossRef] [PubMed]
205. Dai, Z.; Liu, Y.; Guo, J.; Huang, L.; Zhang, X. Yeast Synthetic Biology for High-Value Metabolites. *FEMS Yeast Res.* **2015**, *15*, 1–13. [CrossRef]

Article

Transcriptional Response of Multi-Stress-Tolerant *Saccharomyces cerevisiae* to Sequential Stresses

Ane Catarine Tosi Costa [1], Mariano Russo [2], A. Alberto R. Fernandes [1], James R. Broach [2,*] and Patricia M. B. Fernandes [1,*]

[1] Biotechnology Core, Health Sciences Center, Federal University of Espírito Santo, Vitória 29047-110, Brazil
[2] Department of Biochemistry and Molecular Biology, Penn State University College of Medicine, Hershey, PA 17033, USA
* Correspondence: jbroach@pennstatehealth.psu.edu (J.R.B.); patricia.fernandes@ufes.br (P.M.B.F.)

Abstract: During the fermentation process, yeast cells face different stresses, and their survival and fermentation efficiency depend on their adaptation to these challenging conditions. Yeast cells must tolerate not only a single stress but also multiple simultaneous and sequential stresses. However, the adaptation and cellular response when cells are sequentially stressed are not completely understood. To explore this, we exposed a multi-stress-tolerant strain (BT0510) to different consecutive stresses to globally explore a common response, focusing on the genes induced in both stresses. Gene Ontology, pathway analyses, and common transcription factor motifs identified many processes linked to this common response. A metabolic shift to the pentose phosphate pathway, peroxisome activity, and the oxidative stress response were some of the processes found. The *SYM1*, *STF2*, and *HSP* genes and the transcription factors Adr1 and Usv1 may play a role in this response. This study presents a global view of the transcriptome of a multi-resistance yeast and provides new insights into the response to sequential stresses. The identified response genes can indicate future directions for the genetic engineering of yeast strains, which could improve many fermentation processes, such as those used for bioethanol production and beverages.

Keywords: fermentation process; RNA-Seq; stress response; acquired stress resistance; yeast

Citation: Costa, A.C.T.; Russo, M.; Fernandes, A.A.R.; Broach, J.R.; Fernandes, P.M.B. Transcriptional Response of Multi-Stress-Tolerant *Saccharomyces cerevisiae* to Sequential Stresses. *Fermentation* **2023**, *9*, 195. https://doi.org/10.3390/fermentation9020195

Academic Editor: Timothy Tse

Received: 18 December 2022
Revised: 14 February 2023
Accepted: 15 February 2023
Published: 20 February 2023

1. Introduction

Fermentation is a process in which a complex organic substance is converted into a simpler one usually due to the metabolic activity of microorganisms, such as yeast and bacteria [1]. This process is widely used in industrial settings to produce chemicals, bioethanol, and beverages. During fermentation, yeast cells encounter different stresses that have a significant influence on their microbial physiology. Initially, yeast cells experience osmotic stress caused by high solute concentrations, temperature shock due to exothermic reactions, and oxidative stress. As fermentation progresses, they encounter other stresses, such as starvation and increased ethanol concentration [2]. These environmental changes presented during fermentation processes might cause a reduced growth rate, an extended lag phase, and metabolic cessation, with some yeast surviving and others dying [3].

To survive these stressful environments, yeast cells can maintain their biological homeostasis through transcription regulation, activating genes associated with cell survival and function restoration. The efficiency of this process in a given yeast strain can determine its robustness [4–6]. While the stress response has been extensively studied, most of those studies focused on a single stress response [7–9]. Although such studies have intrinsic and fundamental value, they disregard issues widely observed in industrial environments, such as the stress fluctuations and variations during the fermentation process.

A few studies have explored the effects of multiple and sequential stresses and have shown that yeast subjected to different mild stresses followed by lethal stress exhibit little overlap in the genes involved in the response [10]. Further analyses of yeast cellular memory in response to short, pulsed hyperosmotic stress also demonstrated diverse gene behavior [11]. However, a comprehensive evaluation of the common response to sequential stresses that considers the conditions of industrial environments, focusing on genes induced in all stresses, remains to be thoroughly explored, particularly in yeast strains that already possess a multi-tolerant profile.

A transcriptome analysis can provide insights into the stress response in yeast, leading to a better understanding of cellular metabolism and the related pathways that play a role in the response [12]. In this study, we employed RNA sequencing to identify a cluster of genes that comprise a common response to sequential stresses and the essential pathways, genes, and transcription factors (TFs) associated with that response. We investigated the *Saccharomyces cerevisiae* strain BT0510 due to its high flocculation capacity and high tolerance to ethanol, osmotic, and heat shock stresses, as well as high fermentation rates [13]. A whole-genome analysis of BT0510 revealed substantial genetic differences between it and the *Saccharomyces cerevisiae* reference genome, S288c, as well as distinct genomic endowments when compared with two other cachaça fermentation strains. That analysis highlighted a collection of intragenic SNPs that could influence cellular processes, such as transport and the stress response, and it identified a set of gene loss and gene fragmentations that could directly influence fermentation processes [14]. This current transcriptomic study can reveal how these genomic changes affect the gene expression in this strain, as well as providing baseline knowledge on the response of this multi-tolerant yeast strain to different sequential stresses. This analysis also provides targeted genetic information to enhance stress tolerance in yeast that could improve a variety of biotechnological processes, such as beverage production, baking, and biofuel production.

2. Materials and Methods

2.1. Yeast Strain, Culture Conditions, and Experimental Design

The industrial yeast strain used in this study was previously isolated from a cachaça distillery in the State of Espírito Santo, Brazil, and it is described in Bravim et al. [13]. This strain (URM 6670) was selected for its flocculation ability and tolerance to ethanol, osmotic, and heat shock stresses and has been fully sequenced [14]. URM 6670 is stored at the Federal University of Pernambuco Culture Collection (URM604). In this and previous studies, the strain is designated BT0510.

Yeast cells were grown at 30 °C with aeration in a liquid synthetic complete medium (SC) (yeast nitrogen base without amino acids, drop-out mix, 2% glucose) to the exponential growth phase (OD600 = 1.0) before being exposed to sequential stresses. Stress treatments were selected based on the chronological progression of stress events during fermentation, with ethanol stress increasing towards the end of the fermentation process. Accordingly, the yeast cells were subjected to three different treatments: (1) osmotic stress followed by ethanol stress; (2) oxidative stress followed by ethanol stress; and (3) glucose withdrawal followed by ethanol stress. For each treatment, samples were collected before the first stress (control), after the first stress, and after the second stress (ethanol stress) for RNA extraction. The control was used to identify differentially expressed genes after the first and sequential stresses.

For osmotic stress, cells were incubated in liquid SC containing 1 M sorbitol for 30 min at 28 °C. The cells were subjected to oxidative stress by incubating them in SC containing 0.6 mM hydrogen peroxide for 30 min. Glucose withdrawal was performed by transferring the cells in SC + 2% glucose to SC + 0.05% glucose for 30 min. After each stress, the cells were collected via filtration, washed with deionized water, and reinoculated in SC supplemented with 8% (v/v) ethanol for 30 min at 28 °C. Several studies have demonstrated that, upon exposure to stress for 30 min, cells are capable of initiating a complex signaling network and displaying robust gene expression patterns [15,16]. The experiments were performed in triplicate.

2.2. RNA Preparation, Library Construction, and Sequencing

The total RNA from three biological replicates for each treatment was extracted using a Qiagen RNeasy Mini kit (Qiagen, Hilden, Germany) following the manufacturer's instructions. The quality and quantity of the total RNA samples were determined using a Nanodrop 2000 spectrophotometer (Thermo Scientific, Waltham, MA, USA). The integrity of the RNA was evaluated on an Agilent Bioanalyzer 2100 system using an RNA Nano 6000 Assay Kit (Agilent Technologies, Santa Clara, CA, USA) following the manufacturer's protocol. Poly-A-tailed RNA was captured and reverse-transcribed to complementary DNA (cDNA), which was used for library preparation with the Stranded Total RNA Prep (Illumina, San Diego, CA, USA) following the manufacturer's protocol.

2.3. Reads Mapping, Annotation, and Gene Expression Analysis

After trimming and filtering, paired-end reads were aligned to the reference genome of *S. cerevisiae* S288c via Hisat2 software (v2.0.1). Fragments per kilobases per million reads (FPKM) values were calculated as the metric of gene expression using Cufflinks [17] and were used for a principal component analysis (PCA) and a K-means clustering analysis via iDEP.91 [18]. The 2000 most variable genes were included and mean-normalized in the K-means analysis. The data for K = 4 are shown since higher values of K yielded no further distinct expression patterns. The assembled transcripts generated by Cufflinks were used to obtain the differential gene expression via Cuffdiff [17]. After adjustment using Benjamini and Hochberg's approach for controlling the false discovery rate, p-value < 0.05 and the value of log2 fold change >1 were aggregated to detect differentially expressed genes in both sequential stresses.

2.4. Enrichment Analysis

DAVID 6.8 [19] was used to annotate functions and identify enriched metabolic pathways. The pathways enriched were highlighted in the DAVID Functional Annotation Tool and retrieved from the Kyoto Encyclopedia of Genes and Genomes (KEGG) database. The GO direct category was used to select GO mapping directly annotated by the source database (no parent terms included). GO terms with FDR < 0.05 were considered significantly enriched within the gene set. The enriched KEGG pathways (FDR < 0.05) identified are reported.

2.5. Enriched Promoter Motifs

ShinyGO v0.65 [20] was used to enrich the motifs in the 300 bp region upstream of the genes in the gene list.

2.6. Quantitative Real-Time PCR Validation

Seven representative genes were analyzed in all different sequential stresses using the real-time PCR method to validate the RNA-Seq analysis. The total RNA was extracted from yeast cells using phenol/chloroform and precipitated with 3 M sodium acetate/absolute ethanol. The extracted RNA was treated with DNase I (Thermo Fisher Scientific, Waltham, MA, USA) and reverse-transcribed into cDNA using a High-Capacity cDNA Reverse Transcription Kit (Applied Biosystems, Waltham, MA, USA). PCR reactions were carried out using a 7500 Fast Real-Time PCR System (Applied Biosystems, Waltham, MA, USA). The relative expression levels of the target genes were measured using the $2-\Delta\Delta CT$ method, and ALG9 was used as the reference gene [21]. All tests were performed at least three times. The primers are listed in Table 1.

Table 1. Oligonucleotides used as primers for qRT-PCR analysis.

Target mRNA	Primer Sequence 5′–3′	Amplicon Size (bp)	PCR Efficiency [a] (%)
HSP12	Forward, 5′CGCAGGTAGAAAAGGATTCG3′ Reverse, 5′TCAGCGTTATCCTTGCCTTT3′	194	105
YGP1	Forward, 5′TGACGGTGGTTACTCTTCCA3′ Reverse, 5′GAACGGCAGAACTCAAGGAG3′	49	104
SYM1	Forward, 5′ACGGGTAGCTGTCGATCAAT3′ Reverse, 5′AGGCCACCATTGCTCTTTTA3′	126	97
STF2	Forward, 5′CGGTGAATCTCCAAATCACA3′ Reverse, 5′CACTGGGGGTATTTCACCAT3′	108	99
HSP26	Forward, 5′ATGCTGGCGCTCTTTATGAT3′ Reverse, 5′TTCTAGGGAAACCGAAACCA3′	95	95
SSE2	Forward, 5′CACTGGGGTCAAGGTTCCTA3′ Reverse, 5′GGTAAAGGCACTGGCTCTTG3′	137	94
HSP42	Forward, 5′TGAACGCATTATCCAACCAA3′ Reverse, 5′TTGTCCATAATGGGGATGGT3′	94	100
ALG9	Forward, 5′ACATCGTCGCCCCAATAAAT3′ Reverse, 5′GATTGGCTCCGGTACGTAAA3′	142	93

[a] The PCR efficiency of each primer was evaluated using the dilution series method using a mix of sample cDNAs as the template and based on the slope of the standard curve.

3. Results

3.1. Global Analysis of the Transcriptome

The principal component analysis (PCA) performed on the normalized FPKM data showed that the samples clustered together in each treatment, except for one replicate after the treatment with glucose withdrawal (Figure S1). We identified this outlier and removed it from subsequent analyses. Overall, the clustering indicates a consistent response to different sequential stresses in the yeast cell population.

3.2. Identification of Common Genes in Response to Sequential Stresses

The K-means clustering of the FPKM data identified groups of genes that displayed a distinct expression pattern. Our aim was to identify these stress-induced genes and their corresponding cellular activities that lead to sequential stress adaptation in yeast, namely, genes whose transcript levels increased in abundance in response to both sequential stresses. A cluster of genes with this behavior was identified in all treatments (Figure 1). To select the responsive genes that were differentially expressed, we selected the ones that were upregulated (log2 > 1; adjusted $p < 0.05$) in both stresses. This yielded 746 differentially expressed genes (DEGs) with the pattern of interest when osmotic stress was applied as the first stress, 342 DEGs with oxidative stress treatments, and 588 DEGs in glucose withdrawal experiments. A total of 219 genes were upregulated in all treatments (Figure 2). This list of 219 genes was analyzed using DAVID's Functional Annotation Tool using *Saccharomyces cerevisiae* S288c as a background to functionally annotate and identify enriched pathways in KEGG.

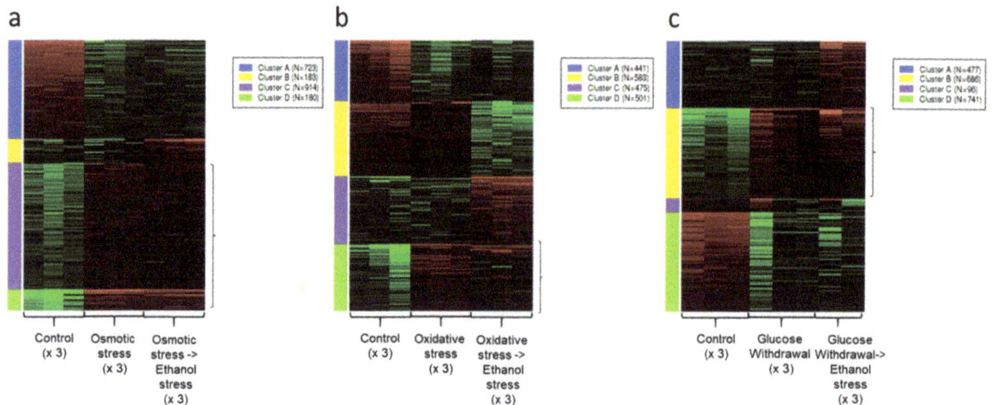

Figure 1. K-means clustering analysis of FPKM data after different sequential stresses. (**a**) Gene clusters after osmotic stress followed by ethanol stress. (**b**) Gene clusters after oxidative stress followed by ethanol stress. (**c**) Gene clusters after glucose withdrawal followed by ethanol stress. For each treatment, four clusters were identified and are color-coded in blue, yellow, purple, and green. The cluster with the gene pattern of interest was highlighted in each treatment. The analysis included the 2000 most variable genes and was mean-normalized. Values below the mean are shown in green, and those above are shown in red. The number of replicates analyzed from each treatment is identified in parentheses.

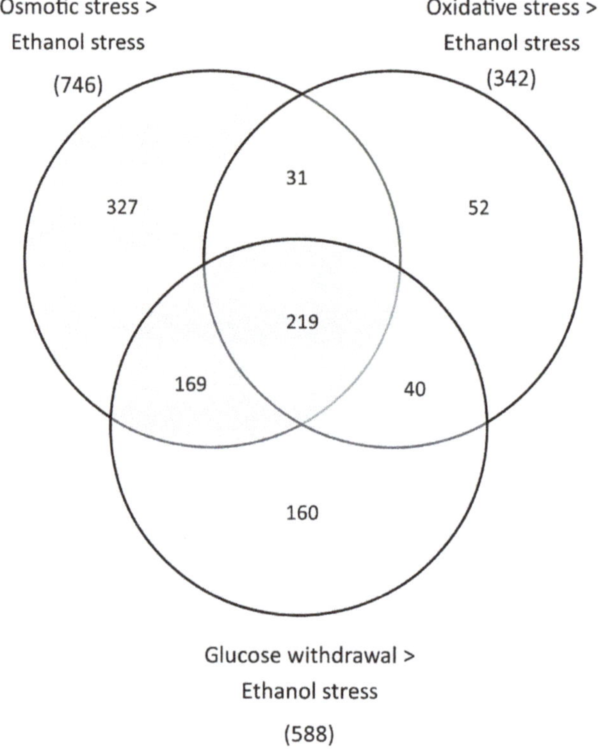

Figure 2. Venn diagram showing overlap between upregulated genes in different sequential stresses.

3.3. Gene Ontology Functional Annotation and Enrichment of Gene Cluster

The 10 most significant terms annotated in all three Gene Ontology (GO) categories are listed in Figure 3. The biological processes included oxidation–reduction (41 genes), metabolism (27 genes), cellular response to oxidative stress (17 genes), carbohydrate metabolism (15 genes), glutathione metabolism (7 genes), negative regulation of gluconeogenesis (5 genes), carbohydrate phosphorylation (5 genes), D-xylose catabolism (4 genes), arabinose catabolism (4 genes), and NADPH regeneration (4 genes). The enriched processes indicate diverse cellular activities, especially those involved in catabolism and defense against oxidative stress. The cellular components included the cytoplasm (113 genes), mitochondria (58 genes), vacuole (14 genes), peroxisome (9 genes), glucose-induced degradation (GID) complex (4 genes), peroxisomal importomer complex (4 genes), and membrane raft (4 genes). Regarding molecular functions, genes were involved in oxidoreductase activity (38 genes); transferase activity (33 genes); catalytic activity (29 genes); hydrolase activity, acting on glycosyl bonds (7 genes); aldehyde dehydrogenase (NAD) activity (6 genes); alditol: NADP+ 1-oxidoreductase activity (5 genes); oxidoreductase activity, acting on the aldehyde or oxo group of donors, NAD, or NADP as acceptor (5 genes); aldo–keto reductase (NADP) activity (4 genes); glutathione transferase activity (4 genes); and oxidoreductase activity, acting on the CH-OH group of donors, NAD, or NADP as acceptor (4 genes). The major molecular function category was oxidoreductase activity, indicating redox-balance-related gene expression alterations under sequential stresses. Molecular functions involving nicotinamide adenine dinucleotide (NAD+), an essential cofactor involved in various cellular processes, were also highlighted.

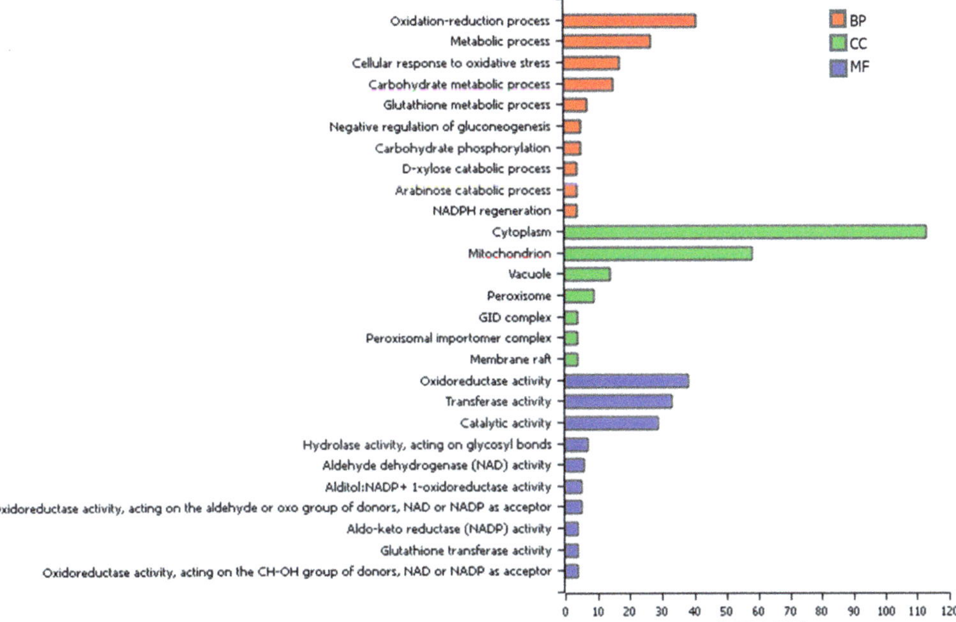

Figure 3. Gene Ontology (GO) enrichment of genes common to the sequential stress response. All the genes were annotated with regard to three categories, namely, biological process (BP), cellular component (CC), and molecular function (MF); these categories are represented on the Y-axis, and the number of genes enriched in each category is indicated on the X-axis. The figure presents the 10 GO terms from each category that contained the largest number of enriched genes.

3.4. KEGG Enrichment Analysis of DEGs

The KEGG pathway database provides molecular-level information about the interactions, reactions, and relation networks of the yeast biological system. A pathway enrichment analysis can reveal metabolic or signal transcription pathways in our gene set that may improve the understanding of yeast's response to sequential stresses. Among the 219 common genes, 79 (36%) were assigned to 18 pathways (Table 2), all of them significantly enriched. This result suggests the activation of several main pathways, including carbon metabolism, glycolysis/gluconeogenesis, the pentose phosphate pathway, amino sugar and nucleotide sugar metabolism, peroxisome, and glutathione metabolism.

Table 2. Significantly enriched KEGG pathways obtained via DAVID 6.8.

Pathways	Count	%	*p*-Value	FDR
Metabolic pathways	48	21.9	6.8×10^{-8}	1.8×10^{-6}
Biosynthesis of secondary metabolites	23	10.5	3.8×10^{-4}	3.6×10^{-3}
Biosynthesis of antibiotics	19	8.7	3.6×10^{-4}	3.6×10^{-3}
Carbon metabolism	16	7.3	6.3×10^{-6}	1.1×10^{-4}
Starch and sucrose metabolism	12	5.5	5.9×10^{-8}	1.8×10^{-6}
Glycolysis/gluconeogenesis	8	3.7	3.9×10^{-3}	1.9×10^{-2}
Peroxisome	7	3.2	1.9×10^{-3}	1.1×10^{-2}
Alanine, aspartate, and glutamate metabolism	6	2.7	3.6×10^{-3}	1.9×10^{-2}
Amino sugar and nucleotide sugar metabolism	6	2.7	4.8×10^{-3}	2.1×10^{-2}
Phenylalanine metabolism	5	2.3	4.1×10^{-4}	3.6×10^{-3}
beta-Alanine metabolism	5	2.3	8.4×10^{-4}	6.4×10^{-3}
Tyrosine metabolism	5	2.3	1.1×10^{-3}	7.6×10^{-3}
Fructose and mannose metabolism	5	2.3	5.7×10^{-3}	2.3×10^{-2}
Glutathione metabolism	5	2.3	9.3×10^{-3}	3.2×10^{-2}
Galactose metabolism	5	2.3	9.3×10^{-3}	3.2×10^{-2}
Pentose phosphate pathway	5	2.3	1.6×10^{-2}	4.7×10^{-2}
Methane metabolism	5	2.3	1.6×10^{-2}	4.7×10^{-2}
Butanoate metabolism	4	1.8	9.7×10^{-3}	3.2×10^{-2}

Sixteen genes were enriched in the carbon metabolism pathway: *SOL4*, *YJL068C*, *AGX1*, *SFA1*, *CTT1*, *CIT3*, *DAK1*, *ZWF1*, *HXK1*, *IDP3*, *MET13*, *GND2*, *GPM2*, *EMI2*, *NQM1*, and *YJL045W*. The induction of *HXK1*, a glucokinase that catalyzes the phosphorylation of glucose at C6 in the first irreversible step of glucose metabolism [22], may indicate the maintenance of internal levels of glucose. Two genes encoding a hexose transporter, HXT5 and YGL104C, were also upregulated upon the sequential stresses. The genes involved in steps 1, 2, and 3 of the oxidative phase of the pentose phosphate pathway (PPP) (*ZWF1*, *SOL4*, and *GND2*) were also overexpressed after the stresses, as well as *NQM1*, a *TAL1* paralog that arose from whole-genome duplication [23]. These results suggest a metabolic shift from glycolysis to the PPP to increase NADPH production. NADPH regeneration by the pentose phosphate shuttle is one fate of glucose in response to stressful environments, along with trehalose synthesis and glycogen storage [24], and genes from the last two processes were also recovered in our analysis.

The genes enriched in the peroxisome pathway were *PXA1*, *POX1*, *AGX1*, *CAT2*, *CTT1*, *SYM1*, and *IDP3*. This gene set attends numerous functions, such as β-oxidation (*POX1*, *PXA1*, and *IDP3*) [25], ethanol metabolism (*SYM1*) [26], and amino acid metabolism (*AGX1*) [27]. The results show stress-induced genes linked to other organelles: the Golgi apparatus (CCC2, EMP46, ATG9, TRX2, GGA1) and vacuole (BXI1, HOR7, PHM7, ATH1, ECM38, SGA1, YJL132W, YNL115C, APE1, PEP4, PRB1, SPO1, RNY1, TXR2, and others). The peroxisome and vacuole are required after oxidative damage and possibly play a role in the response to sequential stresses. The upregulation of five genes (*CUZ1*, *UBC8*, *FYV10*, *GID7*, and *VID28*) that were related to proteasome-mediated ubiquitin-dependent protein and glutathione metabolism (*GTT1*, *GRX2*, *ECM38*, *GLO1*, *ECM4*, *GTO3*, and *TRX2*) may also indicate mechanisms of protein homeostasis and redox balance that may occur in the cell upon sequential stresses.

Five genes encoding heat shock proteins (Hsps) were upregulated upon sequential stresses: *HSP12, HSP26, HSP31, HSP42*, and *SSE2*. These proteins are involved in maintaining organization and resistance during stress conditions, as cytoplasmatic, membrane, and mitochondrial chaperones [28], and they might be required by the cell machinery in the response to sequential stresses. Moreover, several genes involved in transcription regulation were upregulated: *CSR2, SIP4, USV1, XBP1, RPN4*, and *ADR1*. *ROM1*, one of the upstream regulatory factors that modulate Pkc1 activity [29], was also found to be induced in all sequential stresses.

3.5. Transcription Factors Enriched in Clustered Genes

We identified the promoter motifs in the gene list that could be involved in the TFs that coordinate the response to sequential stress. The motifs and TFs are listed in Table 3. Most of the enriched TFs, such as Msn2, Msn4, Usv1, and Gis1, were related to different stress responses [30,31]. Two of them, Usv1 and Adr1, were enriched in the motif analysis and were genes upregulated during both stresses. Accordingly, they could be crucial to a common response to sequential stresses.

Table 3. The enriched motif in the promoter's area (upstream 300 bp) from the gene list. Transcription factors (TFs) marked with (*) were both enriched in the analysis and present in the list of DEGs.

Enriched Motif in Promoter	TF	TF Family
GGGG	Adr1 (*)	C2H2 ZF
TAGGGG	Gis1	C2H2 ZF
AGGGG	Gis1	C2H2 ZF
ACCCC	Gis1	C2H2 ZF
GCGGGG	Mig3	C2H2 ZF
TAT	Mot2	RRM
AAGGGG	Msn2	C2H2 ZF
AGGGG	Msn2	C2H2 ZF
AGGG	Msn2	C2H2 ZF
AGGGG	Msn4	C2H2 ZF
AGGGG	Msn4	C2H2 ZF
AGGGG	Msn4	C2H2 ZF
TCAGGGG	Rei1	C2H2 ZF
TCAGGGG	Rei1	C2H2 ZF
TCAGGGG	Rei1	C2H2 ZF
AGGGG	Rgm1	C2H2 ZF
AGGGG	Rgm1	C2H2 ZF
TCAGGGG	Rgm1	C2H2 ZF
TTAGGGGT	Rph1	C2H2 ZF
ATTTAGGGGG	Rph1	C2H2 ZF
TTAGGGGT	Rph1	C2H2 ZF
TTAGGGGT	Rph1	C2H2 ZF
TCAGGGG	Usv1 (*)	C2H2 ZF
AGGGG	Usv1 (*)	C2H2 ZF
TCAGGGG	Usv1 (*)	C2H2 ZF
ATAGGGG	YER130C	C2H2 ZF
ATAGGGG	YER130C	C2H2 ZF
ATAGGGG	YER130C	C2H2 ZF
ATAGGGG	YER130C	C2H2 ZF
GTGGGGGG	YGR067C	C2H2 ZF

3.6. RNA-Seq Expression Validation via Quantitative Real-Time PCR

The expressions of seven genes (*HSP12, YGP1, SYM1, STF2, HPS26, SSE2*, and *HSP42*), which were shown to exhibit the behavior of interest in all sequential stresses, were measured using RT-qPCR to confirm the transcriptome results. The RT-qPCR results indicated that these genes were upregulated after glucose withdrawal followed by ethanol stress and oxidative stress followed by ethanol stress, as illustrated in Figure 4.

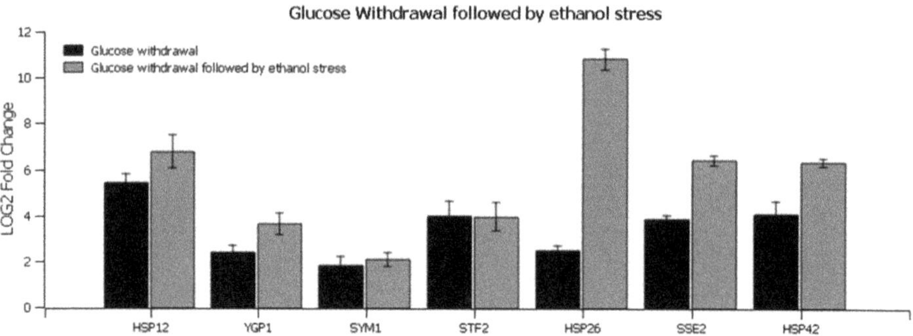

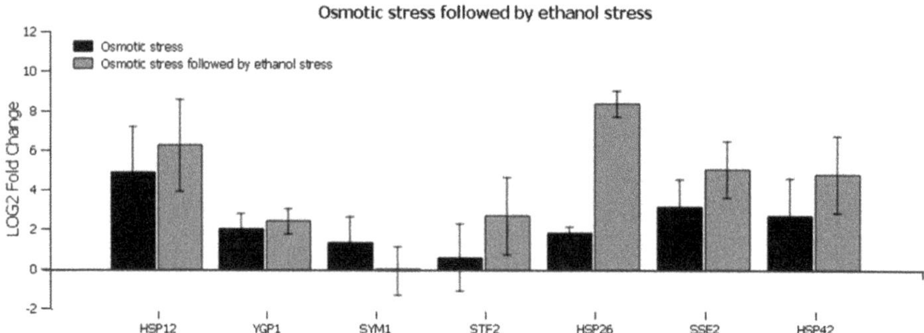

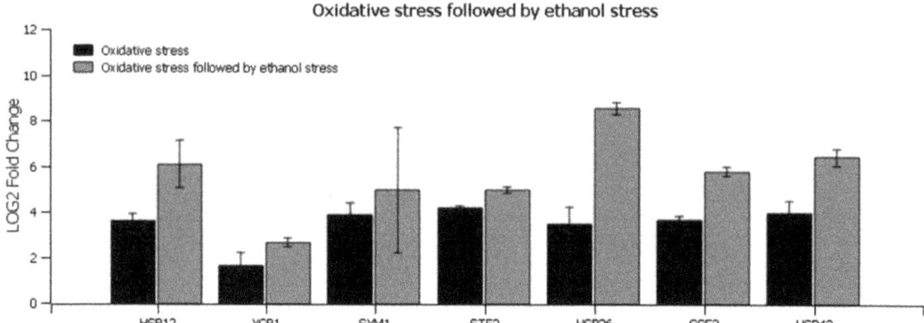

Figure 4. Validation of differentially expressed genes using quantitative real-time PCR (qPCR). Relative expressions of 7 genes were analyzed in all different sequential stresses: glucose withdrawal followed by ethanol stress, osmotic stress followed by ethanol stress, and oxidative stress followed by ethanol stress. The data are expressed as mean (SD) (n = 3).

Among the HPSs analyzed, *HSP26* displayed the highest overexpression, particularly after glucose withdrawal followed by ethanol stress, which may be associated with cellular protection and efficient refolding after sequential stresses. Similar behavior was observed for all HPSs analyzed, indicating a robust activity of this class of genes during the treatments.

YGP1, a gene that plays a role in the response to nutrient limitation and weak acids [32], also demonstrated consistent upregulation across all treatments. Conversely, five genes analyzed showed similar behavior during osmotic stress followed by ethanol stress, while *SYM1* and *STF2* exhibited more heterogeneous expression patterns. Previous studies have linked both *SYM1* and *STF2* to cell recovery and stress tolerance in the BT0510 strain, with the overexpression of *SYM1* also resulting in enhanced ethanol production [33].

4. Discussion

Our RNA-Seq analysis revealed that genes display different behaviors in response to sequential stresses with several distinct patterns. The pattern of interest in this study was that of the genes induced in both stresses regardless of the initial stress applied. We were able to identify genes with this specific pattern in all treatments, and hundreds of genes were common in all sequential stresses applied. Through this data, we were able to identify shared processes and metabolic pathways in all treatments (Figure 5).

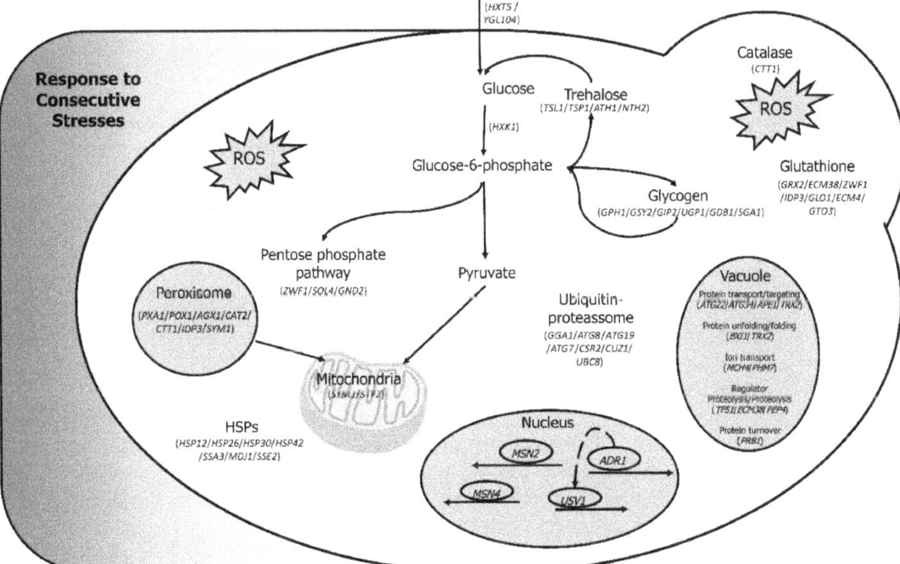

Figure 5. Schematic representation of the yeast response to sequential stresses. Upon sequential stresses, several genes (some of them represented within parenthesis) from different pathways and mechanisms were induced. The transcriptome analysis showed the induction of genes related to carbohydrate metabolism, trehalose syntheses, glycogen storage, and NADPH regeneration by the pentose phosphate pathway. Some organelles also played a role in the response, such as the peroxisome, mitochondria, and vacuole. Several genes that are associated with oxidative stress response and with chaperone functions were also upregulated. Gene motif analysis indicated several TFs coordinating the response, including Msn2 and Msn4, as well as ADRdr1 and Usv1.

A general stress response is important during the bioethanol production process since multiple stresses act together in different combinations depending on the time of the fermentation, the process conditions, and the type of biomass used [34]. Any abrupt change in the yeast cells' environment leads to a universal genomic expression program known as the environmental stress response (ESR) [5,24]. Stressful environments often occur in combinations or sequentially, especially in the fermentation process inside fermentative vats. The ESR likely operates the successive abrupt changes during sequential stresses and the transcriptional reprogramming of metabolic processes. The genes induced in the

ESR are involved in carbohydrate metabolism, metabolite transport, fatty acid metabolism, the maintenance of the cellular redox potential, the detoxification of reactive oxygen species, protein folding and degradation, DNA-damage repair, vacuolar and mitochondrial functions, and others [24]. Many genes that belong to these processes were highlighted in our analysis, as well as two TF motifs from this general response mediated by Msn2 and Msn4. We also identified two other TF motifs in the promoters of genes induced in the response, as well as the genes themselves in the list of activated genes by sequential stresses: Usv1(YPL230W) and Adr1. Both TFs affect the transcriptional regulation of genes involved in growth on non-fermentable carbon sources [30,35].

Usv1 mediates transcriptional changes in response to various stress conditions and alcoholic fermentation processes. It relocates to the nucleus in response to either gluco-neogenic conditions or the presence of salt stress and affects the transcriptional activation of several carbon- or salt-regulated genes. Usv1-mediated transcriptional activation is partially dependent on the Snf1 signaling pathway that controls the activity of several downstream transcription regulators, including Adr1. Two copies of the consensus Adr1 binding site were identified in the promoter of *USV1*, and its transcription is decreased in *adr1Δ* mutants growing in glycerol, indicating that Adr1 plays a role in *USV1* activation [30]. Adr1 encodes a transcriptional activator involved in the expression of genes that are regulated by glucose repression and is one of many genes controlling the fermentation/respiration balance [35,36]. *ADR1* deleterious mutants are unable to switch between fermentative and respiratory metabolism [37]. This gene is also related to the activation of genes involved in ethanol metabolism (e.g., *ALD4*), glycerol metabolism (e.g., *GUT2*), fatty acid utilization (e.g., *POX1* and *FOX2*), and peroxisome biogenesis [38,39]. Koerkamp et al. [40] suggested that Adr1 could act be a globally acting factor in the response to transient oxidative stress, and our transcriptome analysis also indicates a role of Adr1 along with Usv1 in the response to sequential stresses.

The signal transduction that is mediated by the single yeast isozyme of protein kinase C (Pkc1p) is also critical for maintaining the integrity of yeast cells and plays a role in the response to oxidative stress [29,41]. The upstream regulators of Pkc1p include Rom1p and Rom2p, which are GDP/GTP exchange proteins for the small GTP-binding protein Rho1p [42]. The activation of Rho1p triggers the activation of Pkc1 and ensures cellular integrity through multiple mechanisms. Firstly, it activates the MAP kinase cascade via Pkc1p, and secondly, it acts as a regulatory subunit of the 1,3-β-glucan synthase complex involved in cell wall biosynthesis [29]. While previous research has suggested that Rom2p plays a major role in signaling [29], our results indicate that *ROM1* rather than *ROM2* may be involved in this response, as all sequential stresses led to the upregulation of *ROM1*. A cellular process recurrent in our results was the modulation of glucose metabolism, and this modulation is largely studied upon yeast stress. The regulated genes in this response include genes with different functions ranging from intracellular glucose transporters to subsequent catabolism [24]. We noticed the modulation of many genes responsible for the phosphorylation of glucose and the conversion from glucose-1-phosphate to glucose-6-phosphate, as well as genes related to trehalose synthesis, glycogen storage, and NADPH regeneration by the pentose phosphate shuttle (Figure 5).

We observed that sequential stresses promoted the redirection of metabolic flux from glycolysis to the pentose phosphate pathway through the induction of three genes (*ZWF1*, *SOL4*, and *GND2*) that catalyze the first steps of the pentose phosphate pathway. This adjustment was previously described as a response to oxidative stress through the increased production of reducing power for cellular redox systems [43]. The ability to maintain intracellular redox homeostasis is important and is directly associated with stress tolerance. Kitichantaropas et al. [44] showed that, in response to simultaneous multi-stresses mimicking the fermentation process, redox homeostasis appeared to be the primary mechanism required for cell protection, and multiple stress-tolerant strains may have the ability to minimize intracellular ROS levels. Previous studies by Bravim et al. [45] showed that the gene expression pattern in response to high hydrostatic pressure (HHP) displays an oxidative stress response profile in the BT0510 strain and that the overexpression of *STF2*, a gene whose upregulation prevents cellular ROS accumulation [46], led to increased tolerance to HHP stress. *STF2*, as well as *PRX1* and *GND2*, which play a role in the oxidative response [24], was recovered in our gene set. The oxidative damage response and the increase in cellular redox systems could be important to the yeast response to sequential stresses.

The NADPH regenerated by the PPP is an essential cofactor for glutathione- and thioredoxin-dependent enzymes that protect cells against oxidative damage [47]. These proteins are located in the yeast peroxisomes. Peroxisomes are found in most eukaryotic cells and are important in cell metabolism, especially in different catabolic processes, including fatty acid β-oxidation, the glyoxylic shunt, and methanol metabolism, as well as in osmotic stress tolerance [48,49]. Our analysis identified upregulated genes throughout the β-oxidation pathway, including *PXA1* encoding the ATP-binding cassette transporter that transports long-chain fatty acids to the peroxisome; *POX1* responsible for the first step of B-oxidation; and *CAT2* required for transporting acetyl-CoA generated from β-oxidation into the mitochondria, where it enters the TCA cycle [50]. Genes from this pathway are induced through Adr1 triggered by Hog1 MAP kinase upon salt stress, demonstrating that yeast adaptation to environmental stresses involves the dynamic modulation of peroxisomal activity to reshape cellular energy metabolism [51]. Peroxisomal proteins might also play a role in cell survival through the spatial regulation of the redox potential [52]. Our RNA-Seq analysis revealed that the genes related to these organelles were highly responsive to sequential stresses, and the modulation of peroxisomal genes and activities could be promising targets in the biotechnological enhancement of yeast that could lead to multi-stress tolerance.

SYM1, also highlighted in our analysis, could provide another strategy for cellular defense and a target to enhance ethanol fermentation. Even though our analysis showed its heterogeneous expression in response to osmotic stress, the same behavior was not presented in the other sequential stresses. Bravim et al. [33] showed that *SYM1* over-expression enhanced the fermentative capacity of BT0510. Sym1 is homologous to the mammalian peroxisomal membrane protein Mpv17 and was found to be required for both the metabolism and tolerance of ethanol during heat shock. In contrast to the peroxisomal localization of Mpv17p, Sym1 is an integral membrane protein of the inner mitochondrial membrane and was suggested to play an important metabolic role in mitochondrial function during heat shock [26]. We propose that *SYM1* may be part of a global response affecting cellular adaptation, not only in response to heat shock but also to other fermentative stresses.

The vacuole is another organelle enhanced after cell exposure to environmental stresses. A fundamental role of the yeast vacuole is the recycling of biological macromolecules. Proteins, small molecules, and even organelles are degraded and then recycled in the vacuoles [24]. A previous electron microscopic analysis showed that this organelle changes morphology after chronic heat stress, with an accelerated invagination of vacuoles as a stress response, perhaps due to the loss of cell surface proteins as the temperature increases [53]. Our analysis implicates a role of the vacuole in the sequential stress response, with genes associated with protein degradation, protein turnover after oxidative damage, and cytosol-to-vacuolar targeting (CVT) upregulated in all treatments. Gene set enrichment indicates the induction of cellular machinery for protein degradation in both the vacuole and glucose-induced degradation complex. A genome analysis of BT0510 has shown a loss of genes linked to vesicle formation and COPII binding that could alter intracellular transport [14], but we demonstrated that, in the face of sequential stresses, this yeast strain regulates other genes that could mitigate this loss, such as *EMP46* and *ATG9*, both present in our gene set.

Genes encoding protein folding chaperones are also induced by a variety of stressful conditions [24]. *HPS12* encodes a heat shock protein highly expressed after stress and a well-established reporter for the ESR pathway [24,54]. *HSP12* overexpression provided resistance to both freezing storage and heat shock, as well as an increase in intracellular levels of glutathione peroxidase and reductase activity [55,56]. Other genes from the *HSP* family that were also in the gene set analyzed included *HSP31*, *HSP42*, *HSP78*, and *SSE2*, which influence yeast tolerance to different stresses, such as ethanol and high hydrostatic pressure [57]. These genes encode proteins that have varied functions, although most of them operate as chaperones, inhibiting protein aggregation [58,59]. Kitichantaropas et al. [44] suggested that the continuous expression of some *HSP* genes during multi-stress treatments may be involved in the acquisition of the thermotolerant phenotype. They showed that thermotolerant *S. cerevisiae* strains maintained *HSP* gene expression even after long-term exposure to heat stress, suggesting that the continuous expression of these genes may contribute to thermotolerance. Yeasts may face a series of physiological changes upon sequential stresses, and a way to adapt to these stresses could be through protein stabilization by the *HSP* gene family.

5. Conclusions

In this study, an RNA-Seq analysis provided a global overview of the response to sequential stresses in a tolerant strain, presenting new insights into how cells coordinate their metabolic pathways during fermentative stresses. Processes involving NADPH regeneration, peroxisome activity, and genes (such as *SYM1*, *STF2*, and *HSPs*) could be essential for tolerance acquisition, and the genes highlighted in our results could be used to construct more tolerant strains. These responsive genes were shared in all treatments, indicating a common defense strategy to different sequential stresses, an interesting attribute in biotechnological processes involving yeast cells and fermentation. This analysis can improve all fermentation-based processes and enhance the understanding of the response to stress in higher eukaryotes, as *S. cerevisiae* is a recognized study model.

Supplementary Materials: The following supporting information can be downloaded at: https://www.mdpi.com/article/10.3390/fermentation9020195/s1, Figure S1. Principal component analysis (PCA) of RNA-Seq data.

Author Contributions: A.C.T.C., P.M.B.F. and J.R.B. conceived and designed the study. A.C.T.C. and M.R. conducted the experiments. All authors analyzed and discussed the results. A.C.T.C., A.A.R.F., J.R.B. and P.M.B.F. wrote the manuscript. All authors have read and agreed to the published version of the manuscript.

Funding: This research was funded by an NIH grant GM076562 to J.R.B. P.M.B.F. was funded by CNPq and FAPES (research productivity award # 308306/2021 and 269) and A.C.T.C. by CAPES (scholarship # 88882.385022/2009-01 and 88881.187871/2018-01).

Institutional Review Board Statement: Not applicable.

Informed Consent Statement: Not applicable.

Data Availability Statement: The RNA sequencing data obtained in this study are available at NCBI Gene Expression Omnibus (GEO, http://www.ncbi.nlm.nih.gov/geo/ accessed on 8 February 2023) under accession number GSE221570. The yeast strain used in this study (URM 6670) is stored at the Federal University of Pernambuco Culture Collection (URM604).

Acknowledgments: The authors would like to thank Lijun Zhang and Lisa Schneper for assistance with data management and storage.

Conflicts of Interest: The authors declare no conflict of interest.

References

1. Martínez-Espinosa, R.M. Introductory Chapter: A Brief Overview on Fermentation and Challenges for the Next Future. In *New Advances on Fermentation Processes*; Martínez-Espinosa, R.M., Ed.; IntechOpen: Rijeka, Croatia, 2020; p. Ch. 1.
2. Gibson, B.R.; Lawrence, S.J.; Leclaire, J.P.; Powell, C.D.; Smart, K.A. Yeast responses to stresses associated with industrial brewery handling. *FEMS Microbiol. Rev.* **2007**, *31*, 535–569. [CrossRef] [PubMed]
3. Beales, N. Adaptation of Microorganisms to Cold Temperatures, Weak Acid Preservatives, Low pH, and Osmotic Stress: A Review. *Compr Rev. Food Sci. Food Saf.* **2004**, *3*, 1–20. [CrossRef] [PubMed]

4. Attfield, P.V. Stress tolerance: The key to effective strains of industrial baker's yeast. *Nat. Biotechnol.* **1997**, *15*, 1351–1357. [CrossRef] [PubMed]
5. Gasch, A.P.; Spellman, P.T.; Kao, C.M.; Carmel-Harel, O.; Eisen, M.B.; Storz, G.; Botstein, D.; Brown, P.O. Genomic expression programs in the response of yeast cells to environmental changes. *Mol. Biol. Cell* **2000**, *11*, 4241–4257. [CrossRef]
6. Guan, Q.; Haroon, S.; Bravo, D.G.; Will, J.L.; Gasch, A.P. Cellular memory of acquired stress resistance in Saccharomyces cerevisiae. *Genetics* **2012**, *192*, 495–505. [CrossRef]
7. Gutin, J.; Joseph-Strauss, D.; Sadeh, A.; Shalom, E.; Friedman, N. Genetic screen of the yeast environmental stress response dynamics uncovers distinct regulatory phases. *Mol Syst Biol* **2019**, *15*, e8939. [CrossRef]
8. Ikner, A.; Shiozaki, K. Yeast signaling pathways in the oxidative stress response. *Mutat. Res.* **2005**, *569*, 13–27. [CrossRef]
9. Yang, J.; Tavazoie, S. Regulatory and evolutionary adaptation of yeast to acute lethal ethanol stress. *PLoS ONE* **2020**, *15*, e0239528. [CrossRef]
10. Berry, D.B.; Guan, Q.; Hose, J.; Haroon, S.; Gebbia, M.; Heisler, L.E.; Nislow, C.; Giaever, G.; Gasch, A.P. Multiple means to the same end: The genetic basis of acquired stress resistance in yeast. *PLoS Genet* **2011**, *7*, e1002353. [CrossRef]
11. Meriem, Z.B.; Khalil, Y.; Hersen, P.; Fabre, E. Hyperosmotic Stress Response Memory is Modulated by Gene Positioning in Yeast. *Cells* **2019**, *8*, 582. [CrossRef]
12. Taymaz-Nikerel, H.; Cankorur-Cetinkaya, A.; Kirdar, B. Genome-Wide Transcriptional Response of Saccharomyces cerevisiae to Stress-Induced Perturbations. *Front. Bioeng. Biotechnol.* **2016**, *4*, 17. [CrossRef] [PubMed]
13. Bravim, F.; Palhano, F.L.; Fernandes, A.A.; Fernandes, P.M. Biotechnological properties of distillery and laboratory yeasts in response to industrial stresses. *J. Ind Microbiol. Biotechnol.* **2010**, *37*, 1071–1079. [CrossRef] [PubMed]
14. Costa, A.C.T.; Hornick, J.; Antunes, T.F.S.; Santos, A.M.C.; Fernandes, A.A.R.; Broach, J.R.; Fernandes, P.M.B. Complete genome sequence and analysis of a Saccharomyces cerevisiae strain used for sugarcane spirit production. *Braz. J. Microbiol.* **2021**, *52*, 1087–1095. [CrossRef] [PubMed]
15. Bravim, F.; da Silva, L.F.; Souza, D.T.; Lippman, S.I.; Broach, J.R.; Fernandes, A.A.; Fernandes, P.M. High hydrostatic pressure activates transcription factors involved in Saccharomyces cerevisiae stress tolerance. *Curr. Pharm. Biotechnol.* **2012**, *13*, 2712–2720. [CrossRef]
16. Chasman, D.; Ho, Y.H.; Berry, D.B.; Nemec, C.M.; MacGilvray, M.E.; Hose, J.; Merrill, A.E.; Lee, M.V.; Will, J.L.; Coon, J.J.; et al. Pathway connectivity and signaling coordination in the yeast stress-activated signaling network. *Mol. Syst. Biol.* **2014**, *10*, 759. [CrossRef]
17. Trapnell, C.; Williams, B.A.; Pertea, G.; Mortazavi, A.; Kwan, G.; van Baren, M.J.; Salzberg, S.L.; Wold, B.J.; Pachter, L. Transcript assembly and quantification by RNA-Seq reveals unannotated transcripts and isoform switching during cell differentiation. *Nat. Biotechnol.* **2010**, *28*, 511–515. [CrossRef]
18. Ge, S.X.; Son, E.W.; Yao, R. iDEP: An integrated web application for differential expression and pathway analysis of RNA-Seq data. *BMC Bioinform.* **2018**, *19*, 534. [CrossRef]
19. Huang, D.W.; Sherman, B.T.; Lempicki, R.A. Systematic and integrative analysis of large gene lists using DAVID bioinformatics resources. *Nature Protoc.* **2009**, *4*, 44–57. [CrossRef]
20. Ge, S.X.; Jung, D.; Yao, R. ShinyGO: A graphical gene-set enrichment tool for animals and plants. *Bioinformatics* **2019**, *36*, 2628–2629. [CrossRef]
21. Teste, M.A.; Duquenne, M.; Francois, J.M.; Parrou, J.L. Validation of reference genes for quantitative expression analysis by real-time RT-PCR in Saccharomyces cerevisiae. *BMC Mol. Biol.* **2009**, *10*, 99. [CrossRef]
22. Bisson, L.F.; Fraenkel, D.G. Involvement of kinases in glucose and fructose uptake by Saccharomyces cerevisiae. *Proc. Natl. Acad. Sci. USA* **1983**, *80*, 1730–1734. [CrossRef] [PubMed]
23. Huang, H.; Rong, H.; Li, X.; Tong, S.; Zhu, Z.; Niu, L.; Teng, M. The crystal structure and identification of NQM1/YGR043C, a transaldolase from Saccharomyces cerevisiae. *Proteins* **2008**, *73*, 1076–1081. [CrossRef] [PubMed]
24. Gasch, A.P. The environmental stress response: A common yeast response to diverse environmental stresses. In *Yeast Stress Responses*; Hohmann, S., Mager, W.H., Eds.; Springer: Berlin/Heidelberg, Germany, 2003; pp. 11–70.
25. Hiltunen, J.K.; Mursula, A.M.; Rottensteiner, H.; Wierenga, R.K.; Kastaniotis, A.J.; Gurvitz, A. The biochemistry of peroxisomal beta-oxidation in the yeast Saccharomyces cerevisiae. *FEMS Microbiol. Rev.* **2003**, *27*, 35–64. [CrossRef] [PubMed]
26. Trott, A.; Morano, K.A. SYM1 is the stress-induced Saccharomyces cerevisiae ortholog of the mammalian kidney disease gene Mpv17 and is required for ethanol metabolism and tolerance during heat shock. *Eukaryot. Cell* **2004**, *3*, 620–631. [CrossRef]
27. Schlösser, T.; Gätgens, C.; Weber, U.; Stahmann, K.-P. Alanine: Glyoxylate aminotransferase of Saccharomyces cerevisiae–encoding gene AGX1 and metabolic significance. *Yeast* **2004**, *21*, 63–73. [CrossRef]
28. Nash, R.; Weng, S.; Hitz, B.; Balakrishnan, R.; Christie, K.R.; Costanzo, M.C.; Dwight, S.S.; Engel, S.R.; Fisk, D.G.; Hirschman, J.E.; et al. Expanded protein information at SGD: New pages and proteome browser. *Nucleic. Acids Res.* **2007**, *35*, D468–D471. [CrossRef]
29. Heinisch, J.J.; Lorberg, A.; Schmitz, H.P.; Jacoby, J.J. The protein kinase C-mediated MAP kinase pathway involved in the maintenance of cellular integrity in Saccharomyces cerevisiae. *Mol. Microbiol.* **1999**, *32*, 671–680. [CrossRef]
30. Hlynialuk, C.; Schierholtz, R.; Vernooy, A.; van der Merwe, G. Nsf1/Ypl230w participates in transcriptional activation during non-fermentative growth and in response to salt stress in Saccharomyces cerevisiae. *Microbiology* **2008**, *154*, 2482–2491. [CrossRef]

31. Zhang, N.; Wu, J.; Oliver, S.G. Gis1 is required for transcriptional reprogramming of carbon metabolism and the stress response during transition into stationary phase in yeast. *Microbiology* **2009**, *155*, 1690–1698. [CrossRef]

32. Fernandes, A.R.; Mira, N.P.; Vargas, R.C.; Canelhas, I.; Sa-Correia, I. Saccharomyces cerevisiae adaptation to weak acids involves the transcription factor Haa1p and Haa1p-regulated genes. *Biochem. Biophys. Res. Commun.* **2005**, *337*, 95–103. [CrossRef]

33. Bravim, F.; Lippman, S.I.; da Silva, L.F.; Souza, D.T.; Fernandes, A.A.; Masuda, C.A.; Broach, J.R.; Fernandes, P.M. High hydrostatic pressure activates gene expression that leads to ethanol production enhancement in a Saccharomyces cerevisiae distillery strain. *Appl. Microbiol. Biotechnol.* **2013**, *97*, 2093–2107. [CrossRef] [PubMed]

34. Deparis, Q.; Claes, A.; Foulquie-Moreno, M.R.; Thevelein, J.M. Engineering tolerance to industrially relevant stress factors in yeast cell factories. *FEMS Yeast Res.* **2017**, *17*, fox036. [CrossRef] [PubMed]

35. Young, E.T.; Dombek, K.M.; Tachibana, C.; Ideker, T. Multiple pathways are co-regulated by the protein kinase Snf1 and the transcription factors Adr1 and Cat8. *J. Biol. Chem.* **2003**, *278*, 26146–26158. [CrossRef] [PubMed]

36. Peltier, E.; Friedrich, A.; Schacherer, J.; Marullo, P. Quantitative Trait Nucleotides Impacting the Technological Performances of Industrial Saccharomyces cerevisiae Strains. *Front. Genet.* **2019**, *10*, 683. [CrossRef] [PubMed]

37. Denis, C.L.; Young, E.T. Isolation and characterization of the positive regulatory gene ADR1 from Saccharomyces cerevisiae. *Mol. Cell Biol.* **1983**, *3*, 360–370. [CrossRef] [PubMed]

38. Simon, M.; Adam, G.; Rapatz, W.; Spevak, W.; Ruis, H. The Saccharomyces cerevisiae ADR1 gene is a positive regulator of transcription of genes encoding peroxisomal proteins. *Mol. Cell Biol.* **1991**, *11*, 699–704. [CrossRef]

39. Tachibana, C.; Yoo, J.Y.; Tagne, J.B.; Kacherovsky, N.; Lee, T.I.; Young, E.T. Combined global localization analysis and transcriptome data identify genes that are directly coregulated by Adr1 and Cat8. *Mol. Cell Biol.* **2005**, *25*, 2138–2146. [CrossRef]

40. Koerkamp, M.G.; Rep, M.; Bussemaker, H.J.; Hardy, G.P.; Mul, A.; Piekarska, K.; Szigyarto, C.A.; De Mattos, J.M.; Tabak, H.F. Dissection of transient oxidative stress response in Saccharomyces cerevisiae by using DNA microarrays. *Mol. Biol. Cell* **2002**, *13*, 2783–2794. [CrossRef]

41. de la Torre-Ruiz, M.A.; Serrano, L.; Petkova, M.I.; Pujol-Carrion, N. Signalling Oxidative Stress in Saccharomyces cerevisiae. In *Oxidative Stress*; Volodymyr, L., Halyna, M.S., Eds.; IntechOpen: Rijeka, Croatia, 2012.

42. Ozaki, K.; Tanaka, K.; Imamura, H.; Hihara, T.; Kameyama, T.; Nonaka, H.; Hirano, H.; Matsuura, Y.; Takai, Y. Rom1p and Rom2p are GDP/GTP exchange proteins (GEPs) for the Rho1p small GTP binding protein in Saccharomyces cerevisiae. *EMBO J.* **1996**, *15*, 2196–2207. [CrossRef]

43. Grant, C.M. Metabolic reconfiguration is a regulated response to oxidative stress. *J. Biol.* **2008**, *7*, 1. [CrossRef]

44. Kitichantaropas, Y.; Boonchird, C.; Sugiyama, M.; Kaneko, Y.; Harashima, S.; Auesukaree, C. Cellular mechanisms contributing to multiple stress tolerance in Saccharomyces cerevisiae strains with potential use in high-temperature ethanol fermentation. *AMB Express* **2016**, *6*, 107. [CrossRef] [PubMed]

45. Bravim, F.; Mota, M.M.; Fernandes, A.A.; Fernandes, P.M. High hydrostatic pressure leads to free radicals accumulation in yeast cells triggering oxidative stress. *FEMS Yeast Res* **2016**, *16*, fow052. [CrossRef] [PubMed]

46. Lopez-Martinez, G.; Rodriguez-Porrata, B.; Margalef-Catala, M.; Cordero-Otero, R. The STF2p hydrophilin from Saccharomyces cerevisiae is required for dehydration stress tolerance. *PLoS ONE* **2012**, *7*, e33324. [CrossRef] [PubMed]

47. Toledano, M.B.; Delaunay, A.; Biteau, B.; Spector, D.; Azevedo, D. Oxidative stress responses in yeast. In *Yeast Stress Responses*; Hohmann, S., Mager, W.H., Eds.; Springer: Berlin/Heidelberg, Germany, 2003; pp. 241–303.

48. Lopez-Huertas, E.; Charlton, W.L.; Johnson, B.; Graham, I.A.; Baker, A. Stress induces peroxisome biogenesis genes. *EMBO J.* **2000**, *19*, 6770–6777. [CrossRef]

49. Sibirny, A.A. Yeast peroxisomes: Structure, functions and biotechnological opportunities. *FEMS Yeast Res* **2016**, *16*, fow038. [CrossRef]

50. Tang, X.; Lee, J.; Chen, W.N. Engineering the fatty acid metabolic pathway in Saccharomyces cerevisiae for advanced biofuel production. *Metab. Eng. Commun.* **2015**, *2*, 58–66. [CrossRef]

51. Manzanares-Estreder, S.; Espi-Bardisa, J.; Alarcon, B.; Pascual-Ahuir, A.; Proft, M. Multilayered control of peroxisomal activity upon salt stress in Saccharomyces cerevisiae. *Mol. Microbiol.* **2017**, *104*, 851–868. [CrossRef]

52. Jung, S.; Marelli, M.; Rachubinski, R.A.; Goodlett, D.R.; Aitchison, J.D. Dynamic changes in the subcellular distribution of Gpd1p in response to cell stress. *J. Biol. Chem.* **2010**, *285*, 6739–6749. [CrossRef]

53. Ishii, A.; Kawai, M.; Noda, H.; Kato, H.; Takeda, K.; Asakawa, K.; Ichikawa, Y.; Sasanami, T.; Tanaka, K.; Kimura, Y. Accelerated invagination of vacuoles as a stress response in chronically heat-stressed yeasts. *Sci. Rep.* **2018**, *8*, 2644. [CrossRef]

54. Gutin, J.; Sadeh, A.; Rahat, A.; Aharoni, A.; Friedman, N. Condition-specific genetic interaction maps reveal crosstalk between the cAMP/PKA and the HOG MAPK pathways in the activation of the general stress response. *Mol. Syst. Biol.* **2015**, *11*, 829. [CrossRef]

55. Fierro-Risco, J.; Rincon, A.M.; Benitez, T.; Codon, A.C. Overexpression of stress-related genes enhances cell viability and velum formation in Sherry wine yeasts. *Appl. Microbiol. Biotechnol.* **2013**, *97*, 6867–6881. [CrossRef] [PubMed]

56. Pacheco, A.; Pereira, C.; Almeida, M.J.; Sousa, M.J. Small heat-shock protein Hsp12 contributes to yeast tolerance to freezing stress. *Microbiology* **2009**, *155*, 2021–2028. [CrossRef] [PubMed]

57. Stanley, D.; Bandara, A.; Fraser, S.; Chambers, P.J.; Stanley, G.A. The ethanol stress response and ethanol tolerance of Saccharomyces cerevisiae. *J. Appl. Microbiol.* **2010**, *109*, 13–24. [CrossRef] [PubMed]

58. Haslbeck, M.; Braun, N.; Stromer, T.; Richter, B.; Model, N.; Weinkauf, S.; Buchner, J. Hsp42 is the general small heat shock protein in the cytosol of Saccharomyces cerevisiae. *EMBO J.* **2004**, *23*, 638–649. [CrossRef]
59. von Janowsky, B.; Major, T.; Knapp, K.; Voos, W. The Disaggregation Activity of the Mitochondrial ClpB Homolog Hsp78 Maintains Hsp70 Function during Heat Stress. *J. Mol. Biol.* **2006**, *357*, 793–807. [CrossRef]

 *fermentation*

Article

Bioconversion of a Lignocellulosic Hydrolysate to Single Cell Oil for Biofuel Production in a Cost-Efficient Fermentation Process

Zora S. Rerop [†], Nikolaus I. Stellner [†], Petra Graban, Martina Haack, Norbert Mehlmer, Mahmoud Masri * and Thomas B. Brück *

Werner Siemens-Chair of Synthetic Biotechnology, School of Natural Sciences, Technical University of Munich, Lichtenbergstraße 4, 85748 Garching bei München, Germany
* Correspondence: mahmoud.masri@tum.de (M.M.); brueck@tum.de (T.B.B.)
† These authors contributed equally to this work.

Abstract: *Cutaneotrichosporon oleaginosus* is a highly efficient single cell oil producer, which in addition to hexoses and pentoses can metabolize organic acids. In this study, fed-batch cultivation with consumption-based acetic acid feeding was further developed to integrate the transformation of an industrial paper mill lignocellulosic hydrolysate (LCH) into yeast oil. Employing pentose-rich LCH as a carbon source instead of glucose significantly improved both biomass formation and lipid titer, reaching 55.73 ± 5.20 g/L and 42.1 ± 1.7 g/L (75.5% lipid per biomass), respectively. This hybrid approach of using acetic acid and LCH in one process was further optimized to increase the share of bioavailable carbon from LCH using a combination of consumption-based and continuous feeding. Finally, the techno-economic analysis revealed a 26% cost reduction when using LCH instead of commercial glucose. In summary, we developed a process leading to a holistic approach to valorizing a pentose-rich industrial waste by converting it into oleochemicals.

Keywords: fermentation; biotransformations; lipids; single cell oil; biofuel; techno-economic analysis; waste valorization; lignocellulosic hydrolysate; oleaginous yeast

check for updates

Citation: Rerop, Z.S.; Stellner, N.I.; Graban, P.; Haack, M.; Mehlmer, N.; Masri, M.; Brück, T.B. Bioconversion of a Lignocellulosic Hydrolysate to Single Cell Oil for Biofuel Production in a Cost-Efficient Fermentation Process. *Fermentation* 2023, 9, 189. https://doi.org/10.3390/fermentation9020189

Academic Editor: Timothy Tse

Received: 6 February 2023
Revised: 15 February 2023
Accepted: 16 February 2023
Published: 18 February 2023

1. Introduction

The global energy system is responsible for more than 75% of greenhouse gas (GHG) emissions since it depends primarily on fossil fuels [1]. These emissions accelerate global warming, which impacts our environment, threatens our biodiversity, increases the frequency and intensity of extreme weather events, and damages our economy [2]. In response to these challenges, the UN calls for sustainable and innovative short- and long-term strategies to diversify energy sourcing, accelerate the clean energy transition, and achieve net-zero emissions by 2050 [3,4]. This energy transition demands large-scale deployment of renewable energy sources, such as sustainable advanced biofuels. Furthermore, integrating the up-cycling and end-of-life recycling of urban, agricultural, and industrial wastes into supply chains is a value-adding strategy directed toward the implementation within a truly circular economy. According to the EU's long-term measurement, combining all these approaches can reduce net emissions by around 80% [5]. Such sustainable energy concepts would also reduce dependence on limited resources and disrupted supply chains, the latter of which could be observed during the COVID-19 pandemic. Current biofuel production faces climatic, environmental, and social problems. It relies on globally traded feedstock, such as plant oil or used cooking oil (UCO) [6,7]. The large-scale production of plant oil-based biofuel results in a significant increase in problematic monocultures, increased land use, deforestation, and food competition [6,8–10]. In this respect, biofuel production from domestic biomass waste streams could help lower the external energy dependence of communities, reduce greenhouse emissions, provide energy storage capability, and implement an efficient route to a circular treatment of waste [8,11].

Oleaginous microorganisms such as the yeast *Cutaneotrichosporon oleaginosus* (ATCC 20509) can utilize these waste streams to produce single cell oil (SCO) as a starting material for advanced biofuel production. As opposed to growing crops for plant oil production, the cultivation of microorganisms is seasonally independent, and biotechnological processes allow for high area efficiency due to vertical scale-up [12,13]. *C. oleaginosus* is one of the most promising oleaginous microorganisms for the production of SCOs as it has been shown to utilize monomers from cellulose, chitin, lignin, and hemicellulose, the most abundant biopolymers on earth [14–16]. The oleaginous yeast displays an excellent ability to grow to very high cell density and to accumulate plant-oil-like lipids with up to 85% of the dry cell weight [8,17]. Moreover, *C. oleaginosus* has a unique ability to simultaneously uptake hexose and pentose sugars, as well as acetic acid, and not follow a typical diauxic cell growth [17]. Furthermore, it shows a high tolerance to inhibitory compounds, such as furans or lignin-derived compounds, such as coumarat and resorcinol, which are major components of depolymerized lignin [15,18,19]. When employed as a high proportion of the carbon source, acetic acid was reported as suitable for *C. oleaginosus*, resulting in high lipid contents of up to 73% but also low biomass generation of only up to 6 g/L [20,21]. Recently, a fermentation process utilizing acetic acid for the cost-efficient production of SCOs with *C. oleaginosus* at high yields was reported [17]. As opposed to other approaches, pure acetic acid was fed to the culture in a consumption-based manner after adding glucose as a starting sugar. To date, this is the process with the highest outcome of SCO, integrating acetic acid as feedstock and using *C. oleaginosus*. Overall, this unconventional yeast is a promising host for the industrial production of SCO as it is efficient, robust, and capable of metabolizing a diverse range of substrates.

The pulp and paper industry is one of the major producers of waste streams, with a high share of biodegradable carbon [22]. The global production of pulp and paper was 188.9 million metric tons in 2020 [23]. Considering that only 30–40% of wood biomass is recovered as cellulose fibers, huge amounts of residual waste are produced. In Europe alone, the pulp and paper industry produces around 11 million tons of waste every year [24]. One of the main processes of pulp production, besides alkaline kraft pulping, is acidic sulfite pulping combined with steam explosion to hydrolyze the plant cells and lignocellulose compounds and separate the valuable cellulose fibers [25]. The pulping waste stream from this process contains high amounts of sugars from hemicellulose (mainly xylose), breakdown compounds from lignin (lignols and lignans), aliphatic carboxylic acids (mainly acetic acid), and furans (mainly furfural and hydroxymethylfurfural) [26,27]. In most industrial plants, the waste stream is heavily concentrated, and the acetic acid therein is evaporated and separated from the rest [28]. To the current date, the released acetic acid is only collected in exceptional cases, such as by the Lenzing AG in Austria [29]. So far, direct combustion for energy production or passing the material into the wastewater is the common strategy to deal with the highly toxic furans, acetic acid, and lignin-derived compounds [30–33]. For these waste streams, specifically for lignin, xylose, and acetic acid, there are no established recovery or valorization systems. However, different lignocellulosic hydrolysates from agriculture residues, such as wheat straw, corn stover, or switch grass, have been applied to produce bioethanol with bacteria or yeasts [34]. Corn steep liquor has been used as a feedstock for lipid production with genetically modified *Rhodosporidium toruloides*, reaching a top-level lipid titer of 39.5 g/L (0.179 g/g yield per sugar) [35]. Moreover, some pilot and demonstration plants have used hard and soft wood chips to produce second-generation bioethanol [36]. Nonetheless, it is important to note that all these examples utilized a lignocellulosic feedstock with a focus on its glucose content, not the pentose sugars or the lignin breakdown products.

For the first time, this study provides an approach for the up-cycling of a crude lignocellulosic hydrolysate, focusing mainly on the valorization of xylose, acetic acid, and lignin-derived compounds. To the best of our knowledge, the fermentative utilization of a lignocellulosic-rich waste stream directly from a pulping mill and without most of the cellulose and its monomers has never been reported. From this hydrolysate, plant oil-like SCO was produced as a potential domestic feedstock for biofuel or other industrial applications. The current work includes a comprehensive analysis of the hydrolysate, the development of a cost-efficient pretreatment step, and a highly efficient fermentation strategy with the oleaginous yeast *C. oleaginous*. Moreover, the economic viability of the developed process was assessed by a comprehensive techno-economic analysis (TEA).

2. Materials and Methods

2.1. Analysis of Lignocellulosic Hydrolysate

2.1.1. Sugar Analysis

The sugars and short organic acids were analyzed with high-performance liquid chromatography (HPLC). All the samples were filtered with a 10 kDa filter. The Agilent 1260 Infinity II LC system with Diode Array (DA) and Refractive Index (RI) detectors was used. For separation, a column Rezex ROA-organic H+ 8% from Phenomenex was used with a mobile phase of 5 mM H_2SO_4. An isocratic flow of 0.5 mL/min was applied over 60 min with an oven temperature of 70 °C. The detection in the RID was carried out at 40 °C.

2.1.2. Ash

A neutralized and lyophilized lignocellulose hydrolysate of 2 to 4 g was burned at 1000 °C for 3 h to ash. The amount was determined gravimetrically after cooling overnight in a desiccator.

2.1.3. Elemental Analysis

Elemental analysis was carried out with a Euro EA CHNS elemental analyzer (HEKAtech Ltd., Wegberg, Germany). Dynamic spontaneous combustion in a tin boat at approximately 1800 °C was performed with subsequent gas chromatographic separation and was detected using a thermal conductivity detector (TCD).

2.1.4. Quantification of Sulfates

The quantification of sulfates in the hydrolysate was performed chemically with treatment with $CaCO_3$ and $BaCl_2$. The resulting $BaSO_4$ precipitate was quantified gravimetrically.

2.2. Strain and Media Composition

2.2.1. Strain

The oleaginous yeast *Cutaneotrichosporon oleaginosus* ATCC 20509 (DSM-11815), supplied by the Deutsche Sammlung von Mikroorganismen und Zellkultur (DMSZ, Braunschweig, Germany), was used for all cultivation and fermentation experiments.

2.2.2. Pretreatment of the Lignocellulosic Hydrolysate

For the precipitation of sulfates within the hydrolysate, 20 g of $CaCO_3$ was added to 1 L of lignocellulosic hydrolysate (LCH) in a 5 L beaker while constantly stirring. The hydrolysate was mixed until the complete outgassing of CO_2. Next, the LCH was frozen at −20 °C overnight. After defrosting, the LCH was centrifuged at 16000 rcf for 10 min and the supernatant was collected. Twenty grams of KH_2PO_4 was added to the remaining liquid for the removal of excess calcium, and the pH was adjusted with 3 M NaOH to 7. The LCH was centrifuged at 16000 rcf for 10 min and sterile-filtered to be used in the cultivation experiments. All the LCH used for the fermentations was pretreated before its addition to the media.

2.2.3. Pre-Culture

Fifty milliliters of YPD medium (10 g/L yeast extract, 20 g/L peptone, and 20 g/L glucose) in an Erlenmeyer flask containing antibiotics (0.05 g/L kanamycin and 0.1 g/L ampicillin) was inoculated with a single colony of *C. oleaginosus* (ATCC 20509) from a YPD plate. Antibiotics in fermentation pre-cultures were added to prevent contaminations. The flasks were incubated at 28 °C under constant shaking at 120 rpm for 2 days.

2.2.4. Medium Composition for Bioreactor Cultivation

Different media were used for either nitrogen limitation or consumption-based cultivation with acetic acid. The base medium was composed of 0.9 g/L Na_2HPO_4, 2.4 g/L KH_2PO_4, 4.5 g/L $CH_3COO\cdot Na$, 2 g/L $MgSO_4\cdot 7H_2O$, 0.5 $CaCl_2\cdot 2H_2O$, 0.00055 mg/L $ZnSO_4\cdot 7H_2O$, 0.024 mg/L $MnCl_2\cdot 6H_2O$, 0.025 mg/L $CuSO_4 \cdot 5H_2O$, 0.027 mg/L $C_6H_8O_7\cdot Fe\cdot H_3N$, 1 g/L urea, 3 g/L peptone, and 2 g/L yeast extract. For nitrogen limitation, the medium above was adjusted to 1.74 g/L urea, 0.5 g/L yeast extract, and no peptone and $CH_3COO\cdot Na$. The C/N ratio was calculated based on the elemental carbon and nitrogen content, using the information provided by the suppliers of the chemicals. As carbon sources, either glucose, pretreated LCH, or a mix of xylose, glucose, and acetic acid (XGA) was used.

2.3. Fermentation

2.3.1. Fermentation in 1.3 L Bioreactor

The fermentations were carried out at pH 6.5, at 28 °C, and with a dissolved oxygen content of 50% in a fed-batch process. For the nitrogen-limited and consumption-based acetic acid approaches, a scale of 1 L maximal working volume in the DASGIP® system (Eppendorf AG, Hamburg, Germany) with a total reactor volume of 1.3 L was utilized. In the 500 mL starting medium, the sugar content of 3% from either LCH, glucose, xylose, or XGA was used. The nitrogen content was adjusted to accomplish a C/N ratio of 15 at the start and 65 at the end of the feeding. Pretreated pure and concentrated LCH or XGA with a carbon source content of 260 g/L was continuously fed, starting after 12 h, at a rate of 10 mL/h and 5 mL/h between 36 and 60 h. For non-limited consumption-based (cb) acetic acid feeding, a feed of 90% acetic acid solution was used for the pH regulation. At the same time, the acetic acid functioned as feed for the cultivation. For the combined (co) feeding, the cb-feed was combined with a continuous feed, with the medium mentioned above and a C/N ratio of 16 at the start and a maximum of 37 at the end of the fermentation.

2.3.2. Fermentation in 350 mL Bioreactor

The feeding strategy optimization was performed in a maximal working volume of 250 mL in the DasBOX® system (Eppendorf AG, Hamburg, Germany) with a 350 mL total reactor volume. Fifty percent acetic acid was used as consumption-based feed, controlled by the pH change. In the combined feeding condition, LCH was added to 50% (*v/v*) acetic acid to a final concentration of either 10% (*v/v*) and 50% (*v/v*) (50:10 and 50:50 mix of acetic acid:LCH). In the case of continuous feeding, the feeding rates were either 0.5 mL/h or 1 mL/h, starting from 12 h after inoculation, by maintaining a consumption-based acetic acid feed with 50% acetic acid.

2.4. Monitoring of Fermentation

2.4.1. OD_{600} Measurement

Cell growth was monitored by measuring the optical density at 600 nm (OD_{600}). The samples were diluted in the respective cultivation media to an OD_{600} between 0.1 and 1.

2.4.2. Dry Weight

The dry weight of the substrate solutions and biomass samples was determined gravimetrically. To measure the dry cell weight, 4 mL of fermentation culture was transferred to pre-weighed reaction tubes, centrifuged (4500 rct, 20 min), and washed two times with an equal amount of purified water or 50% ethanol in the case of lipid-rich cells. Alternatively, 0.5 mL of a lipid-rich culture was filtered through a pre-weighed 0.2 µm filter paper and washed three times with 2 mL water. The samples were frozen and lyophilized. For each biological replicate, the technical duplicates at least were measured. Growth curve fitting was performed by the Gompertz function, as shown in Equation (1); here, A is the upper asymptote, k_G is the growth-rate coefficient, and T_i is the time of inflection [37].

$$W(t) = A \times \exp(-\exp(-k_G(t - T_i))) \tag{1}$$

2.4.3. Lipid Content

For the lipid content analysis, the cells from the fermentation were centrifuged and washed two times with 50% ethanol and resolved in water. The cells were disrupted mechanically with a High-Pressure Homogenizer Type HPL6 from Maximator. Triplicates of the 7 mL disrupted cell solution were frozen and lyophilized. The chloroform-methanol lipid extraction was carried out after modifying with the Bligh and Dyer method [38]. Briefly, 100–200 mg biomass was weighed in a glass tube, and 4 mL Cl_3CH: methanol (2:1) and and 1 mL H_2O (0.58% NaOH) were added. After 60 min shaking at 120 rpm, it was centrifuged for 10 min at 2000 rcf, and the bottom layer was transferred to a new glass tube. An additional 3 mL Cl_3CH: methanol (2:1) was added to the upper phase in the first tube and quickly mixed. After centrifugation, the bottom layer was combined with the lower phase from the first extraction step, and 2 mL of H_2O (0.58% NaOH): methanol (1:1) was added. After mixing and centrifugation, the bottom phase was transferred to a fresh pre-weighed glass tube, and the solvents were evaporated under a nitrogen flow. The lipid amount was determined gravimetrically.

2.4.4. Fatty Acid Profile

The fatty acid profile was measured through gas chromatography. Therefore, fatty acid methyl esterification (FAME) of the lyophilized cells was performed; the biomass was not washed, to avoid the loss of very fat cells. Three to ten milligrams of the biomass was weighed in glass vials; all further steps were automated with the Multi-Purpose Sampler MPS robotic from Gerstel. An internal standard of 10 g/L C19 TAG in toluene was used for quantification. First, 490 μL toluene and 10 μL internal standard were added and mixed for 1 min at 1000 rpm, followed by the addition of 1 mL 0.5 M sodium methoxide in methanol, and the solution was heated to 80 °C and shaken at 750 rpm for 20 min. After cooling to 5 °C, 1 mL of 5% HCl in methanol was added, and the mixture was heated to 80 °C while shaking at 800 rpm for 20 min and cooled to 5 °C afterwards. Four hundred microliters of de-ionized H_2O was added and mixed for 30 s at 1000 rpm before 1 mL hexane was added. Extraction was performed by shaking three times for 20 s at 2000 rpm in a quickMix device. The samples were centrifuged for 3 min at 1000 rpm and cooled at 5 °C before a 200 μL sample of the organic phase was transferred to a 1.5 mL vial for chromatography. Gas chromatograph flame ionization detection (GC-FID) was used to quantify the fatty acids. GC-MS was carried out to identify the acids, with the TRACE™ Ultra Gas Chromatograph from Thermo Scientific coupled to a Thermo DSQ™ II mass spectrometer and a Triplus™ Autosampler injector in positive ion mode. A Stabilwax® fused silica capillary column (30 m × 0.25 mm, film thickness 0.25 μm) was used for separation. The temperature profile for the analysis was set to an initial column temperature of 50 °C, increasing at a rate of 4 °C/min up to a final temperature of 250 °C. Hydrogen was used as a carrier gas at a constant flow rate of 35 mL/min. Standardization was performed with the FAMEs Marine Oil Standard (20 components, from C14:0 to C24:1).

2.4.5. Statistical Analysis

For statistical analysis of the significant differences in the analysis outcomes, first an ANOVA test was performed. When the ANOVA test showed a test value of $p < 0.001$, and thereby indicated a significant difference, in the next step a Duncan test was applied. The Duncan tests were performed with a test value of $p < 0.01$, and they detected differences and similarities between the test conditions by grouping the results.

2.5. Techno-Economic Analysis

To estimate the total capital investment and operation cost for the production of oil from LCH using *C. oleaginosus*, a techno-economic analysis (TEA) was performed. The process was designed with a consequential approach in silico, according to previous laboratory results and available data from the literature [39]. SuperPro Designer (SPD) version 10 (Intelligen, Inc., Scotch Plains, NJ, USA) was used, and the functions describing the biochemical processes were integrated [39]. In the process simulation, a production plant was created with a lipid production capacity of 0.81 metric tons per hour (t/h). The feedstock requirements and the chemicals needed for the LCH pretreatment were estimated based on the media components used in the current work. The mathematical equations and parameters for simulating yeast biomass generation, lipid formation, and enzymatic hydrolysis were based on the results generated in this work or previous experiments [17]. To calculate energy balance, equipment sizing, and purchasing prices, the values from the SPD database were completed with publicly available current prices and data from the literature. The in silico plant featured several operations for LCH pretreatment, fermentation, and downstream processing, including recovery of SCO and recycling of waste streams and side products. The list of modules used in the SPD for the different fermentation conditions (glucose and LCH with acetic acid co-feeding and combination of continuous and consumption-based feeding) is included in the supplements. Lipid productivity was deduced from the lipid titers from the fermentations performed in the 1 L DASGIP® system after 71 h.

$$\text{Lipid productivity (g/L/h)} = \text{lipid titer (g/L)}/\text{time (h)} \qquad (2)$$

The lipid productivities were calculated for three conditions: glucose with acetic acid co-feeding, LCH with acetic acid co-feeding, and the combination of continuous feed and consumption-based acetic acid co-feeding. The equipment purchase cost (PC) was calculated with an internal SPD function based on the size of the equipment. The installation cost factors relative to the PC were included with values between 1.1 and 1.3 [40]. The cost factors for storage (4%), piping (9%), and site development (5%) were included [41]. The indirect capital costs consisted of insurance (5%), field expenses (5%), construction (10%), contingency (10%), and other expenses (10%). Capital interest was included as 6% of the total investment. Utility and labor costs, as well as waste disposal, were estimated based on the current prices in Germany [42]. The costs for consumables and raw materials were set according to current market prices. Maintenance and insurance were included as faculty-dependent costs at 3% and 0.7% [41].

3. Results

3.1. Analysis of the Lignocellulosic Hydrolysate

The lignocellulosic hydrolysate (LCH), coming from the spent liquor of the pulping process, was comprehensively analyzed to gain a better understanding of its composition and to set up a pretreatment strategy. The pH was 1.7, with a dry mass of 247.7 ± 13.9 g/L and a high sulfate content of 19.4 ± 2.0 g/L. HPLC analysis was used to quantify the sugar, organic acids, and furans. It showed that the primary sugar of the hydrolysate was xylose, with 77.05 ± 2.46 g/L, and smaller amounts of other sugars, such as glucose (11.51 ± 0.39 g/L), mannose (8.21 ± 0.84 g/L), galactose (6.47 ± 0.21 g/L), and others (12.37 ± 1.10 g/L), resulting in a total sugar content of 115.60 ± 4.99 g/L. The organic acids had a total concentration of 16.83 ± 1.43 g/L, acetic acid being the major one with 12.34 ± 1.20 g/L. A considerable amount of hydroxymethylfurfural (HMF), with 4.53 ± 0.59 g/L, and furfural, with 0.68 ± 0.04 g/L, was detected. The hydrolysate is a waste stream from the industrial production of cellulose fibers using hardwood as a source material in an acidic sulfite pulping process. The chemical hydrolysis of lignocellulose results in the detected sugar monomers, organic acids, and breakdown compounds from lignin. The total ash was 0.70 ± 0.01 g/L, with low concentrations of phosphorus (0.035 g/L) and no detectable nitrogen. The rest of the material derives from lignols and lignans as well as other plant metabolites, resulting in a total amount of 89.95 $\pm$ 22.97 g/L. Table S1 gives a summarized overview of the composition.

3.2. Pretreatment of the Lignocellulosic Hydrolysate

To transform the waste stream LCH into a suitable carbon source for yeast fermentation, the challenges of low pH and the lack of essential nutrients needed to be addressed with a suitable pretreatment strategy. However, the formation of insoluble particles was observed after the addition of the other medium compounds such as buffering salts, vitamins, nitrogen, and trace elements, as well as during the cultivation. To address this problem, some simple pretreatment steps were taken. As a result of the chemical hydrolysis from the industrial pulping process, the LCH has a high sulfate content and a highly acidic pH. To reduce the amount of residual sulfate, overliming was performed using $CaCO_3$. Nevertheless, particle formation could still be observed. Titration of the overlimed LCH with $H_2PO_4^-$ or HPO_4^{2-} salts was empirically found to prevent this particle formation. As this process still results in an acidic pH, a final neutralization with NaOH was performed after the KH_2PO_4 treatment. The details of the different tested pretreatment conditions are provided in the Table S2. The reduction in the overall bioavailable carbon by the pretreatment (159.85 $\pm$ 2.80 g/L) was negligible compared to the starting material (162.05 $\pm$ 1.44 g/L). This qualifies the LCH pretreatment as a suitable step in the preparation of the waste stream as a carbon source for bioreactor fermentations. This optimized pretreatment strategy was applied to all the LCH used for fermentation in this study.

3.3. Nitrogen-Limited Fermentation

Fermentation under nutrient limitation, such as with phosphate or nitrogen, is a common strategy to trigger lipid accumulation in oleaginous yeast [43]. Therefore, nitrogen-limited fermentation was performed using LCH as a carbon source (Figure 1a). To understand the effect of the lignocellulose-derived compound on the yeast metabolism, a further control medium was composed. The control medium contained only the main sugars from the LCH in their respective ratios: xylose (8.28%), glucose (1.03%), and acetic acid (0.69%), and it is therefore abbreviated with xylose, glucose, and acetic acid (XGA). The fermentations resulted in a biomass formation of 7.02 $\pm$ 0.88 g/L and 16.65 $\pm$ 0.24 g/L for LCH and XGA, respectively, showing better growth performance on the XGA medium.

3.4. Acetic Acid-Based Fermentation with Commercial Sugars as Carbon Source

The successful production of SCO from refined sugars with acetic acid-based fermentation was shown previously [17]. In this work, LCH was integrated as a carbon source for fed-batch fermentation with consumption-based acetic acid feeding. As controls, glucose and xylose, as well as the XGA mix, as starting sugars were tested. The analysis showed that the concentration of acetic acid remained constant due to the consumption-based feeding; the sugars were consumed within the first 40 h after inoculation, as shown in Figure 1b–d. The biomass increased further after the full consumption of the sugar and reached 35.11 $\pm$ 1.11 g/L, 39.95 $\pm$ 4.59 g/L, and 39.00 $\pm$ 0.76 g/L for glucose, xylose, and XGA by 71 h, with lipid titers of 18.5 $\pm$ 3.6 g/L, 23.6 $\pm$ 0.5 g/L, and 21.6 $\pm$ 2.8 g/L for glucose, xylose, and XGA, respectively (Figure 2a). The same applied to the carbon-to-lipid conversion yield of lipid carbon from substrate carbon, which was 0.204 $\pm$ 0.003 g/g, 0.230 $\pm$ 0.001 g/g, and 0.235 $\pm$ 0.001 g/g for glucose, xylose, and XGA, respectively (Figure 2b). The fatty acid profile of the oil is shown in Figure 2c; the most abundant fatty acid was C18:1 with around 54%, followed by C16:0 (~24%), 18:0 (~15%), and C18:2 (~6%) and by traces of C18:3, C16:1, and C22:0 with less than 0.3% each.

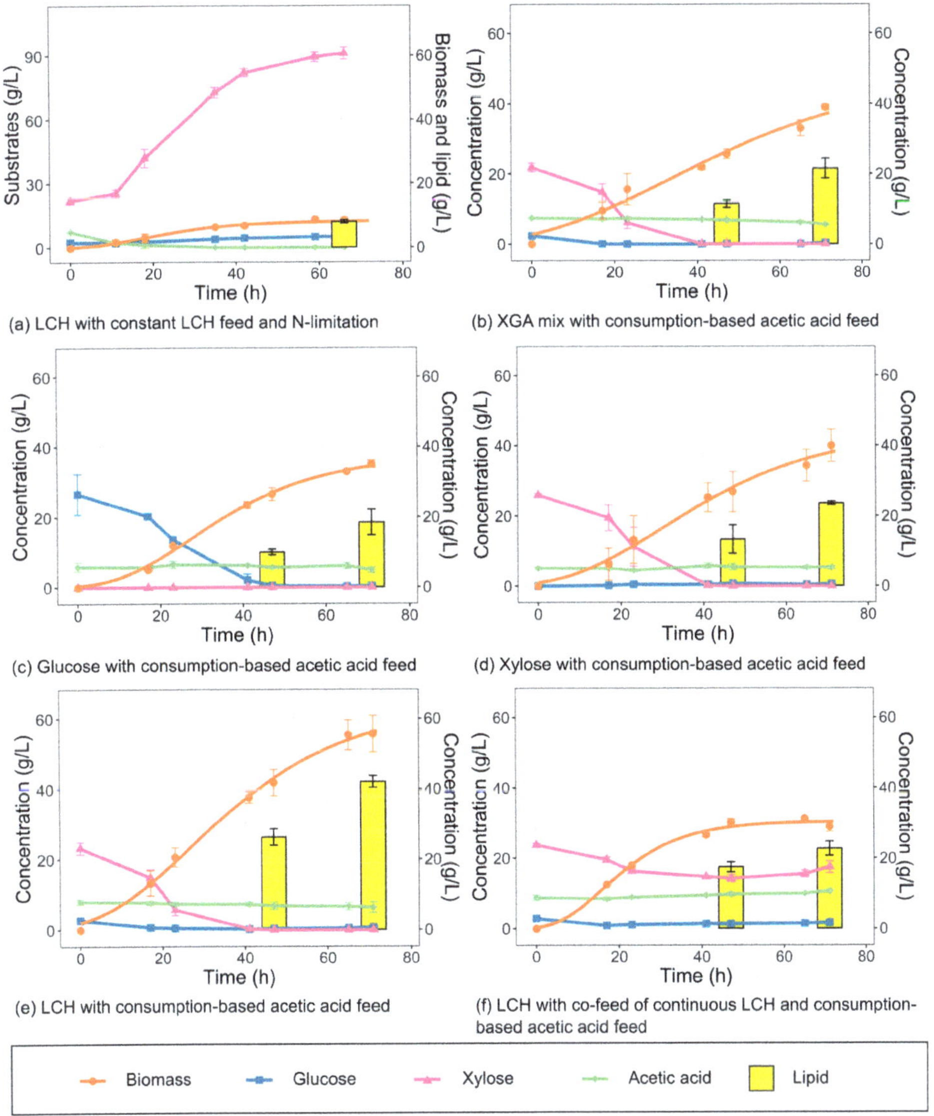

Figure 1. Biomass, substrate—namely glucose, xylose, and acetic acid—concentration, and lipid titer for different fermentation conditions. (**a**) LCH with constant feed and N-limitation; (**b**) XGA mix with consumption-based acetic acid feed; (**c**) Glucose with consumption-based acetic acid feed; (**d**) Xylose with consumption-based acetic acid feed; (**e**) LCH with consumption-based acetic acid feed; (**f**) LCH with co-feed of continuous LCH and consumption-based acetic acid feed. Fermentations were performed in triplicates (except for glucose and xylose, which were performed in duplicates). Error bars display two times the standard deviation. LCH—lignocellulosic hydrolysate, XGA—xylose, glucose, and acetic acid mix as model substrate.

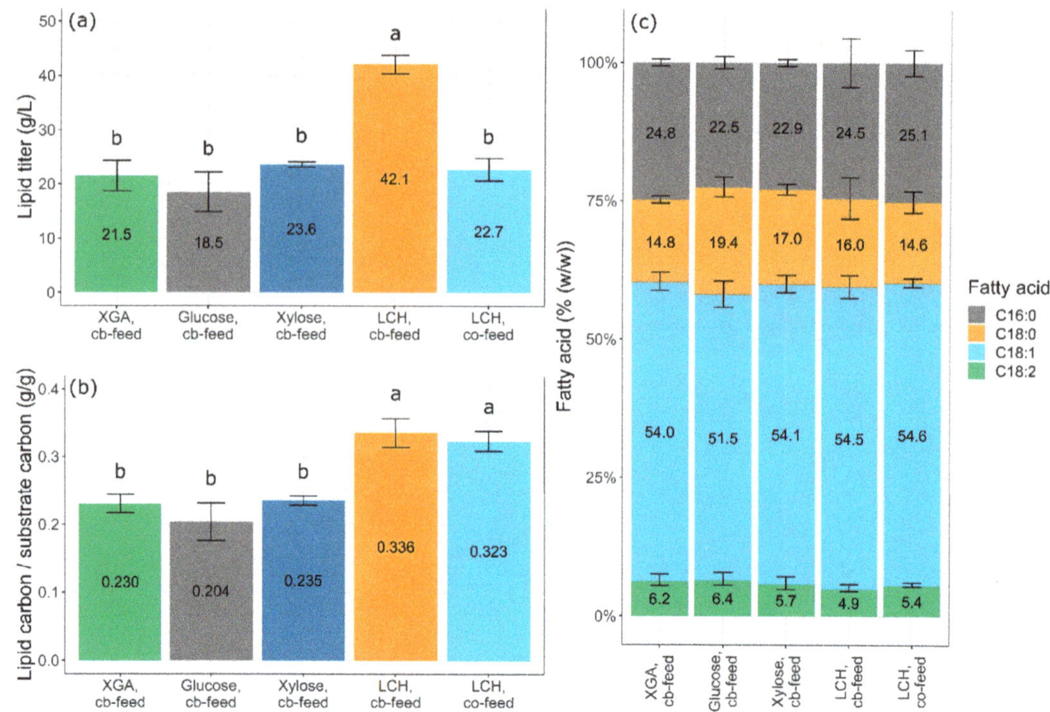

Figure 2. Lipid analysis of the four most important fermentation conditions. (**a**) Lipid titers after 71 h fermentation in 1 L scale. (**b**) Carbon conversion yield from substrate carbon to lipid carbon. (**c**) fatty acid profile of the main fatty acids quantified with GC-FID. Error bars display two times standard deviation; Duncan tests were performed with $p < 0.01$ to determine statistical groups a and b for lipid titer and carbon conversion, Duncan results for the fatty acid profile are shown in the Table S3. cb-feed—consumption-based acetic acid feed; LCH—lignocellulosic hydrolysate; LCH co-feed—lignocellulosic hydrolysate as starting carbon, and constant feed with cb-feed.

3.5. Acetic Acid-Based Fermentation of Lignocellulosic Hydrolysate at Different Starting Concentrations

The LCH used in this study is the spent liquor from acidic hardwood pulping, containing sugars, mainly xylose, as well as acetic acid. LCH further contains furans and lignin breakdown products that can have an inhibitory effect on microorganisms. To assess the optimal balance between carbon supply and the toxic level of inhibitory compounds, fermentations with three different starting concentrations of LCH were performed. The amount of hydrolysate corresponded to the concentrations of 3%, 5%, and 7% of the bioavailable carbon in the starting medium. The fermentation with the highest LCH concentration of 7% showed a slower overall growth with a growth-rate coefficient of 1.21 compared with 1.33 for 3%. The dry biomass formations were 39.64 ± 6.30 g/L and 51.92 ± 0.18 g/L for 7% and 5%, compared to 55.73 ± 5.20 g/L in the case of 3%. Accordingly, an increase in the concentration to 5% resulted in a slightly longer lag phase and a slightly lower biomass accumulation after three days of fermentation. The lipid titer was lower for 7% LCH (19.83 ± 1.38 g/L) and 5% (25.75 ± 2.02 g/L) and much higher for 3% (42.1 ± 1.7 g/L). The starting carbon source concentration of 3% resulted in the shortest lag phase, the best growth rate, and the overall biomass formation, as shown in Figure S1.

3.6. Acetic Acid-Based Fermentation on Lignocellulosic Hydrolysate with Optimized Conditions

The optimized conditions for the fed-batch fermentation of *C. oleaginosus* on LCH were carried out in three biological replicates. The starting carbon content of the hydrolysate corresponding to 3% sugar and acetic acid was used. A high biomass formation of 55.73 ± 5.20 g/L was achieved after 71 h, which is a seven-fold increase compared to the nitrogen-limited fermentation of LCH. Moreover, the biomass formation was about 50% higher with LCH as a carbon source compared to glucose or xylose under equivalent reaction conditions (Figure 1b,c). However, the sugar consumption showed a similar profile to that of the fermentation processes using sole sugars. The lipid titer after 71 h was, with 42.1 ± 1.7 g/L (lipid productivity 0.593 g/L/h), twice as high as that from the glucose fermentation (18.5 ± 3.6 g/L, productivity 0.261 g/L/h) (Figure 2a). The carbon conversion was 0.336 g/g lipid carbon from substrate carbon (Figure 2b) with 76.05 ± 9.04% lipid per dry biomass. The fatty acid profile did not show significant changes compared to the lipid profile of the fermentation on xylose or XGA and slight changes compared to the fermentation on glucose (Figure 2c and Table S3).

Furthermore, lignin breakdown products were shown to be metabolized by *C. oleaginosus*, in a range of 91.6% of the initial compound amount (Figure S2 and Table S4).

3.7. Screening of Different Strategies for Feeding Lignocellulosic Hydrolysate

In several fermentation approaches, the combined feeding of a mix of LCH and acetic were tested. The biomass of the cells fed with a 50:10 (50% acetic acid, 10% LCH, 40% purified water) and a 50:50 (50% acetic acid, 50% LCH) mix of acetic acid:LCH were 51.5 ± 3.7 g/L and 59.6 ± 1.2 g/L. Those solely fed with acetic acid after 65 h (50.8 ± 0.1 g/L), with 50:50 co-feeding, resulted in the highest biomass (Figure 3). The lipid titers after 65 h were 28.4 ± 0.4 g/L and 30.4 ± 1.4 g/L for the 50:10 and 50:50 feeding, respectively, while 25.2 ± 3.1 g/L was measured for the control settings (Figure 3). As an alternative feeding strategy, the pure pretreated LCH was continuously fed into the reactor throughout the fermentation process at two feeding rates: 0.5 mL/h or 1 mL/h, in addition to the consumption-based feeding of acetic acid. The dry biomass accumulated after 65 h was 50.4 ± 2.2 g/L and 50.4 ± 4.6 g/L for 0.5 mL/h and 1 mL/h, respectively. The maximum lipid titers were 0.5 ± 2.4 g/L for 0.5 mL/h and 26.3 ± 2.7 g/L for 1 mL/h and were therefore above the control (25.2 ± 3.1 g/L). The control and the two best feeding strategies are visualized in Figure 3; a comparison of all five strategies can be found in the Figure S3.

To select the setup with the highest LCH turnover, the share of LCH in the overall consumed carbon was calculated. The strategy using 1 mL/h of continuous feeding resulted in the highest percentage of LCH consumed (21.3 ± 2.9%), followed by 50:50 acetic acid:LCH (12.5%) and continuous feed with 0.5 mL/h (10.4 ± 1.2%). For the differences between the duplicates continuously fed with LCH results and the different consumed quantities of acetic acid, see Table S6. As the share of LCH using 50:50 acetic acid:LCH stayed constant over the fermentation, no standard deviation could be calculated. In addition, the carbon conversion rate of fed carbon to lipid was calculated for all conditions. In comparison, the setups using continuous feed had the highest carbon conversion yield, with 0.198 ± 0.002 g/g for 0.5 mL/h and 0.188 ± 0.0002 g/g for 1 mL/h. The control showed a conversion yield of 0.164 ± 0.0002 g/g, and co-feeding with the acetic acid:LCH mix resulted in 0.167 ± 0.0001 g/g, for a ratio of 50:10 and 0.167 ± 0.0005 g/g for 50:50.

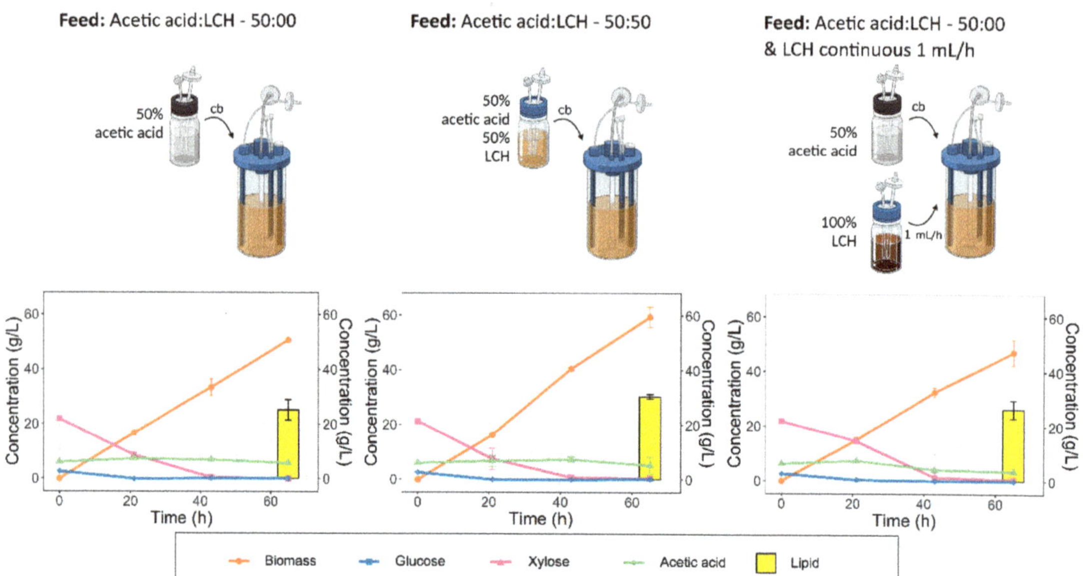

Figure 3. Comparison of the different feeding strategies in 0.25 L scale in the DASbox® system. Below each schema, the biomass accumulation, substrate concentration, and total lipid titer after 65 h are shown. On the left side, the consumption-based acetic acid feed with a 50% acetic acid solution is displayed (acetic acid:LCH 50:00), in the middle the consumption-based feed mixture of LCH and acetic acid (acetic acid:LCH—50:50), and on the right side the combination of continuous feeding with LCH at 1 mL/h and consumption-based acetic acid feeding (acetic acid:LCH 50:00 and LCH continuous). Fermentations were performed in duplicates at least. Error bars display two times standard deviation. LCH—Lignocellulosic hydrolysate, cb—consumption-based acetic acid feed.

3.8. Fermentation with Co-Feeding of Acetic Acid and Lignocellulosic Hydrolysate

Using 1 mL/h continuous feed of LCH (LCH co-feed) led to the highest share of LCH consumed of the overall carbon source and had the second highest carbon conversion yield among all the tested conditions. Therefore, this condition was selected for the 1 L DASGIP® system. The slight scale-up also allowed a better direct comparison of the fermentation output with the other conditions tested in the DASGIP® system. A consumption-based feed with 90% acetic acid was used in combination with a continuous LCH feed of 3.3 mL/h. The biomass accumulation reached 29.0 ± 5.87 g/L after 71 h, with a lipid titer of 22.7 ± 1.95 g/L (lipid productivity 0.320 g/L/h). HPLC analysis showed that the sugars and acetic acid were evenly consumed throughout the fermentation, and only xylose accumulated slightly from 15.5 ± 1.11 g/L at 65 h to 17.44 ± 1.85 g/L at 71 h, as shown in Figure 1f. The xylose accumulation could presumably be solved by using a slightly lower feeding rate, resulting in a steady supply of sugar and fewer inhibitory compounds. When compared to the other conditions tested in the DASGIP® system, biomass accumulation with LCH co-feed was only approximately 17% less than that with glucose as the starting sugar. The lipid titers were comparable to the titers reached with glucose or xylose, with the LCH co-feed performing only 4% worse than the titers reached with xylose.

3.9. Techno-Economic Analysis

Aside from the experimental results in this study, a consequential TEA was performed to assess the economic feasibility of a commercial-scale production plant using LCH as an industrially relevant feedstock. Two different modes of LCH fermentation from the 1 L scale DASGIP® experiments were implemented and compared to the setup with glucose as a carbon source, as shown in Figure S4. Firstly, there was the consumption-based acetic acid feeding using LCH as a starting carbon source, and secondly, there was the combination of consumption-based acetic acid and the continuous feeding of LCH. Both fermentation modes were compared to an operation setup, where glucose was used as the starting carbon source for consumption-based acetic acid feeding. According to the simulation, the capital expenditure (CAPEX) was lowest for the LCH cb-feed at USD 30.1 M. In comparison, the CAPEX in the glucose cb-feed was USD 42.5 M and USD 41.8 M for the LCH co-feed. The increased capital cost was mainly caused by the different amounts and the vessel volumes of the fermentation vessels in the respective model. Different vessel combinations had to be used for the different fermentation strategies because of differences in the dwell times to simulate comparable lipid productivities. Other than the vessel amounts and the volumes, the equipment costs were the same. Depreciation was calculated as 10% over ten years, resulting in USD 4.2 M, USD 3.0 M, and USD 4.2 M for glucose cb-feed, LCH cb-feed, and LCH co-feed, respectively. The quantity and volumes of the vessels also resulted in differences in the operating costs, especially the costs for power (glucose cb-feed: USD 4.8 M, LCH cb-feed: USD 2.3 M, LCH co-feed: USD 4.0 M), cooling water (glucose cb-feed: USD 1.0 M, LCH cb-feed: USD 0.5 M, LCH co-feed: USD 0.9 M), labor cost (glucose cb-feed: USD 2.2 M, LCH cb-feed: USD 1.5 M, LCH co-feed: USD 1.8 M), maintenance (glucose cb-feed: USD 0.8 M, LCH cb-feed: USD 0.6 M, LCH co-feed: USD 0.8 M). Raw materials were a major expense, dominated by the acetic acid price. For the main analysis, an acetic acid price of 600 USD/t was used, resulting in a raw material cost of USD 8.5 M for glucose cb-feed, USD 7.7 M for LCH cb-feed, and USD 7.2 for LCH co-feed. The final costs for the yeast oil calculated from the in silico model were 3985 USD/t (glucose cb-feed), 2938 USD/t (LCH cb-feed), and 3576 USD/t (LCH co-feed), respectively (Figure 4a).

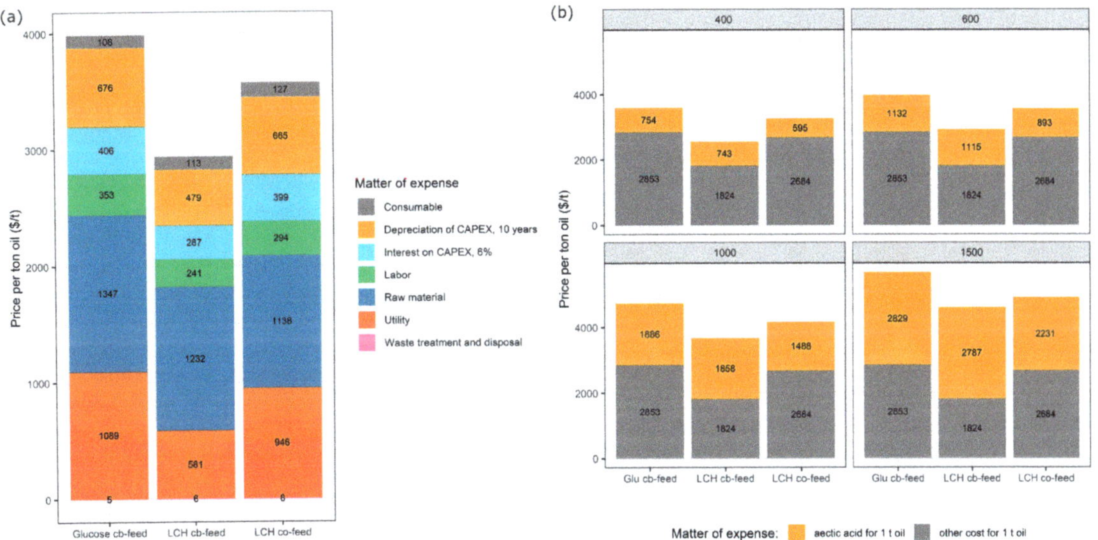

Figure 4 (a) Annual production cost of yeast oil in USD/Mt for the three fermentation strategies analyzed with the TEA. The respective amount of feedstock was set to produce yeast oil at a rate of 0.81 Mt/h (LCH cb-feed: 1 Mt/h, glucose cb-feed: 0.151 Mt/h, LCH co-feed: 2.1 Mt/h). (b) Annual production costs in a sensitivity analysis of the influence of the acetic acid prices 400, 600, 1000, and 1500 USD/t. LCH—lignocellulosic hydrolysate; cb-feed—consumption-based acetic acid feed; co-feed—lignocellulosic hydrolysate as starting carbon and constant feed with cb-feed.

In addition, a sensitivity analysis was performed considering different acetic acid prices, as it accounted for the major part of the raw material costs (glucose cb-feed: 84%, LCH cb-feed: 90%, LCH co-feed: 78%). Acetic acid prices in Western Europe fluctuated between 400 USD/t and 1500 USD/t between 2012 and 2023; so, the cost of yeast oil per t is largely dependent on the current acetic acid price [44,45]. Therefore, acetic acid costs of 400 USD/t, 600 USD/t, 1000 USD/t, and 1500 USD/t were assumed for the sensitivity analysis. The scenario using LCH cb-feed responded strongest to the price changes, as shown in Figure 4b. The price decrease to 400 USD/t led to a 12.6% decrease in the cost to 2567 USD/t oil. The rise to 1000 USD/t led to a 25.3% increase in production costs (3682 USD/t), and an elevated acetic acid price of 1500 USD/t resulted in a 56.9% increase (4611 USD/t). These tendencies were similar for the glucose cb-feed and LCH co-feed. Overall though, LCH cb-feed still showed the cheapest production costs for a decrease or increase in acetic acid prices.

4. Discussion

The lignocellulosic hydrolysate from acidic hardwood pulping mainly consisted of xylose, glucose, and other sugars, as well as acetic acid and lignin breakdown products, as expected and described by the literature [26,27]. This composition makes LCH an ideal carbon source for *C. oleaginosus*, which has been shown to efficiently metabolize both hexose and pentose sugars as well as acetic acid [17]. Waste streams with a high content of xylose and lignin-derived compounds are of special research interest because only a few commercial valorization strategies are available [46,47].

Due to the low amounts of nitrogen, phosphorus, and other elements, such as sulfur, magnesium, or calcium, additional nutrient supplementation is required before using the hydrolysate in a fermentation medium. Without pretreatment the LCH formed insoluble particles when combined with the other media components, which was presumably the result of sulfate and phosphate salt formation.

The pretreatment strategy established in this study produced a neutral LCH that did not form insoluble material when buffering salts or trace elements were added. As the sugar content was maintained, it could be efficiently applied as a carbon source for *C. oleaginosus*.

Primarily, the conventional nitrogen-limited fermentation approach was applied for the comparison of the LCH-containing medium with the XGA model. The reduced growth in LCH conditions was presumably caused by the high concentration of inhibitory furans and phenolic compounds in the medium. However, the contained phenolic and other lignin-derived compounds, did not inhibit growth completely. Furthermore, *C. oleaginosus* has already been shown before to metabolize some of the lignin breakdown compounds [16]. Generally, nitrogen limitation induces lipid formation but may restrict biomass formation at the same time, as nitrogen is required for protein synthesis and other metabolic processes [48]. Due to these drawbacks, nutrient limitation strategies are difficult for the production of SCO with the high yields required for commercial production.

The successful production of SCO with *C. oleaginosus* from refined sugars with acetic acid-based fermentation was shown previously. This approach combines a non-limiting sugar-containing starting medium with a consumption-based acetic acid feeding strategy [17]. For a commercially more attractive process, the LCH as a waste product was applied as a starting carbon source in this study, in place of costly sugars. This was further advanced in the fed-batch fermentations of this study. Xylose and glucose as sole carbon sources showed a similar growth behavior to the fermentation using XGA, which indicates a good and simultaneous uptake of both glucose and xylose. This is a considerable advantage of *C. oleaginosus* as a fermentation strain because the full potential of the biomass as a carbon source can be exploited. Moreover, the acetic acid can theoretically be directly channeled into the metabolism of lipid synthesis by acetyl-CoA synthetase. Therefore, a more efficient method of lipid biosynthesis in *C. oleaginosus*, compared to nutrient limitation conditions, can be employed [17]. Fortunately, acetic acid is part of the pulp and paper waste streams and hence represents an industrially more promising carbon source than refined sugars, as it is a byproduct of hardwood pulping alongside the LCH [29,49].

The fatty acid profile of the SCO produced showed a typical composition of oleaginous yeasts, mainly consisting of C16 to C18 fatty acids, which is favorable for biodiesels [50]. Bio-based diesel from plant oils is used in the form of fatty acid esters. One main factor in evaluating the suitability of the fatty acids for biodiesel is the cetane number. This number indicates the combustion speed; it increases with chain length and decreases with saturation. The minimal overall values for use as biodiesel are 47 (U.S. American specification) and 51 (European specification). Examples of cetane ranges of biodiesels are 48–67 (soy oil) and 60–63 (palm oil) [51,52]. Among the unsaturated fatty acids, oleic acid has the greatest cetane number of 56 and is most prominent in the fatty acid profile of *C. oleaginosus* [53]. The cetane numbers of the other dominating fatty acids are 75 (palmitic acid) and 76 (stearic acid) [53]. As C16:0, C18:0, and C18:1 compose around 93% of the fatty acids, based on its composition the oil of *C. oleaginosus* fits well as a raw material for biodiesel.

The starting carbon source concentration of 3% resulted in the shortest lag phase, the best growth rate, and the overall biomass formation, demonstrating that the combination of monomeric sugars and acetic acid in the LCH can be metabolized efficiently and simultaneously by *C. oleaginosus*. Higher concentrations resulted in higher concentrations of inhibitory furans and lignin breakdown products, explaining the longer lag phase and decreased growth rates. Therefore, the starting carbon concentration of 3% was used for all further experiments.

The optimized fermentation strategy of *C. oleaginosus* on LCH with acetic acid consumption-based feeding showed the best performance in biomass accumulation, lipid titer, and lipid yield, with minor changes in the lipid profile. The increased performance of LCH as a substrate might be attributed to the contained lignol and lignan compounds that can partly be metabolized by *C. oleaginosus* [16]. In summary, this combination of the waste stream and oleaginous yeast is promising for a sustainable new process of SCO production. It combines the application of a waste stream rich in xylose, lignols, and lignans with a fermentation strategy resulting in high lipid yields. The lipid titer of 42.1 ± 1.7 g/L with a wild-type *C. oleaginosus* from this study is higher than the top yields from the literature of 39.5 ± 0.49 g/L with a genetically modified *R. toruloides* [35]. In the measured lipid carbon conversion, both sugars and acetic acid were considered but not the lignols and lignans from the hydrolysate. Nevertheless, the conversion yield of 0.336 g/g per carbon source is close to the theoretical maximum for xylose of 0.34 g/g [54]. Furthermore, efficient metabolization of the lignin breakdown products was shown, which can explain the high conversion yield and demonstrates the potential of *C. oleaginosus* for the valorization of waste streams from the pulp and paper industry.

Consumption-based acetic acid feeding results in a high lipid productivity of over 80%, as shown in the literature [17]. However, compared to acetic acid or commercially available sugars, LCH is a cheap feedstock as it is a waste product with limited options for value creation. To increase LCH uptake by the oleaginous yeast relative to the consumption of acetic acid and to better balance the two waste streams, different fermentation modes were compared. For the overall evaluation of the feeding strategies, the focus was set on the share of the consumed LCH and the lipid titer. With regard to this, the two best feeding strategies were cb-feed with 50:50 acetic acid:LCH as well as a co-feed with a continuous feed of 1 mL/h LCH. Providing more LCH during the fermentation process, these strategies could significantly increase the share of LCH. At the same time, the starting concentration of inhibitory compounds was not increased, which prevents an elongated lag phase, as observed for higher LCH concentrations.

The consequential TEA revealed a price reduction of 26% using LCH instead of glucose. Looking forward, the price points for the LCH processes could potentially be reduced by the further optimization of the fermentation processes and the resulting lipid conversion yields. All the calculated price regimes are in the current price range for organic palm oil (2500–3000 USD/t) [55]. Nevertheless, the consequential TEA, as applied in this study, focuses mainly on the changes in the input and output conditions relative to each other [56]. The calculated prices per t should therefore be considered as an estimate of magnitude. Other plant oils used for biofuel production, such as canola oil, are priced well below 1000 USD/t [57]. However, utilizing an industrial waste stream circumvents the competition for edible oils, thus enhancing the sustainability aspect of the LCH-based oil production processes presented in this study. Furthermore, vertical fermentation setups are seasonally independent compared to their agricultural counterparts, and in contrast to conventional agriculture, do not require fertilizers and pesticides [58]. Moreover, fermentative SCO production is not heavily affected by climate change and represents an area-efficient alternative to plant oils.

A main factor for the production cost of the SCO is the price of acetic acid, as it is one of the main carbon sources in this process. The high recent fluctuation in acetic acid prices can be explained by the limited number of suppliers, namely the USA and China, which supply the majority of the acetic acid, with the market shares of 17% and 55%, respectively [59]. Therefore, problems or even disruptions in the supply chains of these two countries have a profound impact on the pricing as the availability can be drastically decreased [44,60]. Local or even on-site production and supply of acetic acid would be the best route to avoid such dependencies and price fluctuations. In that context, up to 5% of wood weight is acetate esters, which could potentially be separated and used directly from pulp and paper waste streams as an acetic acid source [27,61].

In Europe alone, the pulp and paper industry produces around 11 million tons of waste every year [24]. Wood is mainly composed of 40–44% cellulose, 18–32% lignin, and 15–35% hemicelluloses [27,61]. The hemicellulose consists of xylan with 10–35% and acetate esters with 3–5% of the total wood [62,63]. Extrapolating these numbers to the actual biomass volume results in waste material of 85–151, 47–151, and 14–24 million metric tons of lignin, xylan, and acetate, respectively. Even a small proportion of these waste amounts would be enough to supply a biotechnological plant, as described here.

In the TEA, our lab-scale experiments for the biotechnological valorization of an industrial waste stream using *C. oleaginosus* were validated in an in silico industrial fermentation plant model. It was performed with a focus on the comparison of different conditions and raw material costs. Compared to glucose, the cost efficiency of LCH was demonstrated for the production of SCO on an industrially relevant scale. SCO can be used as a starting material for similar applications to those of plant-based oils in the biofuel sector as well as for the production of polymeric materials, lubricants, and other oleochemicals [64,65]. Such resource-efficient approaches could lead to more independent, circular, and sustainable economies in the future.

5. Conclusions

Single cell oils are a promising alternative to plant-based oils for the production of sustainable biofuels. This study provides a novel strategy to produce SCOs with the oleaginous yeast *C. oleaginosus*, utilizing lignocellulosic hydrolysate, a biogenic waste stream from the pulp and paper industry. To this end, fermentations were performed, and a maximal lipid titer of 42.1 ± 1.7 g/L (75.5% lipid per biomass) and a carbon conversion to lipid of 0.336 g/g were reached with optimized conditions. Furthermore, feeding strategies were tested to enhance the share of LCH in the total amount of fed carbon, where acetic acid has the main share. The most effective feeding strategy utilized a mix of acetic acid and LCH as the main carbon sources, both of which account for the main part of the waste stream from the pulp and paper industry. In summary, our data indicate that the xylose-rich LCH is a highly suitable substrate for the efficient bioconversion to SCO. By adding continuous feeding to the established consumption-based approach, the LCH share of consumed carbon was increased. The economic advantages of the use of LCH over glucose were shown in a techno-economic analysis. The predicted prices for the SCO produced moved in a range (2900 USD/t) that was competitive with sustainably produced organic plant oils (1000–3000 USD/t) [55,57]. At the same time, vertical fermentation setups can be run in a seasonally independent manner, without the use of fertilizers and pesticides, and offer options for land-efficient scale-up. Our further research will continue in the direction of maximizing the reuse of industrial waste streams in an economically feasible way to close the remaining gaps for a circular bioeconomy and to contribute to the environmental challenges the world faces today. In conclusion, we present a sustainable and economical strategy to produce SCOs as an alternative platform to produce advanced biofuels, lubricants, and other oleochemical products.

6. Patents

Thomas B. Brueck, Mahmoud Masri, Nikolaus I. Stellner, and Zora S. Rerop have a European patent field with the Technical University of Munich involving the methodology described in this work.

Supplementary Materials: The following supporting information can be downloaded at: https://www.mdpi.com/article/10.3390/fermentation9020189/s1, Table S1: composition of the LCH, Table S2: pretreatment strategies, Table S3: sugar and organic acid composition of LCH before and after pretreatment, Figure S1: comparison between three different starting concentration of LCH, Table S4: Duncan groups of the statistical analysis of the fatty acid profile, Figure S2: GC-MS chromatograms of fermentation medium before and after fermentation, Table S5: share of LCH from the overall consumed carbon for the different feeding strategies, Figure S3: comparison of all five different feeding strategies as well as respective carbon conversion rates, Table S6: GC-FID relative quantification of LCH-derived lignin breakdown products before and after the fermentation, Figure S4: Super Pro Designer model for the TEA of SCO production on the basis of LCH and acetic acid.

Author Contributions: Conceptualization, T.B.B., M.M., N.I.S. and Z.S.R.; methodology, M.H., P.G., M.M., N.I.S. and Z.S.R.; software, M.M., N.I.S. and Z.S.R.; validation, T.B.B., M.M. and N.M.; investigation, P.G., M.M., N.I.S. and Z.S.R.; resources, T.B.B.; data curation, P.G., M.M., N.I.S. and Z.S.R.; writing—original draft preparation, N.I.S. and Z.S.R.; writing—review and editing, T.B.B. and M.M.; visualization, N.I.S. and Z.S.R.; supervision, T.B.B., N.M. and M.M.; project administration, M.M.; funding acquisition, T.B.B. All authors have read and agreed to the published version of the manuscript.

Funding: This research was funded by: The German Federal Ministry of Education and Research (Bundesministerium für Bildung und Forschung), grant number: FKZ 031B0662B; European Union's Horizon Europe research and innovation program, grant number: 101059786; Werner Siemens Foundation.

Institutional Review Board Statement: Not applicable.

Informed Consent Statement: Not applicable.

Data Availability Statement: The data presented in this study are available on request from the corresponding author. The data are not publicly available due to privacy protection.

Acknowledgments: We would like to thank Jeremias Widmann for his outstandingly diligent and dependable help with the sample preparation and analysis throughout our work. We also very much appreciate the expert knowledge and advice of Uwe Arnold, managing director of AHP Group GmbH, which led to the optimization of our techno-economic analysis setup. Finally, we express our thanks to Gülnaz Celik for her reliable support and advice in the fermentations performed for this study.

Conflicts of Interest: Thomas B. Brueck and Mahmoud Masri are board members at Global Sustainable Transformation GmbH.

References

1. Ge, M.; Friedrich, J.; Vigna, L. *5 Facts about Country & Sector GHG Emissions*; World Resources Institute: Washington, DC, USA, 2022; pp. 1–9.
2. Pörtner, H.-O.; Roberts, D.C.; Adams, H.; Adelekan, I.; Adler, C.; Adrian, R.; Aldunce, P.; Ali, E.; Begum, R.A.; Friedl, B.B.; et al. *2022: Technical Summary*; Cambridge University Press. Cambridge, UK, New York, NY, USA, 2022, ISBN 9781009157896.

3.   United Nations. "17 Sustainable Development Goals (SDGs)", Sustainable Development Goals (SDGs), 2022. Available online: https://sdgs.un.org/goals (accessed on 29 September 2022).

4.   European Commission. *REPowerEU Plan*; European Commission: Brussels, Belgium, 2022.

5.   European Commission. *A Strategic Long-Term Vision for a Prosperous, Modern, Competitive & Climate-Neutral Economy EU Vision for 2050—"A Clean Planet for All"*; European Commission: Brussels, Belgium, 2018.

6.   Flach, B.; Lieberz, S.; Bolla, S. *EU Biofuels Annual 2019*; USDA: Washington, DC, USA, 2019.

7.   Irena & Ace. *Renewable Energy Outlook for ASEAN*; Irena & Ace: Masdar City, Abu Dhabi, 2022; ISBN 9789295111288.

8.   Abeln, F.; Chuck, C.J. The history, state of the art and future prospects for oleaginous yeast research. *Microb. Cell Fact.* **2021**, *20*, 221. [CrossRef] [PubMed]

9.   Huang, C.; Chen, X.-F.; Xiong, L.; Chen, X.-D.; Ma, L.L.; Chen, Y. Single cell oil production from low-cost substrates: The possibility and potential of its industrialization. *Biotechnol. Adv.* **2013**, *31*, 129–139. [CrossRef] [PubMed]

10.   Agricultural Market Information Company (AMI). *Report on Global Market Supply: 2020/2021*; Agricultural Market Information Company (AMI): Las Vegas, NV, USA, 2020.

11.   Brandenburg, J. Lipid Production from Lignocellulosic Material by Oleaginous Yeasts. Doctoral Thesis, Swedish University of Agricultural Sciences Uppsala, Uppsala, Sweden, 2021.

12.   Shaigani, P.; Awad, D.; Redai, V.; Fuchs, M.; Haack, M.; Mehlmer, N.; Brueck, T. Oleaginous yeasts- substrate preference and lipid productivity: A view on the performance of microbial lipid producers. *Microb. Cell Fact.* **2021**, *20*, 220. [CrossRef] [PubMed]

13.   Vasconcelos, B.; Teixeira, J.C.; Dragone, G.; Teixeira, J.A. Oleaginous yeasts for sustainable lipid production—From biodiesel to surf boards, a wide range of "green" applications. *Appl. Microbiol. Biotechnol.* **2019**, *103*, 3651–3667. [CrossRef] [PubMed]

14.   Banwell, M.G.; Pollard, B.; Liu, X.; Connal, L.A. Exploiting Nature's Most Abundant Polymers: Developing New Pathways for the Conversion of Cellulose, Hemicellulose, Lignin and Chitin into Platform Molecules (and Beyond). *Chem.-Asian J.* **2021**, *16*, 604–620. [CrossRef] [PubMed]

15.   Lee, S.; Monnappa, A.K.; Mitchell, R.J. Biological activities of lignin hydrolysate-related compounds. *BMB Rep.* **2012**, *45*, 265–274. [CrossRef] [PubMed]

16.   Yaguchi, A.; Robinson, A.; Mihealsick, E.; Blenner, M. Metabolism of aromatics by Trichosporon oleaginosus while remaining oleaginous. *Microb. Cell Fact.* **2017**, *16*, 206. [CrossRef]

17.   Masri, M.; Garbe, D.; Mehlmer, N.; Brück, T. A sustainable, high-performance process for the economic production of waste-free microbial oils that can replace plant-based equivalents. *Energy Environ. Sci.* **2019**, *12*, 2717–2732. [CrossRef]

18.   Yaguchi, A.; Franaszek, N.; O'Neill, K.; Lee, S.; Sitepu, I.; Boundy-Mills, K.; Blenner, M. Identification of oleaginous yeasts that metabolize aromatic compounds. *J. Ind. Microbiol. Biotechnol.* **2020**, *47*, 801–813. [CrossRef]

19.   Yu, X.; Zheng, Y.; Dorgan, K.M.; Chen, S. Oil production by oleaginous yeasts using the hydrolysate from pretreatment of wheat straw with dilute sulfuric acid. *Bioresour. Technol.* **2011**, *102*, 6134–6140. [CrossRef]

20.   Gong, Z.; Shen, H.; Zhou, W.; Wang, Y.; Yang, X.; Zhao, Z.K. Efficient conversion of acetate into lipids by the oleaginous yeast *Cryptococcus curvatus*. *Biotechnol. Biofuels* **2015**, *8*, 189. [CrossRef]

21.   Gong, Z.; Zhou, W.; Shen, H.; Yang, Z.; Wang, G.; Zuo, Z.; Hou, Y.; Zhao, Z.K. Co-fermentation of acetate and sugars facilitating microbial lipid production on acetate-rich biomass hydrolysates. *Bioresour. Technol.* **2016**, *207*, 102–108. [CrossRef] [PubMed]

22.   Saxena, G.; Bharagava, R.N. *Bioremediation of Industrial Waste for Environmental Safety*; Springer: Singapore, 2020; Volume I, ISBN 9789811318900.

23.   Tiseo, I. Annual production of plastics worldwide from 1950 to 2020 (in million metric tons). *Statista* **2022**, *2020*, 22–24.

24.   European Commission. New market niches for the Pulp and Paper Industry waste based on circular economy approaches. In *H2020-Eu*; European Commission: Brussels, Belgium, 2017; pp. 1–4.

25.   Gottumukkala, L.D.; Haigh, K.; Collard, F.X.; van Rensburg, E.; Görgens, J. Opportunities and prospects of biorefinery-based valorisation of pulp and paper sludge. *Bioresour. Technol.* **2016**, *215*, 37–49. [CrossRef] [PubMed]

26.   Luo, H.; Liu, Z.; Xie, F.; Bilal, M.; Peng, F. Lignocellulosic biomass to biobutanol: Toxic effects and response mechanism of the combined stress of lignin-derived phenolic acids and phenolic aldehydes to Clostridium acetobutylicum. *Ind. Crops Prod.* **2021**, *170*, 113722. [CrossRef]

27.   Jönsson, L.J.; Martín, C. Pretreatment of lignocellulose: Formation of inhibitory by-products and strategies for minimizing their effects. *Bioresour. Technol.* **2016**, *199*, 103–112. [CrossRef] [PubMed]

28.   Koutinas, A.A.; Vlysidis, A.; Pleissner, D.; Kopsahelis, N.; Garcia, I.L.; Kookos, I.K.; Papanikolaou, S.; Kwan, T.H.; Lin, C.S.K. Valorization of industrial waste and by-product streams via fermentation for the production of chemicals and biopolymers. *Chem. Soc. Rev.* **2014**, *43*, 2587–2627. [CrossRef] [PubMed]

29.   Sixta, H. Lenzinger AG Zellstoffherstellung und Recycling von Roh- und Hilfsstoffen nach dem Lenzinger Mg- Bisulfitverfahren. *Lenzing. Ber.* **1986**, *61*, 5–11.

30.   Singh, A.K.; Bilal, M.; Iqbal, H.M.N.; Meyer, A.S.; Raj, A. Bioremediation of lignin derivatives and phenolics in wastewater with lignin modifying enzymes: Status, opportunities and challenges. *Sci. Total Environ.* **2021**, *777*, 145988. [CrossRef]

31.   Koul, B.; Yakoob, M.; Shah, M.P. Agricultural waste management strategies for environmental sustainability. *Environ. Res.* **2021**, *206*, 112285. [CrossRef]

32.   Nitsos, C.K.; Matis, K.A.; Triantafyllidis, K.S. Optimization of hydrothermal pretreatment of lignocellulosic biomass in the bioethanol production process. *ChemSusChem* **2013**, *6*, 110–122. [CrossRef]

33. Almeida, J.R.M.; Wiman, M.; Heer, D.; Brink, D.P.; Sauer, U.; Hahn-Hägerdal, B.; Lidén, G.; Gorwa-Grauslund, M.F. Physiological and Molecular Characterization of Yeast Cultures Pre-Adapted for Fermentation of Lignocellulosic Hydrolysate. *Fermentation* **2023**, *9*, 72. [CrossRef]

34. Duwe, A.; Tippkötter, N.; Ulber, R. Lignocellulose-biorefinery: Ethanol-focused. *Adv. Biochem. Eng. Biotechnol.* **2017**, *166*, 177–215.

35. Díaz, T.; Fillet, S.; Campoy, S.; Vázquez, R.; Viña, J.; Murillo, J.; Adrio, J.L. Combining evolutionary and metabolic engineering in Rhodosporidium toruloides for lipid production with non-detoxified wheat straw hydrolysates. *Appl. Microbiol. Biotechnol.* **2018**, *102*, 3287–3300. [CrossRef] [PubMed]

36. Nigam, J.N. Ethanol production from hardwood spent sulfite liquor using an adapted strain of Pichia stipitis. *J. Ind. Microbiol. Biotechnol.* **2001**, *26*, 145–150. [CrossRef] [PubMed]

37. Tjørve, K.M.C.; Tjørve, E. The use of Gompertz models in growth analyses, and new Gompertz-model approach: An addition to the Unified-Richards family. *PLoS ONE* **2017**, *12*, e0178691. [CrossRef] [PubMed]

38. Bligh, E.G.; Dyer, W.J. A rapid method of total lipid extraction and purification. *Can. J. Biochem. Physiol.* **1959**, *37*, 911–917. [CrossRef] [PubMed]

39. Intelligen Inc. *SuperPro® Designer V10*; Intelligen, Inc.: Medford, MA, USA, 2019.

40. Walas, S.M. *Chemical Process Equipment: Selection and Design*; Butterworth-Heinemann Reed Publ. Inc.: Oxford, UK, 1990; pp. 1–755.

41. Humbird, D.; Davis, R.; Tao, L.; Kinchin, C.; Hsu, D.; Aden, A. *Process Design and Economics for Biochemical Conversion of Lignocellulosic Biomass to Ethanol*; National Renewable Energy Lab.: Golden, CO, USA, 2011; pp. 153–170.

42. Eurostat Industriestrompreise in Deutschland in den Jahren 2001 bis 2021. *Statista*. 2022. Available online: https://de.statista.com/statistik/daten/studie/155964/umfrage/entwicklung-der-industriestrompreise-in-deutschland-seit-1995/ (accessed on 18 October 2022).

43. Ageitos, J.M.; Vallejo, J.A.; Veiga-Crespo, P.; Villa, T.G. Oily yeasts as oleaginous cell factories. *Appl. Microbiol. Biotechnol.* **2011**, *90*, 1219–1227. [CrossRef] [PubMed]

44. Bann, W. *Acetic Acid/Vinyl Acetate—The Waiting Is The Hardest Part*; ResourceWise: Charlotte, NC, USA, 2021.

45. Alibaba.com. Acetic Acid. 2022. Available online: https://www.alibaba.com/trade/search?fsb=y&IndexArea=product_en&CatId=&tab=all&SearchText=acetic+acid (accessed on 18 October 2022).

46. Broda, M.; Yelle, D.J. Bioethanol Production from Lignocellulosic Biomass—Challenges and Solutions. *Molecules* **2022**, *27*, 8717. [CrossRef]

47. Sharma, S.; Tsai, M.L.; Sharma, V.; Sun, P.P.; Nargotra, P.; Bajaj, B.K.; Chen, C.W.; Dong, C. Di Environment Friendly Pretreatment Approaches for the Bioconversion of Lignocellulosic Biomass into Biofuels and Value-Added Products. *Environments* **2023**, *10*, 6. [CrossRef]

48. Chopra, J.; Sen, R. Process optimization involving critical evaluation of oxygen transfer, oxygen uptake and nitrogen limitation for enhanced biomass and lipid production by oleaginous yeast for biofuel application. *Bioprocess Biosyst. Eng.* **2018**, *41*, 1103–1113. [CrossRef] [PubMed]

49. Schneider, H. Selective removal of acetic acid from hardwood-spent sulfite liquor using a mutant yeast. *Enzym. Microb. Technol.* **1996**, *19*, 94–98. [CrossRef]

50. Singh, S.; Pandey, D.; Saravanabhupathy, S.; Daverey, A.; Dutta, K.; Arunachalam, K. Liquid wastes as a renewable feedstock for yeast biodiesel production: Opportunities and challenges. *Environ. Res.* **2022**, *207*, 112100. [CrossRef] [PubMed]

51. Van Gerpen, J. Cetane Number Testing of Biodiesel. In *Third Liquid Fuel Conference: Liquid Fuel and Industrial Products from Renewable Resources*; American Society of Agricultural Engineers: St. Joseph, MI, USA, 1996; pp. 197–206.

52. Zahan, K.A.; Kano, M. Biodiesel production from palm oil, its by-products, and mill effluent: A review. *Energies* **2018**, *11*, 2132. [CrossRef]

53. Tamilselvan, P.; Sassykova, L.; Bhaskar, K.; Subramanian, S.; Gomathi, K.; Prabhahar, M.; Prakash, S. Effect of Saturated Fatty Acid Composition of Biodiesel on Oxides of Nitrogen and Smoke Emissions. *J. Chem. Technol. Metall.* **2023**, *58*, 167–177.

54. Papanikolaou, S.; Aggelis, G. Lipids of oleaginous yeasts. Part I: Biochemistry of single cell oil production. *Eur. J. Lipid. Sci. Technol.* **2011**, *113*, 1031–1051. [CrossRef]

55. Alibaba_Palm-Oil. 2022. Available online: https://www.alibaba.com/product-detail/Best-Quality-cheap-organic-refined-Crude_60360488487.html?spm=a2700.galleryofferlist.normal_offer.d_title.56f1cfd5ZqfiRs&s=p (accessed on 18 October 2022).

56. Brander, M.; Tipper, R.; Hutchison, C.; Davis, G. Consequential and attributional approaches to LCA: A Guide to policy makers with specific reference to greenhouse gas LCA of biofuels. *Econ. Press* **2008**, *44*, 1–14.

57. Alibaba2_Canola-Oil. 2022. Available online: https://www.alibaba.com/trade/search?fsb=y&IndexArea=product_en&CatId=&SearchText=canola+oil&selectedTab=product_en (accessed on 20 October 2022).

58. Ullmann, J.; Grimm, D. Algae and their potential for a future bioeconomy, landless food production, and the socio-economic impact of an algae industry. *Org. Agric.* **2021**, *11*, 261–267. [CrossRef]

59. Acetic Acid. *Chemical Economics Handbook*; IHS Markit, S&P Global: London, UK, 2021. Available online: https://ihsmarkit.com/products/acetic-acid-chemical-economics-handbook.html (accessed on 5 April 2022).

60. Helmlinger, E. Celanese Corporation Berichtet Ergebnisse für das Erste Quartal 2021. 2021. Available online: https://www.celanese.de/News-And-Media/2021/April/ergebnisse-fur-das-erste-quartal-2021 (accessed on 17 February 2023).

61. Doelle, K.; Bajrami, B. Sodium Hydroxide and Calcium Hydroxide Hybrid Oxygen Bleaching with System. *IOP Conf. Ser. Mater. Sci. Eng.* **2018**, *301*, 12136. [CrossRef]
62. Patil, R.; Genco, J.; Pendse, H.; Van Heiningen, A. Process for producing acetic acid in hardwood kraft pulp mills. *Tappi J.* **2017**, *16*, 287–300. [CrossRef]
63. Kobetičová, K.; Nábělková, J. Effect of wood hemicellulose composition on binding interactions with caffeine. *Buildings* **2021**, *11*, 515. [CrossRef]
64. Sivasamy, A.; Cheah, K.Y.; Fornasiero, P.; Kemausuor, F.; Zinoviev, S.; Miertus, S. Catalytic applications in the production of biodiesel from vegetable oils. *ChemSusChem* **2009**, *2*, 278–300. [CrossRef] [PubMed]
65. Nurchi, C.; Buonvino, S.; Arciero, I.; Melino, S. Sustainable Vegetable Oil-Based Biomaterials: Synthesis and Biomedical Applications. *Int. J. Mol. Sci.* **2023**, *24*, 2153. [CrossRef] [PubMed]

 *fermentation*

Article

High-Gravity Fermentation for Bioethanol Production from Industrial Spent Black Cherry Brine Supplemented with Whey

Javier Ricardo Gómez Cardozo, Jean-Baptiste Beigbeder, Julia Maria de Madeiros Dantas and Jean-Michel Lavoie *

Biomass Technology Laboratory, Department of Chemical Engineering and Biotechnology Engineering, Université de Sherbrooke, 2500 Boul. de l'Université, Sherbrooke, QC J1K 2R1, Canada
* Correspondence: jean-michel.lavoie2@usherbrooke.ca

Abstract: By-products from different industries could represent an available source of carbon and nitrogen which could be used for bioethanol production using conventional *Saccharomyces cerevisiae* yeast. Spent cherry brine and whey are acid food by-products which have a high organic matter content and toxic compounds, and their discharges represent significant environmental and economic challenges. In this study, different combinations of urea, yeast concentrations, and whey as a nutrient source were tested for bioethanol production scale-up using 96-well microplates as well as 7.5 L to 100 L bioreactors. For bioethanol production in vials, the addition of urea allowed increasing the bioethanol yield by about 10%. Bioethanol production in the 7.5 L and 100 L bioreactors was 73.2 g·L^{-1} and 103.5 g·L^{-1} with a sugar consumption of 81.5% and 94.8%, respectively, using spent cherry brine diluted into whey (200 g·L^{-1} of total sugars) supplemented with 0.5 g·L^{-1} urea and 0.5 g·L^{-1} yeast at 30 °C and a pH of 5.0 after 96 h of fermentation for both systems. The results allow these by-products to be considered low-economic-value alternatives for fuel- or food-grade bioethanol production.

Keywords: bioethanol; fermentation; yeast; cherry brine; whey; industrial by-products

Citation: Gómez Cardozo, J.R.; Beigbeder, J.-B.; Dantas, J.M.d.M.; Lavoie, J.-M. High-Gravity Fermentation for Bioethanol Production from Industrial Spent Black Cherry Brine Supplemented with Whey. *Fermentation* 2023, 9, 170. https://doi.org/10.3390/fermentation9020170

Academic Editor: Timothy Tse

Received: 16 January 2023
Revised: 6 February 2023
Accepted: 7 February 2023
Published: 14 February 2023

1. Introduction

The global biofuel market is led by biodiesel and bioethanol production, in which the production of the latter corresponds to about 74% of the biofuel industry [1]. Bioethanol is a commodity used in many industries, such as medicine, pharmacy and biofuel without forgetting as an additive for gasoline to cut down the emissions of pollutants. In this sense, this compound has been important in the race against reducing our dependence on fossil fuels and greenhouse gas emissions. Bioethanol is obtained via the fermentation of sugars present in several sources, such as sugarcane, corn, sugar beet, wheat and sweet sorghum, as well as agricultural, industrial and forest by-products [2,3].

The valorization of industrial by-products for bioethanol production has gained more interest at the industrial level to reduce overall production process costs. Among them, molasses [4], whey [5], hydrolyzed lignocellulosic materials [6] and by-products from different industries [7] have been evaluated to obtain this commodity.

Black cherry (*Prunus serotina*) fruits are used for human consumption in many forms, such as fresh, canned, juiced and dried, but also as a flavoring for soft drinks and in jams and jellies. Nonetheless, black cherry brine is an acid solution produced as a preservation medium for cherry fruits before being used in food products, especially in the ice cream industry. Its composition depends on the formulation of the producers and cherry type. Typically, cherries, water, sugars (4.0–11.0% *w/v*) and starches are present in this medium, as well as natural flavors, colorants and preservatives, such as citric acid, sodium benzoate, potassium sorbate, calcium chloride and sulfites, used to avoid microbial contamination [8]. Once the fruits are removed, the solution remains as a by-product that must be treated before its disposal due to its high biological (DBO) and chemical oxygen demand (DCO), toxicity and acidity.

Whey, on the other hand, is the main by-product of the cheese industry. It is estimated that, for every 1 kg of cheese produced, 9 kg of whey is generated [9]. This makes it an abundant by-product, containing an important organic (lactose, 3.8–5.0% *w/v*, and proteins, 0.8–1.0% *w/v*) and inorganic (0.5–0.7% *w/v*) load [10]. However, it is estimated that about 50% of the total whey production (189 million tons/year) is considered as a waste and is disposed of without any treatment [11]. Therefore, an alternative would be to use it as a supplement for fermentation processes, and although it is true that yeasts do not usually metabolize lactose, the nutrients present in whey can be used in the enzymatic processes of these microorganisms.

This study focuses on the recovery of these two industrial residues (spent cherry brine, hereafter simply called cherry juice, and whey from a local company, located in the Quebec Province, Canada). Fermentations were carried out using cherry juice as a carbon source and whey as a nutritional supplement, aiming to improve the sugar consumption present in cherry juice and increase the yield in ethanol production. For this, growth tests were carried out in microplates, after which the process was scaled up to bioreactors with capacities of 7.5 L and 100 L, respectively.

To our knowledge, this is the first time that whey and cherry juice were combined for fermentation, which is an alternative use of such by-products for the sustainable production of bioethanol.

2. Materials and Methods

2.1. Cherry Juice and Whey Characterization

Both cherry juice and whey were collected from a local sugar refinery located in Coaticook (Québec, Canada) and stored at 4 °C and −20 °C, respectively, until further use. The cherry juice (Black Cheery Halves, $_{MD}$Cibona^{TM}, Lot# C21236) was the liquid part used to preserve the black cherry fruit before being used in various Laiterie de Coaticook Ltée products. This preserving liquid was mainly composed of sugar, water, citric acid, sodium benzoate, $CaCl_2$, natural and/or artificial flavor, color and sulfites, according to the producer. The whey was generated from cheese production by the same company. The cherry juice and whey were initially characterized at the Biomass Technology Laboratory analytical facilities to identify the concentrations of compounds such as sugars and organic acids, as well as for an elemental analysis to quantify the Carbon (C), Hydrogen (H), Nitrogen (N) and Oxygen (O) content (see Table 1).

2.2. Yeast Inoculum, Culture Media and Growth Curves in Microplates

Yeast inoculum was prepared by rehydrating *Saccharomyces cerevisiae* dry yeast cells (DistilaMax$^®$ HT, Lallemand Biofuels & Distilled Spirits, Canada) with tap water for 15 min at 30 °C and 140 rpm using a shaking incubator (Infors-HT Inc., Bottmingen, Switzerland).

The composition of the media used to evaluate yeast growth was cherry juice at 60 g·L^{-1} of total fermentable sugars (6.0°Brix) diluted in tap water or whey, supplemented with urea, Crystal Reagent, ACS, 99.0–100.5%, Anachemia, Mississauga, ON, Canada, Inc. (0.0 g L^{-1}, 0.5 g·L^{-1}, and 1.0 g·L^{-1}), at two initial yeast concentrations (0.5 g·L^{-1} and 1.0 g·L^{-1}). Different yeast suspensions were prepared to obtain the desired initial concentration for each experiment. The initial pH of the medium was adjusted to 5.0 with NaOH 1.0 N. Fermentations were performed by adding 200 μL of the culture medium inoculated according to the initial yeast concentration in a 96-well microplate, and each condition was evaluated in triplicate. The plates were then shaken in a thermostat microplate reader (BioTek EPOCH2) at 30 °C for 49 cycles (about 24 h in total) of double orbital shaking (1 min at 282 rpm) followed by optical density measurements at 600 nm (OD$_{600}$). The data collected were converted to biomass using a calibration curve (see Figure S1). All the experiments were conducted in triplicate, and the results represent the Mean (M) ± Standard Error (STD).

Table 1. Characterization of cherry juice and whey used in this study.

Parameter	Cherry Juice	Whey
pH	3.3	4.0
Density (g·mL^{-1})	1.2 ± 0.1	1.1 ± 0.1
Humidity (%wt)	61.6 ± 0.2	93.7 ± 0.0
Total Solid (%wt)	38.4 ± 0.2	6.3 ± 0.0
Ashes (%wt)	0.1 ± 0.0	0.6 ± 0.0
Total Protein (g·L^{-1})	-	0.5 ± 0.1
Sugars (g·L^{-1})		
Sucrose	248.0 ± 0.8	-
Lactose	-	41.9 ± 0.1
Glucose	105.6 ± 0.3	-
Fructose	99.4 ± 1.1	-
Total sugars	453.0 ± 2.2	41.9 ± 0.1
Other Compounds (g·L^{-1})		
Lactic Acid	-	33.1 ± 0.5
Citric Acid	3.9 ± 0.1	-
Sodium Benzoate	1.2 ± 0.1	-
Elemental Analysis (%wt)		
Carbon	17.0 ± 0.0	2.4 ± 0.5
Hydrogen	2.6 ± 0.2	0.3 ± 0.1
Nitrogen	<LOD	0.1 ± 0.0
Oxygen	22.0 ± 1.1	2.2 ± 0.3

LOD: Limit of detection.

2.3. Culture Media and Fermentation in Vials

The culture media employed in all fermentations was composed of cherry juice that was added to whey to reach 200 g·L^{-1} (around 20 °Bx) of total fermentable sugars (sucrose, glucose and fructose). The initial and final sugar concentrations of the mixture of cherry juice and whey were measured using a DIONEX ICS-500+ ion chromatography system (see analytical procedures). The initial pH of the medium was adjusted to 5.0 with NaOH 1.0 N. Carbon dioxide (CO_2) production indicated the metabolism of the sugars present in the medium. To monitor fermentations, a gravimetric method was used to measure the amount of gas released according to the weight difference, using an accurate balance (0.001 g). The final ethanol concentration was quantified via HPLC (see analytical procedures).

The fermentations were carried out in 50 mL vials with rubber septum stoppers and aluminum rings, with a working volume of 30 mL using the culture media described above. The initial yeast concentration used was 0.5 g·L^{-1}. Urea was used as a nitrogen source at two concentrations (0.0 g·L^{-1} and 0.5 g·L^{-1}) to evaluate the effects on the ethanol yield. Fermentation ran under anaerobic conditions (by flushing the vials with $N_{2(g)}$ for 1.0 min) for 170 h at 30 °C and 140 rpm.

2.4. Bioethanol Production Using Cherry Juice and Whey in Bioreactors

Bioethanol production in the bioreactors was evaluated using the best conditions from screening. Based on the experimental results obtained in 50 mL vials, yeast and total reducing sugar concentrations were used to keep constant fermentation media conditions in a 7.5 L and in a 100 L batch bioreactor with a working volume of 5.0 L and 50.0 L, respectively. The 7.5 L batch bioreactor (Infors-HT Inc., Bottmingen, Switzerland) was operated at 30 °C and was stirred at 200 rpm using 2 Rushton turbine impellers. The 100 L bioreactor, a 0.5 m tall vertical cylinder made from stainless with 0.5 m of internal diameter, equipped with a 2000 W heating element to control the internal temperature of the system and an agitation system consisting of a 2-blade inclined impeller powered by a 12 V DC electric motor, was operated at 30 °C and 70 rpm. Samples were taken in triplicate, and the data reported were the M ± STD of the samples analyzed from each batch.

2.5. Analytical Procedures

Brix degrees (°Brix), or the total soluble solid content of cherry juice media, were measured using a digital pocket refractometer (ATAGO, PAL-BX/RI, Bellevue, WA, USA). Samples were tested in triplicate, and the data reported were the Mean (M) ± Standard Error (STD) of each one.

The determination of moisture and dry matter was carried out according to the mass loss, for which around 5.0 g of cherry juice and whey were weighed in previously weighed aluminum dishes, and the samples were dried at 105 °C in an oven (Fisher Sci 100L, Pittsburg, PA, USA) for 12 h or until a constant weight was reached. The dry matter obtained after moisture determination was used to quantify the CHNS/O content with an elemental analyzer (Flash 2000 OEA, Thermo Fisher Scientific, Toronto, ON, Canada), as well as the ash content, which was determined according to the NREL/TP-510-42622 method [12]. Finally, the total protein content was determined following the method described by Bradford [13]. In all cases, samples were tested in triplicate, and the data reported were the M ± STD for each parameter.

High-performance liquid chromatography (HPLC) was used to quantify ethanol and lactic acid. Samples were diluted, filtered and injected into the chromatographic system (Agilent 1100 series equipped with a G1362A Refractive Index Detector) (Agilent Technologies Inc., Colorado Springs, CO, USA). The system was operated at 40 °C with an isocratic elution method (2.5 mM). The HPLC set-up also had a G1322A Degasser and a G1311A Quaternary Pump. A G1313A Autosampler injected 40 µL of the sample, and the column used was ROA-Organic Acid H$^+$ (8%) at 65 °C. The elution was performed at a constant flow of 0.600 mL·min^{-1} of a 2.5 mM H_2SO_4 solution. A calibration curve from 10 ppm to 1000 ppm was performed using the following standards: L-lactic 99% (Alfa Aesar, Tewkbury, MA, USA), ethanol 99% (Sigma Aldrich, Oakville, ON, Canada) and glycerol 99% (Sigma Aldrich, Oakville, ON, Canada).

The DIONEX ICS-500+ ion chromatography system allowed the quantification of the total fermentable sugars (sucrose, glucose and fructose), citric acid and sodium benzoate. The system was equipped with a KOH eluent generator to provide a proper eluent concentration. A 200 mM KOH post-injection with a Dionex GP 50 gradient pump was implemented to ensure signal stability. A Dionex CarboPac Sa10-4 µM column was used. The oven was set to 45 °C, and the electrochemical detector was at 30 °C. The injection volume was 0.4 µL, and elution was made with an aqueous solution of KOH at 1.25 mL·min^{-1} at the following concentrations: 1 mM for 12 min, 10 mM for 5 min and 1 mM for 1 min. For sugars, the calibration curve involved a concentration of standards varying from 10 ppm to 1000 ppm and involved the following standards: sucrose 99%, glucose 99% and fructose 99%, which were all purchased from Sigma-Aldrich. For citric acid and sodium benzoate, the chromatograms of the standard samples, citric acid 99% (Sigma-Aldrich) and sodium benzoate 99% (Sigma-Aldrich) with a concentration of 1000 ppm, were taken and compared with the chromatograms of the raw cherry juice and whey samples.

2.6. Ethanol Yield, Sugar and Productivity

The fermentation performance was evaluated using the ethanol yield (Y_{EtOH}) based on the theoretical ethanol production and using the sugar uptake according to Equations (1) and (2). Ethanol productivity was calculated as the grams of ethanol produced per day per liter of liquid volume of the bioreactor.

$$Y_{EtOH}(\%) = \frac{EtOH}{0.51 \cdot S_0} \times 100 \tag{1}$$

$$Sugar\ Consumption(\%) = \frac{S_0 - S_f}{S_0} \times 100 \tag{2}$$

where:

EtOH: Ethanol concentration (g·L^{-1});

$0.51 \cdot S_0$: Theoretical ethanol concentration from hexoses (g·L^{-1});

S_0: Initial fermentable sugars (Sucrose as C6, Glucose and Fructose (g·L^{-1});

S_f: Final fermentable sugars (Sucrose as C6, Glucose and Fructose) (g·L^{-1}).

3. Results and Discussion

3.1. Yeast Growth in Cherry Juice Diluted into Whey

In order to evaluate the yeast growth using cherry juice as a substrate, some tests were carried out to determine the possible toxic effects of the medium on *Saccharomyces cerevisiae* in a microplate system. In addition, the media were supplemented with different urea concentrations because it was initially known that cherry juice has no nitrogen sources, as shown in Table 1. Moreover, tests were carried out by diluting the cherry juice in whey with and without the addition of a nitrogen source.

After 12 h of fermentation, it was observed that the media that used whey as a diluent reached the maximum values for the conditions evaluated, whereas in the media in which the diluent was tap water, slower growth was observed (Figure 1). Increasing the initial yeast load from 0.5 to 1.0 g·L^{-1} improved the growth kinetics of the cherry juice media diluted with water and supplemented with urea. In addition, growth inhibition was observed in the media in which water was used as diluent and no nitrogen source was added, due to the lack of nutrients that are not supplemented only with the addition of urea. It is known that yeasts in their fermentation processes use nitrogen for their growth [14], and the presence of citric acid and sodium benzoate in cherry juice could inhibit its growth [15] even when the medium is supplemented with urea. Finally, the addition of urea does not significantly affect yeast growth in whey-supplemented media. For instance, at 0.5 g·L^{-1} of yeast, CJWU0.0 followed the same trend as CJWU0.5 and CJWU1.0, which means that whey provided enough nutrients to the media without adding urea. However, when using 1.0 g·L^{-1} of yeast, it was observed that the inhibition was lower, and the growth was similar for all the combinations tested, except for CJU0.0, in which strong growth inhibition was observed due to the high toxicity of the medium.

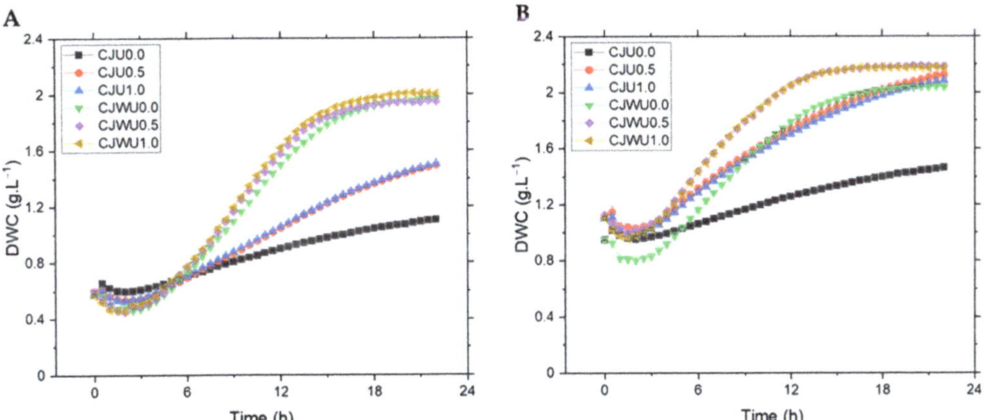

Figure 1. Yeast growth curves in the different fermentation media tested: Cherry Juice diluted in Tap Water (CJ) at 6.0°Brix, Cherry Juice diluted in Whey (CJW) at 6.0°Brix; Urea (0.0 g·L^{-1}, 0.5 g·L^{-1} and 1.0 g·L^{-1}) (U0.0, U0.5 and U1.0) and initial yeast concentrations (**A**) 0.5 g·L^{-1} and (**B**) 1.0 g·L^{-1}. All the tests were performed at pH 5.0 (e.g., CJU0.5. Cherry Juice diluted in Tap Water supplemented with 0.5 g·L^{-1} urea. CJWU0.5: Cherry Juice diluted in Whey supplemented with 0.5 g·L^{-1} Urea).

The results also showed that whey contained the nutritional value necessary for the yeast uptake of sugars present in the cherry juice for its growth despite the yeast-inhibiting compounds, such as citric acid and sodium benzoate, which were present in the juice

(Table 1). For example, citric acid has been reported to be a yeast growth inhibitor, reducing its growth rate between 64% and 88% when compared to media without the addition of this compound [15]. On the other hand, a study conducted by Yardimci et al. reported that the yeast growth decreased from 84% to 35% when the concentration of sodium benzoate was duplicated from 25 to 50 mM in YPDA media [16]. However, in our study, in the medium that was diluted with whey, lag, exponential and stationary stages were observed, describing the typical microbial growth in such processes. Studies on the valorization of whey have demonstrated the mineral and protein content of this dairy industry by-product [10,11], which explains why whey-containing media were less toxic than those that only used urea.

The yeast used in this study had a strong tolerance to high-gravity media, and the fact that diluting the cherry juice in whey decreased the inhibitory effect of the toxic compounds it contained led to the use of high-gravity media using these by-products at a weight ratio of cherry juice/whey of 1.11 to reach approximately 200 g·L^{-1} (20° Brix) of total fermentable sugars and to evaluate ethanol production, as described in the following sections.

3.2. Fermentation and CO_2 Production

The growth kinetics results obtained in the last section allowed us to establish the initial fermentation conditions in order to prepare the yeast inoculum for a series of new fermentation trials. In this context, the initial concentration of sugar was fixed to 20° Brix to simulate industrial fementation and to maximize the ethanol productivity. To delineate the effects of urea in yeast growth and ethanol production, the CO_2 production and ethanol yield were measured, as discussed in the following section.

CO_2 production shows the fermentative activity of yeast in different media, which can be directly correlated with ethanol production, as presented in Section 2.3. In this study, the results showed a direct relationship between CO_2 production and ethanol production, demonstrating the metabolic activity of the yeast on the sugars present in the media (Figure 2). The cherry juice incubated in the absence of nutrients and/or whey (CJ20U0.0) generated very limited production of CO_2 (20.2 $\pm$ 1.7 g·L^{-1}) and ethanol yield (18.6 $\pm$ 1.8%). This behavior was caused by yeast-inhibiting compounds present in the cherry juice along with the production of ethanol and glycerol, two other compounds that are toxic for yeast, which limited the consumption of fermentable sugars present in the medium, such as fructose, decreasing the yields in the fermentation process. However, the supplementation of 0.5 g·L^{-1} of urea in the same media (CJ20U0.5) showed significant effects on the CO_2 production kinetics, reaching up to 72.2 $\pm$ 0.8 g·L^{-1} of CO_2.

As expected, the cherry juice diluted in whey presented high fermentation responses. The addition of a nitrogen source in whey-containing media increased yields by approximately 11%, from 77.3 $\pm$ 1.5% in the media supplemented with whey (CJW20U0.0) to 87.1 $\pm$ 4.2% in the media supplemented with whey plus 0.5 g·L^{-1} of urea (CJW20U0.5). The addition of urea in high-gravity media such as 20.8° Brix has been reported to improve ethanol production yields [17] using higher concentrations of urea (2.75 g·L^{-1}). However, a high urera content is not recommended because it might favor the production of carcinogenic compounds, such as ethyl carbamate [18]. Considering that, even if the whey provides certain minerals and nutrients to enhance yeast growth (see Table S1), the addition of a nitrogen source favors yeast metabolism for ethanol production. Finally, it should be noted that, although the fermentation processes in the whey-diluted media were carried out up to 170 h, it was observed that, after 120 h, the process reached its maximum value.

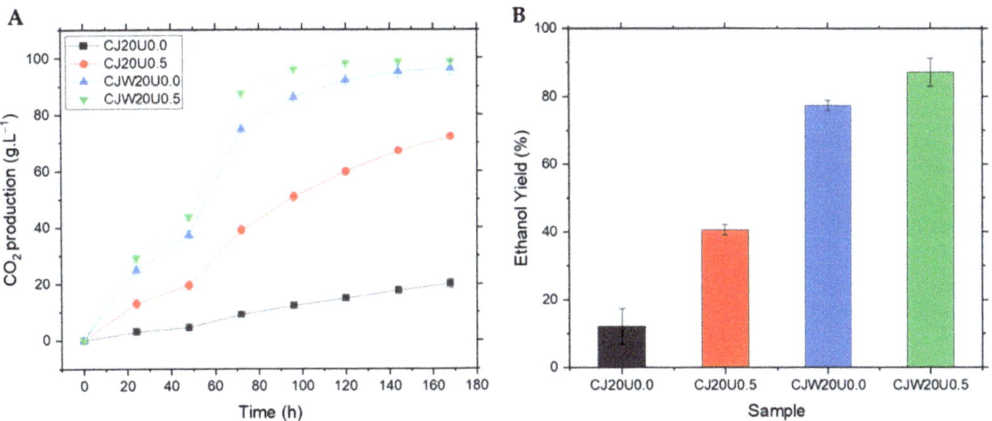

Figure 2. (**A**) CO_2 production in the different fermentation media tested: Cherry Juice diluted in Tap Water (CJ20) at 20°Brix; Cherry Juice diluted in Whey (CJW20) at 20°Brix; Urea (0.0 g·L^{-1}, and 0.5 g·L^{-1}) (U0.0, and U0.5) and initial yeast concentration of 0.5 g·L^{-1} at 30 °C and pH 5.0. (**B**) Ethanol yield in the different fermentation media tested at 170 h of incubation (e.g., CJ20U0.5: Cherry Juice diluted in Tap Water at 20°Brix supplemented with 0.5 g·L^{-1} Urea. CJ20WU0.5: Cherry Juice diluted in Whey at 20°Brix supplemented with 0.5 g·L^{-1} urea).

In this section, the results we found are comparable to those presented by Park and Bakalinsky [8] who reported an ethanol yield of 92%. However, they used a spent cherry brine pretreated with alkaline and acidic processes to detoxify it prior to supplementing it with different nitrogen sources, such as urea and diammonium phosphate (DAP). In that sense, this would result in more stages, which would require more energy and increase the costs of the fermentation process.

3.3. Bioethanol Production Using Bioreactors

Once the laboratory-scale fermentations were completed, the conditions were used for scaling-up ethanol production using 7.5 L and 100 L bioreactors. The yeast growth took place inside the bioreactors under anaerobic conditions. Liquid samples were taken every 24 h to evaluate the sugar consumption and ethanol production for 120 h of incubation. Samples were stored at −20 °C until analysis.

After 24 h of fermentation, sucrose was quickly hydrolyzed into glucose and fructose molecules, and ethanol production started in both reactors (Figure 3). Glucose was fully consumed after 72 h in both systems, and fructose degradation began after 24 h. The maximum ethanol production was obtained after 96 h of fermentation, reaching values up to 73.2 ± 6.0 g·L^{-1} and 103.5 ± 2.3 g·L^{-1} in the 7.5 L and 100 L bioreactors, respectively. However, the sugar profiles were different due to the operating conditions used in each case because these bioreactors did not have the same geometrical configuration and operating capacity. Nonetheless, scaling-up was performed based just on the optimal conditions for ethanol production from the yeast used in this study (pH and temperature). This had a major impact on the yields because, after 72 h in 7.5 L bioreactor, the yeast was not able to degrade fructose remaining in the media. Factors such as agitation [19] and ethanol concentration can affect the sugar consumption. Thus, a comparison could be performed, as shown in Table 2.

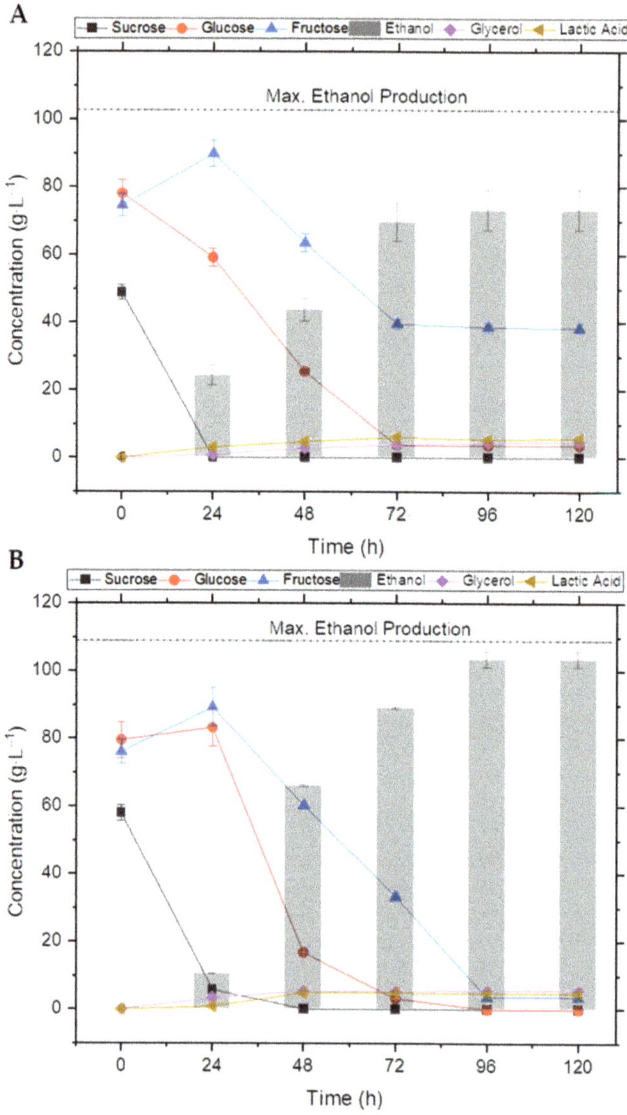

Figure 3. Variations in sugars, ethanol, glycerol and lactic acid concentration during fermentation scale-up trials for volumes of 7.5 L (**A**) and 100 L (**B**) in a batch reactor using an initial yeast concentration of 0.5 g·L^{-1}, an initial urea concentration of 0.5 g·L^{-1}, cherry juice diluted into whey at 200 g·L^{-1} (20°Brix) at 30 °C, a pH of 5.0, and 200 rpm and 70 rpm, respectively.

This difference in system behavior has been reported in a previous study using sugar beet molasses, in which using the same bioreactor systems reached different ethanol yields [20]. From the literature, it is known that, when these scaling processes are performed, it is important to maintain certain parameters, both dimensional configurations [22] and mass transfer, such as power input [21], to minimize variations in the microenvironments where biological processes are performed. Nonetheless, the ethanol concentration and ethanol yield are comparable with those found in the literature, as we can see in Table 2. On the other hand, the selected operational conditions allowed the yeast to uptake most of the sugar available in media composed by cherry juice diluted into whey. Likewise, the concentration of by-products such as glycerol and lactic acid was similar for both scales, indicating that the stress generated by the scale changes did not affect the yeast metabolism, always favoring ethanol production.

Table 2. Comparison of fermentation performance and by-products measured for different substrates: Cherry Juice diluted into Whey (200 g·L^{-1} of total sugars) (this study). Sugar Beet Molasses (170 g·L^{-1} of total sugar) [20] *. Hydrolyzed Raw Cassava Starch (200 g·L^{-1}) [21] **.

Substrates/Parameters	Cherry Juice/Whey		Sugar Beet Molasses *		Hydrolyzed Raw Cassava Starch **	
	Bioreactor					
	7.5-L	100-L	7.5-L	100-L	5.0-L	200-L
	Fermentation					
Ethanol (g·L^{-1})	73.2 ± 6.0	103.5 ± 2.3	79.6 ± 0.7	63.2 ± 1.1	81.9 ± 1.9	80.9 ± 0.5
Ethanol Yield (%)	71.2 ± 5.8	93.8 ± 1.8	99.5 ± 0.9	78.9 ± 0.4	75.3 ± 1.3	74.4 ± 0.3
Sugar Consumption (%)	81.5 ± 0.3	98.4 ± 0.1	92.2 ± 0.8	100.0 ± 0.0	-	-
	By-products					
Glycerol (g·L^{-1})	4.8 ± 0.2	5.6 ± 0.1	8.7 ± 0.2	6.5 ± 0.2	11.4 ± 0.1	11.0 ± 0.0
Lactic Acid (g·L^{-1})	5.7 ± 0.2	4.7 ± 0.6	3.8 ± 0.1	3.4. ± 0.2	0.5 ± 0.0	0.5 ± 0.0
Acetic Acid (g·L^{-1})	0.0 ± 0.0	0.0 ± 0.0	0.8 ± 0.1	0.2 ± 0.0	0.0 ± 0.0	0.2 ± 0.0

Finally, if the results found in vials are compared with the 100 L bioreactor, it can be observed that they are very similar and, in that sense, the process should be scalable with reproducible results. On the other hand, in the case of the 7.5 L reactor, it is possible that the agitation speed of the system had a negative impact on the fermentation process, as we have already mentioned, and that the inhibition of fructose consumption decreased the ethanol yields.

4. Conclusions

The outcomes of the present work show the potential of cherry juice as a carbon source and whey as a nutrient source for bioethanol production using *Saccharomyces cerevisiae*. The urea addition improved the ethanol yield by up to 10%. Because urea is a cheap product, it is an excellent option that improves the yield and would not increase the costs of the processes. The sugar consumption (between 81.5% and 98.4%) during the fermentation process and the ethanol yield (between 71.2% and 93.8%), showed that a major part of the sugars were metabolized to ethanol under the operational conditions evaluated,. Other metabolites, such as glycerol and lactic acid, reached low values that were good for future distillation processes, and they demonstrated, once again, that the selected operational conditions are ideal for ethanol production using cherry juice supplemented with whey. Finally, cherry juice by itself is a medium that inhibits yeast growth, limiting yields in fermentation processes, and the addition of urea is not sufficient to diminish this toxic effect. Therefore, detoxification processes are necessary if it is to be used as the sole source of carbon for fermentation processes.

Supplementary Materials: The following supporting information can be downloaded at https://www.mdpi.com/article/10.3390/fermentation9020170/s1: Figure S1: Dry Cell Weight (DWC) versus Optical Density at 600 nm (OD600) of *Saccharomyces cerevisiae*. Table S1: Analysis of minerals present in whey.

Author Contributions: J.R.G.C.: conceptualization, methodology, investigation, formal analysis and writing—review and editing; J.-B.B.: conceptualization, methodology, investigation, formal analysis and writing—review and editing; J.M.d.M.D.: conceptualization, methodology, investigation, formal analysis and writing—review and editing; J.-M.L.: supervision, writing—review and editing, and funding acquisition. All authors have read and agreed to the published version of the manuscript.

Funding: This research was funded by the Conseil de Recherches en Sciences Naturelles et en Génie du Canada (ALLRP 555816–20) and the authors would also like to acknowledge the financial contribution from our industrial partners Gestion P.A.S and Laiterie de Coaticook Ltée.

Institutional Review Board Statement: Not applicable.

Informed Consent Statement: Not applicable.

Data Availability Statement: Not applicable.

Acknowledgments: The authors would like to thank Jean Provencher and Philippe Robert (Gestion P.A.S Inc.) for providing the industrial cherry juice, whey and all valuable information regarding their production. Further acknowledgement goes to the Analytical Chemistry Laboratory of the Biomass Technology Laboratory and to Thierry Ghislain and Maxime Lessard for their support in the analysis of the samples.

Conflicts of Interest: The authors declare no conflict of interest.

References

1. Roukas, T.; Kotzekidou, P. Rotary Biofilm Reactor: A New Tool for Long-Term Bioethanol Production from Non-Sterilized Beet Molasses by *Saccharomyces cerevisiae* in Repeated-Batch Fermentation. *J. Clean. Prod.* **2020**, *257*, 120519. [CrossRef]
2. Camargos, C.V.; Moraes, V.D.; de Oliveira, L.M.; Guidini, C.Z.; Ribeiro, E.J.; Santos, L.D. High Gravity and Very High Gravity Fermentation of Sugarcane Molasses by Flocculating *Saccharomyces cerevisiae*: Experimental Investigation and Kinetic Modeling. *Appl. Biochem. Biotechnol.* **2021**, *193*, 807–821. [CrossRef] [PubMed]
3. Ambaye, T.G.; Vaccari, M.; Bonilla-Petriciolet, A.; Prasad, S.; van Hullebusch, E.D.; Rtimi, S. Emerging Technologies for Biofuel Production: A Critical Review on Recent Progress, Challenges and Perspectives. *J. Environ. Manag.* **2021**, *290*, 112627. [CrossRef] [PubMed]
4. Cruz, M.L.; de Resende, M.M.; Ribeiro, E.J. Improvement of Ethanol Production in Fed-Batch Fermentation Using a Mixture of Sugarcane Juice and Molasse under Very High-Gravity Conditions. *Bioprocess Biosyst. Eng.* **2021**, *44*, 617–625. [CrossRef] [PubMed]
5. Pasotti, L.; Zucca, S.; Casanova, M.; Micoli, G.; Cusella De Angelis, M.G.; Magni, P. Fermentation of Lactose to Ethanol in Cheese Whey Permeate and Concentrated Permeate by Engineered *Escherichia coli*. *BMC Biotechnol.* **2017**, *17*, 48. [CrossRef] [PubMed]
6. Carrillo-Nieves, D.; Saldarriaga-Hernandez, S.; Gutiérrez-Soto, G.; Rostro-Alanis, M.; Hernández-Luna, C.; Alvarez, A.J.; Iqbal, H.M.N.; Parra-Saldívar, R. Biotransformation of Agro-Industrial Waste to Produce Lignocellulolytic Enzymes and Bioethanol with a Zero Waste. *Biomass Convers. Biorefinery* **2020**, *12*, 253–264. [CrossRef]
7. Bhadana, B.; Chauhan, M. Bioethanol Production Using *Saccharomyces cerevisiae* with Different Perspectives: Substrates, Growth Variables, Inhibitor Reduction and Immobilization. *Ferment. Technol.* **2016**, *5*, 2–5. [CrossRef]
8. Park, H.; Bakalinsky, A.T. Ethanol Production from Spent Cherry Brine. *J. Ind. Microbiol. Biotechnol.* **1997**, *19*, 12–17. [CrossRef] [PubMed]
9. Hadiyanto; Ariyanti, D.; Aini, A.P.; Pinundi, D.S. Optimization of Ethanol Production from Whey through Fed-Batch Fermentation Using *Kluyveromyces marxianus*. *Energy Procedia* **2014**, *47*, 108–112. [CrossRef]
10. Tsakali, E.; Petrotos, K.; D'Alessandr, A.G.; Goulas, P. A Review on Whey Composition and the Methods Used for Its Utilization for Food and Pharmaceutical Products. In Proceedings of the 6th International Conference on Simulation and Modelling in the Food and Bio-Industry 2010, FOODSIM 2010, Braganca, Portugal, 24–26 June 2010; pp. 195–201.
11. Pires, A.F.; Marnotes, N.G.; Rubio, O.D.; Garcia, A.C.; Pereira, C.D. Dairy By-Products: A Review on the Valorization of Whey and Second Cheese Whey. *Foods* **2021**, *10*, 1067. [CrossRef] [PubMed]
12. Sluiter, A.; Hames, B.; Ruiz, R.; Scarlata, C.; Sluiter, J.; Templeton, D. *Determination of Ash in Biomass*; NREL Laboratory Analytical Procedure (LAP); NREL: Golden, CO, USA, 2008.
13. Bradford, M. A Rapid and Sensitive Method for the Quantitation of Microgram Quantities of Protein Utilizing the Principle of Protein-Dye Binding. *Anal. Biochem.* **1976**, *72*, 248–254. [CrossRef] [PubMed]
14. Marx, S.; Brandling, J.; Van Der Gryp, P. Marx Ethanol Production from Tropical Sugar Beet Juice. *Afr. J. Biotechnol.* **2012**, *11*, 11709–11720. [CrossRef]
15. Nielsen, M.K.; Arneborg, N. The Effect of Citric Acid and PH on Growth and Metabolism of Anaerobic *Saccharomyces cerevisiae* and *Zygosaccharomyces bailii* Cultures. *Food Microbiol.* **2007**, *24*, 101–105. [CrossRef]
16. Yardimci, B.K.; Sahin, S.C.; Sever, N.I.; Ozek, N.S. Biochemical Effects of Sodium Benzoate, Potassium Sorbate and Sodium Nitrite on Food Spoilage Yeast *Saccharomyces cerevisiae*. *Biologia (Bratisl)* **2022**, *77*, 547–557. [CrossRef]
17. Sriputorn, B.; Laopaiboon, P.; Phukoetphim, N.; Uppatcha, N.; Phuphalai, W.; Laopaiboon, L. Very High Gravity Ethanol Fermentation from Sweet Sorghum Stem Juice Using a Stirred Tank Bioreactor Coupled with a Column Bioreactor. *J. Biotechnol.* **2021**, *332*, 1–10. [CrossRef]

18. Wongsurakul, P.; Termtanun, M.; Kiatkittipong, W.; Lim, J.W.; Kiatkittipong, K.; Pavasant, P.; Kumakiri, I.; Assabumrungrat, S. Comprehensive Review on Potential Contamination in Fuel Ethanol Production with Proposed Specific Guideline Criteria. *Energies* **2022**, *15*, 2986. [CrossRef]

19. Pinheiro, Á.D.T.; Barros, E.M.; Rocha, L.A.; Ponte, V.M.D.R.; de Macedo, A.C.; Rocha, M.V.P.; Gonçalves, L.R.B. Optimization and Scale-up of Ethanol Production by a Flocculent Yeast Using Cashew Apple Juice as Feedstock. *Braz. J. Chem. Eng.* **2020**, *37*, 629–641. [CrossRef]

20. Beigbeder, J.-B.; de Medeiros Dantas, J.M.; Lavoie, J.-M. Optimization of Yeast, Sugar and Nutrient Concentrations for High Ethanol Production Rate Using Industrial Sugar Beet Molasses and Response Surface Methodology. *Fermentation* **2021**, *7*, 86. [CrossRef]

21. Krajang, M.; Malairuang, K.; Sukna, J.; Rattanapradit, K.; Chamsart, S. Single-Step Ethanol Production from Raw Cassava Starch Using a Combination of Raw Starch Hydrolysis and Fermentation, Scale-up from 5-L Laboratory and 200-L Pilot Plant to 3000-L Industrial Fermenters. *Biotechnol. Biofuels* **2021**, *14*, 68. [CrossRef] [PubMed]

22. Imamoglu, E.; Sukan, F.V. Scale-up and Kinetic Modeling for Bioethanol Production. *Bioresour. Technol.* **2013**, *144*, 311–320. [CrossRef] [PubMed]

fermentation

MDPI

Article

Qualitative Screening of Yeast Biodiversity for Hydrolytic Enzymes Isolated from the Gastrointestinal Tract of a Coprophage *"Gymnopleurus sturmi"* and Dung of Ruminants

Touijer Hanane [1,2], Benchemsi Najoua [3], Hamdi Salsabil [4], Janati Idrissi Abdellatif [2], Bousta Dalila [1], Irfan Ahmad [5], Sayyad Ali Raza BukharI [6], Muhammad Irfan [6,*], Lijing Chen [7,*] and Bekkari Hicham [1,2]

1   Laboratory of Biotechnology, Faculty of Sciences Dhar El Mahraz, Sidi Mohamed Ben Abdellah University, P.O. Box 1796, Fez-Atlas 30003, Morocco
2   Biodiversity, Bioenergy & Environment Research Group (BBE), Faculty of Sciences Dhar El Mahraz, Sidi Mohamed Ben Abdellah University, P.O. Box 1796, Fez-Atlas 30003, Morocco
3   Laboratory of Ecology and Environment, Faculty of Sciences & Techniques Saiss, Sidi Mohamed Ben Abdellah University, P.O. Box 1796, Fez 30000, Morocco
4   Institut Pasteur Casablanca, Casablanca 20250, Morocco
5   Department of Clinical Laboratory Sciences, College of Applied Medical Sciences, King Khalid University, Abha 61421, Saudi Arabia
6   Department of Biotechnology, Faculty of Science, University of Sargodha, Sargodha 40100, Pakistan
7   Key Laboratory of Agriculture Biotechnology, College of Biosciences and Biotechnology, Shenyang Agricultural University, Shenyang 110866, China
*   Correspondence: irfan.ashraf@uos.edu.pk (M.I.); chenlijing1997@126.com (L.C.)

Citation: Hanane, T.; Najoua, B.; Salsabil, H.; Abdellatif, J.I.; Dalila, B.; Ahmad, I.; BukharI, S.A.R.; Irfan, M.; Chen, L.; Hicham, B. Qualitative Screening of Yeast Biodiversity for Hydrolytic Enzymes Isolated from the Gastrointestinal Tract of a Coprophage *"Gymnopleurus sturmi"* and Dung of Ruminants. *Fermentation* **2022**, *8*, 692. https://doi.org/10.3390/fermentation8120692

Academic Editor: Timothy Tse

Received: 22 October 2022
Accepted: 27 November 2022
Published: 30 November 2022

Abstract: In this study, thirty yeast strains isolated from the gut of coprophagous *"Gymnopleurus sturmi"* and twenty-four from the dung of ruminants were shown to be producers of cellulases. Cellulolytic yeast isolates could also produce other hydrolytic enzymes such as pectinase, lipase, β-glucosidase, catalase, inulinase, urease, gelatinase, and protease. The oroduction of amylase was present in only one isolate of dung of ruminants. On the other hand, the production of tannase was absent in these isolates. All the yeasts isolated from two sources could utilize various carbon sources, including sorbitol, sucrose, and raffinose, and withstand high concentrations of glucose (300 g/L), salt (100 g/L), and exogenous ethanol. They could grow in a wide pH range of 3 to 11. The growth was stable up to a temperature of 40 °C for isolates from the gut of coprophage and 37 °C for the yeast from the dung of ruminants. These activities and growing conditions were similar to the diet of coprophagous insects and the composition of ruminant manure, likely because the adaptation and distribution of these microorganisms depend on the phenology and trophic preferences of these insects.

Keywords: insect; yeast; enzymes; cellulase

1. Introduction

Yeasts are unicellular eukaryotic microorganisms with the ability to multiply rapidly because they are less demanding in nutrients. They are easily implemented in several farms, research, and industrial applications [1,2]. Indeed, they represent the largest group of exploited microorganisms compared to prokaryotes [3]. Yeasts play an essential role in recycling organic matter by drawing their energy from external carbon sources [4]. They can also cohabit in various environments where the living conditions are specific, such as acidic environments, tree bark, digestive tract, and insect dung [1,5–8]. Yeasts are widely used in food fermentation industry and making more progress in other industries [2,9]. They represent an important source to be exploited to develop new and very extensive biotechnological processes as they have a highly developed enzymatic system [10]. Indeed, yeasts can produce enzymes such as cellulases, ligninase, xylanase, pectinase, amylase, glucoamylase, and lipase (Table 1) [11] for applications in various industrial processes, which require the use of specific thermostable enzymes [12]. Several types of research have been reported on the ability of yeast strains to hydrolyze plant-derived substrates to provide an alternative to chemical hydrolysis for bioenergy production.

In this context, we have established a collection of yeasts isolated from various ecological niches; the digestive tract of a coprophagous beetle *"G. sturmi"* and ruminant dung, a food source for this insect. The choice of these sources is based on the fact that coprophagous beetles constitute a group of beetles well-adapted to grazed ecosystems [13], as they derive their food resources from the dung of large mammals [14]. This dung could be of carbohydrate origins, or lipid- or phenolic-dominant. The degradation of these compounds can be done through a system of enzymes secreted by the microorganisms colonizing the digestive tract of the animals or the microorganisms specific to the dung, in particular the cellulases to degrade the cellulose not transformed by the animals. These cellulases could be used for industrial applications due to their great biotechnological potential, and recycling of cellulosic biomass [15,16]. This research is focused on the isolation and screening of various yeast isolates from the gut and dung of coprophagous insect *Gymnopleurus sturmi* for biotechnological applications.

Table 1. Technological applications of yeast isolates.

Enzymes	Yeasts	Applications	References
Cellulases	*Aureobasidium pullulans* 98	Food, chemical, textile, paper, and biofuel industries	[17,18]
Pectinase	*Kluveromyces marxianus* *Metschnikowia pulcherrima*	Wine, cider, and fruit juice industries	[19,20]
Lipase	*Aureobasidium pullulans* HN2.3	Food, wastewater treatment, cosmetics, pharmaceuticals, leather processing, and biofuel industries	[21]
ß-glucosidase	*Guehomyces pullulans* 17-1	Food industry	[22]
Catalase	*Saccharomyces cerevisiae*	Food industry	[23]
Inulinase	*Pichia guilliermondii*	Food, beverage, and biofuel industries	[24]
Urease	*Cryptococcus gattii*	Diagnostic kits, beverages, and animal feed	[25]
Gelatinase	*Trichosporon pullulans*	Food, chemical, and medical industries	[26]
Protease	*Rhodotorula mucilaginosa* L7	Clinical applications, food, beverage, and leather processing industries	[27]
Amylase	*Aureobasidium pullulans* N13dSaccharomyces cerevisiae	Food, textile, paper, detergent industry, medical, and pharmaceutical industries	[28,29]
Tannase	*Kluyveromyces marxianus*	Feed, food, beverages, brewing, pharmaceutical, chemical, cosmetic, and leather industries	[30]

2. Materials and Methods

2.1. Sampling

The coprophagous organisms and ruminant dung used for this study were collected in June 2015 and 2016 from two different regions of Fez, Morocco; Fez sais (33°54′14″ N, 4°59′55″ W), at 609 m of altitude, Ain Aicha (34°29′59″ N, 4°42′01″ W), at 246 m of altitude, Ain Aicha (34°29′59″ N, 4°42′01″ W), at 246 m of altitude, and Ain Beda (33°57′07.5″ N, 4°50′59.9″ W), at 579 m of altitude.

The isolation of the yeasts was carried out from the gastrointestinal tract of the coprophagous *"Gymnopleurus sturmi"* and the ruminate dung on Yeast Pepton Glucose (YPG) culture medium, comprised of Peptone (20 g/L), yeast extract (10 g/L), glucose (20 g/L), agar (20 g/L), and with two antibiotics, ampicillin (60 µg/mL) and kanamycin (60 µg/mL), used to prevent bacterial growth.

2.2. Qualitative Screening of Isolates for Hydrolytic Enzymes

The different yeast isolates were tested for their ability to produce hydrolytic enzymes.

2.2.1. Oxidation of Phenolic Substrates

The determination of the capacity of all the isolates to utilize and degrade phenolic substrates was carried out on the M9 culture medium, comprising of (g/L) Na_2HPO_4 (6 g), KH_2PO_4 (3 g), NH_4Cl (1 g), and $NaCl$ (0.5 g) with 0.1% of catechol or pyrogallol. The positive result was visualized by a black coloration around the isolate, indicating assimilation of these substrates.

2.2.2. Cellulase Activity

This activity was determined using carboxymethylcellulose (CMC) agar medium as a source of carbon. The M9 buffer consists of (g/L) Na_2HPO_4 (6 g), KH_2PO_4 (3 g), NH_4Cl (1 g), and $NaCl$ (0.5 g). The medium was autoclaved for 15 min at 120 °C. Then, 1 mL of 0.1 M $CaCl_2$ and 1 mL of $MgSO_4$ at 1 M were added. The cellulase medium was composed of carboxymethyl cellulose (CMC) or cellulose fiber (10 g), yeast extract (0.2 g), agar (20 g), and the M9 buffer. The final medium was autoclaved for 15 min at 110 °C [31]. The yeast was incubated on the prepared medium for 72 h at 30 °C. After that 1%, Congo red was flooded, which selectively binds to the polymers of cellulose. After 15 to 20 min, the dishes were washed several times with $NaCl$ (1 M). The presence of a light halo around the colonies indicates the degradation of the CMC in the medium by the cellulase secreted from the isolates that were called "cellulase +" [32].

2.2.3. Amylase Activity

Yeast isolates with amylase activity were detected by cultivation on a starch agar medium. This culture media was comprised (g/L) of peptone (2 g), $MgSO_4 7H_2O$ (0.5 g), $CaCl_2$ (0.2 g) $NaCl$ (0.5 g), and starch (1 g). Isolates possessing the amylase enzyme were able to degrade the starch present in the agar, thus developing a clear zone around the colonies. The evidence of starch degradation by the secreted enzyme was confirmed after adding Lugol and rinsing with distilled water [33].

2.2.4. Inulinase Activity

Inulinase production by the isolated yeasts was detected on a medium containing inulin (2 g), yeast extract (0.5 g), KH_2PO_4 (3 g), agar (20 g), and distilled water (1 L). Inulin was used as the only carbon source in this medium; yeast growth after 4 days of incubation at 30 °C showed the presence of inulinase activity [34].

2.2.5. β-Glucosidase Activity

The β-glucosidase activity was carried out in a medium containing (g/L) yeast extract (1 g), peptone (1 g), ferric ammonium citrate (0.01 g), esculin (0.3 g), and agar (20 g). The Petri dishes inoculated with yeast cultures were incubated at 30 °C for 24 to 48 h. The presence of enzymatic activity was visualized as a dark halo surrounding yeast growth [35].

2.2.6. Pectinase Activity

To check the pectinase activity of the isolates, yeast cultures were grown on a medium based on (g/L) pectin (2.5 g), yeast extract (5 g), $(NH_4)_2SO_4$ at 10% (5 mL), 1 M $MgSO_4$ (0.5 mL), 50% glycerol (5 mL), and agar (20 g). The pH of the culture medium was adjusted to 8.0. The medium was autoclaved at 120 °C for 15 min [36]. The pectinase secretion was detected by adding lugol (selectively binding to the pectin polymers) and a successive rinsing of the dishes with distilled water after 10–15 min. "The pectinase +" isolates were characterized by the presence of a clear halo around colonies [12].

2.2.7. Lipase Activity

Lipase activity was carried out on a culture medium based on olive oil and rhodamine B. This method was based on the emission of fluorescence by the colonies, which was due to the interaction of rhodamine B with the fatty acids released during the enzymatic hydrolysis of the olive oil. The medium consisted of (g/L) $NaHPO_4$ (12 g), KH_2PO_4 (2 g), $MgSO_4$ $7H_2O$ (0.3 g), $CaCl_2$ (0.25 g), (NH_4) SO_4 (2 g), and pH was adjusted to 7.0. After sterilization, the medium was mixed with olive oil (31.25 mL) and rhodamine B 10 mL (1 mg/mL). Colonies with lipase activity develop a fluorescence after exposure of the Petri dishes to UV radiation (350 nm) [37].

2.2.8. Gelatinase Activity

The detection of gelatinase activity was carried out on a gelatin-based culture medium: (g/L) gelatin (15 g), peptone (4 g), yeast extract (1 g), meat extract (1 g), agar (15 g). While the medium showed an opaque appearance on the proteinaceous substrate, having a clear zone around the colonies indicates the hydrolysis of gelatin substrate [38].

2.2.9. Urease Activity

The medium Christensen was used to reveal urease activity in the yeasts studied. This enzyme converts urea to ammonia, which increases the pH that changes the color indicator. After 2 days of incubation at 30 °C, a pink to violet coloration in the culture media indicated positive results [39].

2.2.10. Protease Activity

Protease production was determined according to the method of Strauss et al. [40] by plating yeast colonies on a medium containing (g/L) yeast extract (0.5 g), $NaNO_3$ (1 g), K_2HPO_4 (2 g), KCl (1 g), $MgSO_4$ (0.5 g), agar (20 g), and milk powder (10 g). Incubation was carried out at 30 °C for 7 days. The proteolytic activity was revealed by the presence of a clear zone around the colony.

2.2.11. Catalase Activity

The catalase activity was evaluated using the method described by Whittenbury [41] by directly adding 3% (v/v) hydrogen peroxide to a 48 h yeast culture. Catalase activity was evidenced by the presence of oxygen bubbles.

2.2.12. Tannase Activity

The tannase activity in the different yeast isolates was detected on YPG media supplemented with 10 mL of a tannic acid (20%) solution. The yeasts that have a Tannase activity showed a clear halo, reflecting the decomposition of tannic acid, gallic acid, and glucose.

2.3. Study of Physiological Characteristics of Cellulolytic Isolates

2.3.1. Thermotolerance

The thermotolerance of the isolate temperatures was evaluated on YPG agar and M9-CMC medium, after incubation for 48 h at different temperatures: 30, 37, 40, 42, 44, 46, and 48 °C.

2.3.2. pH Tolerance

The pH tolerance of the isolates was carried out on a YPG agar medium at different pH values: 3, 4, 5, 6, 8, 9, 10, 11, and incubated at 30 °C for 48 h.

2.3.3. Utilization of Carbon Sources

The digestion of carbon sources by these isolates was carried out on a minimal medium of M9 agar supplemented with 1% of each source, CMC [42], cellobiose, cellulose, sorbitol, sucrose, raffinose, maltose, ribose, xylose, galactose, arabinose, fructose, casein, mannitol, lactose, dextrin, glycine, and glycerol. The growth of isolates was noted after incubation at 30 °C for 48 h.

2.3.4. Glucose Tolerance of Yeasts

The growth of the isolates at different glucose concentrations—50, 100, 150, 200, 250, and 300 (g/L)—was evaluated to investigate the yeast's tolerance. The isolates were incubated on a YPG agar medium, properly supplemented with different concentrations of glucose. Next, the culture plates were incubated at 30 °C for 48 h.

2.3.5. Tolerance to Ethanol

We tested the isolate's tolerance for exogenous ethanol on YPG agar medium that was supplemented with different concentrations of ethanol: 1, 5, 6, 8, 10, and 12%. Then, the plate was incubated at 30 °C for 48 h.

2.4. Molecular Identification

Cellulolytic isolates were cultured in test tubes containing 5 mL of YPG medium. Incubation was carried out at 30 °C for 24 h. DNA extraction from yeast isolates was carried out by the classic method [43]. 1.5 mL of yeast suspension were centrifuged for 10 min at 12,000 rpm, and the pellet was suspended in 200 µL of lysis buffer (2% Triton X-100, 1% SDS, 100 mM NaCl, 10 mM Tris-HCl at pH 8, 1 mM EDTA at pH 8). The mixture was placed at -20 °C for 2 min and then at 95 °C for 1 min. This step was repeated twice to create a thermal shock and promote cell bursting. Next, the DNA was extracted with phenol/chloroform. The amplification by PCR of the sequence ITS1 (TCCGTAGGTGAAC-CTGCGG) ITS4 (TCCTCCGCTTATTGATATGC) was carried out according to the method of White et al. [44]. The reaction was carried out in a final volume of 25 µL, containing primers (final concentration 10 pm, each), dNTPs (200 µM final concentration), $MgCl_2$ (1.5 mM final concentration), 10 µL 1× PCR buffer, Taq polymerase (0.5 U), pure H_2O, and extracted DNA. PCR conditions were as follows: initial denaturation at 95 °C for 2 min and 30 cycles of denaturation at 95 °C for 45 s, primer annealing at 55 °C for 1 min, and extension at 72 °C for 1 min. A final extension was completed at 72 °C for 10 min. Sanger sequencing was performed at the pastor institute (Casablanca-Morocco) using an ABI PRISM 3130XL Genetic Analyzer, Applied Biosystems. Preliminary identifications were performed based on sequence assembly and by searching in the NCBI ITS RefSeq Fungi database using command line interface.

2.5. Multiple Sequence Alignment and Phylogenetic Tree

ITS sequences from all the isolates were used to perform BLASTN, and the top 5 hits were retrieved to find close relatives to the isolates. Sequences from all isolates were subjected to multiple sequence alignment using MUSCLE v3.8.31 with default parameters. A phylogenetic tree was constructed using FastTree v2.1.10 with –nt –gtr parameters. The phylogenetic tree was visualized with ggtree v3.4.0 in RStudio and R v4.2.0.

3. Results and Discussion

3.1. Distribution of Isolated Yeasts

The yeasts isolated from the gut of coprophagous "*Gymnopleurus sturmi*" revealed a large load of 79×10^8 CFU/g and 65.24×10^8 CFU/g in June 2015 and 2016, respectively. The ruminant manure contained 16.8×10^6 CFU/g and 11×10^6 CFU /g in 2015 and 2016, respectively (Table 2). In addition, yeast biodiversity was different among the four samples. The collection consists of 96 isolates, of which 57.29% were isolated from gut *Gymnopleurus sturmi* (GGS) and 42.71% were isolated from the dung of ruminants (DR).

Table 2. Percentage of isolates screened for cellulase activity.

Isolation Medium	Sampling Period	Total Counts of Yeasts (CFU/g)	Number of Isolates
Isolates from GGS	2015	7.9×10^9	24
	2016	6.52×10^9	31
Isolates from DR	2015	16.8×10^6	19
	2016	11×10^6	22

3.2. Oxidation of Phenolic Substrates

The isolated yeasts were tested to demonstrate their degradation potential of phenolic lignin by-product compounds (pyrogallol and catechol), essential components of the plant cell wall. Results showed (Table 3) that 40% of the isolates from GGS and 41.5% of the isolates from DR were able to degrade these compounds. These yeasts were isolated from media with a high concentration of phenolic products after the degradation of plants rich in ligninolytic products by the mammals and the assimilation of these products by the clean microflora of the dung and the microflora of the coprophages feeding this dung [45]. Indeed, the phenolic compounds are always liable to be degraded by the enzymes produced by microorganisms [46–49]. Other studies have shown that fungi such as *Aspergillus terreus* [49] and yeasts such as *Candida tropicalis* [50] have a great ability to degrade phenolic products.

Table 3. Percentage of isolates capable of degrading lignin by-products.

Isolation Medium	Positive Isolates	Total Number of Isolates
Isolates from GGS	22 (40%)	55
Isolates from DR	17 (41.5%)	41

3.3. Screening of Cellulolytic Isolates

One of the main objectives of this study was to determine the ability of isolated yeasts to produce cellulase enzymes. Therefore, two different substrates (CMC and cellulose fiber) were used for the screening of cellulolytic isolates. Only isolates that showed a positive result in both substrates were considered cellulolytic. This has enabled us to build a library of isolates capable of producing our enzyme of interest from which 30 isolates (54.54%) were collected from GGS and 24 isolates (58.54%) from DR. The degradation time of two substrates varies between isolates (Figure 1A,B). For the GGS isolates, the hydrolysis zone of CMC varied between 6 to 25 mm, and the hydrolysis of the cellulose fiber varied between 3 to 15 mm (Figure 1A). For the DR isolates, the zone of hydrolysis for CMC was noted from 2 to 16 mm and 3 to 7 mm for cellulose fiber (Figure 1B). Thongekkaew and Kongsanthia [18] showed that 45 yeast isolates, obtained from various samples (soil, tree bark, and insect excrement), can hydrolyze cellulose.

3.4. Hydrolytic Enzymes

Yeasts have the potential to contribute to the development of biotechnological processes with specific applications [51,52]. To characterize the different isolates biochemically, different substrates were tested. The choice of these substrates was related to the composition of the dung [45] and the food requirement of the insect studied [53], which contains a large number of complex substrates. The qualitative tests revealing the production of hydrolytic enzymes in the cellulosic isolates tested made it possible to select efficient isolates. Figure 2 shows that all DR isolates were capable of producing the β-glucosidase and catalase. For GGS, 93.33% of isolates were β-glucosidase-positive and 96.67% were catalase-positive. In addition, the hydrolysis of olive oil by the lipase was noted in 86, 67% of the GGS isolates and 95.83% of the DR isolates. On the other hand, the considerable production of pectinase was detected in 80% GGS isolates and 50% DR isolates. Protease was also more prevalent in 87% of GGS and 75% of DR isolates. In addition, 63.33% of GGS, and 29.17% of DR isolates were positive for Inulinase. Gelatinase synthesis was detected for many of the 56.67% GGS and 4% DR isolates. Most isolates had urease activity, of which 50% were isolated from GGS and 66.67% from DR. Tannase activity was absent in all isolates considered in this study. The amylase activity was present only in one isolate of DR out of a total of 24 isolates, which was 4% (Figure 2). In previous studies, (Table 4) [11,27,40,41,54], the isolated various strains exhibited the ability to produce interesting enzymes, with interesting technological applications (Table 4).

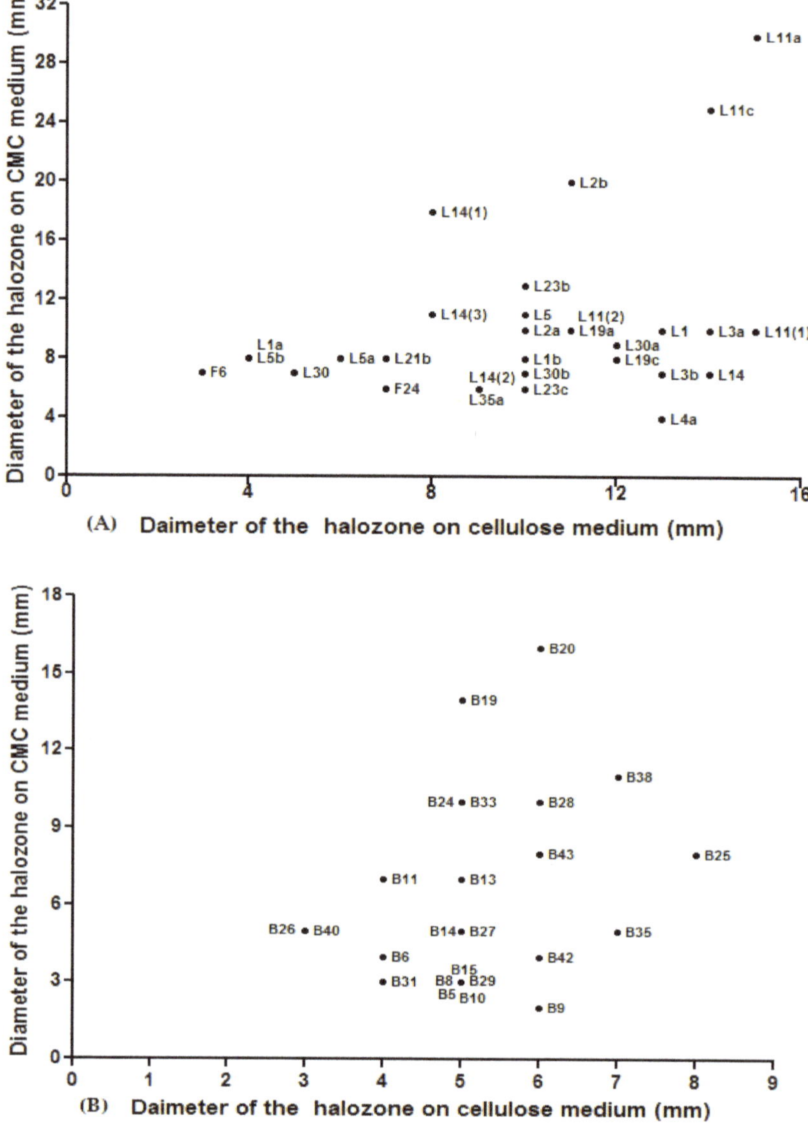

Figure 1. Screening of cellulase activities carried out for the strains isolated (**A**) from the gut of *Gymnopleurus Sturmi* (GGS), (**B**) and from the dung of ruminants (DR).

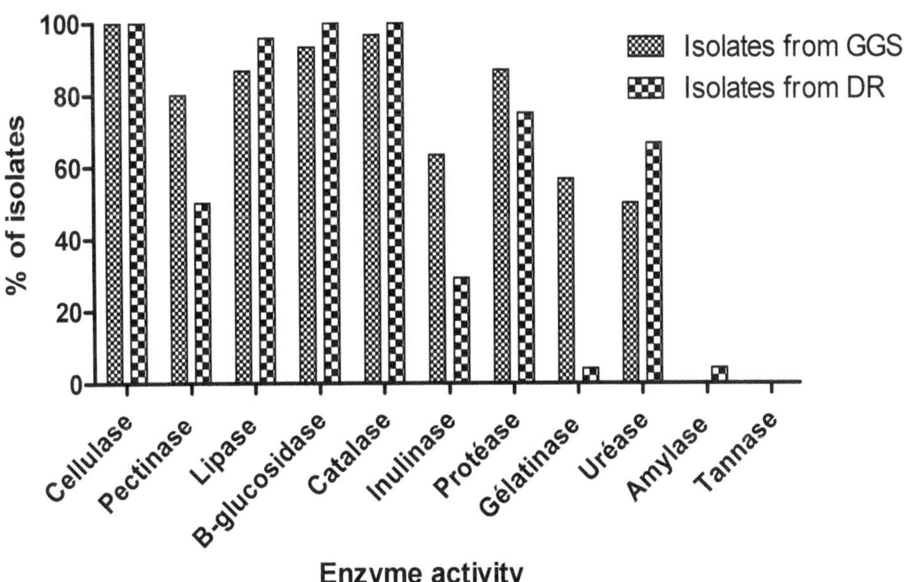

Figure 2. Qualitative tests of enzymatic activities of yeasts isolated from GGS and DR.

Table 4. Enzymatic activity of isolated yeast a: Isolates from GGS, b: Isolates from DR, 1: Bautista-G et al. [54], 2: Strauss et al. [40], 3: Lilao et al. [27], 4: Romo-Sánchez et al. [11], 5: Hernandez et al. [41].

Enzyme	Current Study		1	2	3	4	5
	a	b					
Cellulase	30	24	-	11	-	38	-
Pectinase	24	12	-	9	62	4	23
Lipase	26	23	4	-	134	-	6
β-glucosidase	28	24	10	-	142	13	-
Catalase	29	24	16	-	-	-	41
Inulinase	19	7	-	-	-	-	-
Urease	15	16	-	-	-	-	-
Gelatinase	17	1	-	-	-	-	-
Protease	26	18	6	10	49	-	-
Amylase	0	1	-	9	-	-	-
Tannase	0	0	-	-	-	-	-
Number of isolates	30	24	30		216	42	83

3.5. Study of Physiological Characteristics

3.5.1. Temperature

The effect of temperature on the yeast growth showed that the growth was stable up to a temperature of 40 °C in GGS isolates and 37 °C for the DR isolates (Figure 3). The percentage of isolates capable of tolerating these temperatures decrease with increasing temperature (Figure 3). Aissam et al. [55] reported an optimum temperature of 37 °C for yeast strains. Babavalian et al. [56] showed that most yeasts isolated from soil cannot withstand a temperature above 37 °C, while complete inhibition of their growth is observed at 45 °C.

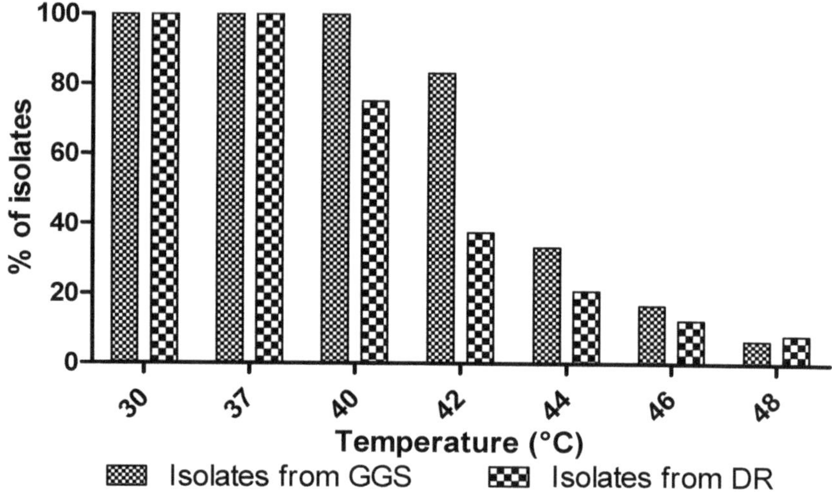

Figure 3. Effect of temperature on the growth of yeast isolates.

3.5.2. Exogenous Ethanol

All cellulosic isolates from DR were able to support up to 8% of ethanol and 52% of isolates were able to grow in a 12% ethanol culture medium. However, isolates from the GGS only supported a 5% concentration of exogenous ethanol. While increasing the concentration of ethanol, the percentage of these isolates, which grow in the medium, decreases (Figure 4). Arguably, the DR isolates support a higher concentration of ethanol than GGS isolates. This type of behavior has also been observed in strains *S. cerevisiae* C2 and TA, *Saccharomyces cerevisiae* K2, *Saccharomyces rosinii* S1 and S2, *Rhodotorula minuta* S3, and *Saccharomyces exiguus* K1 [57].

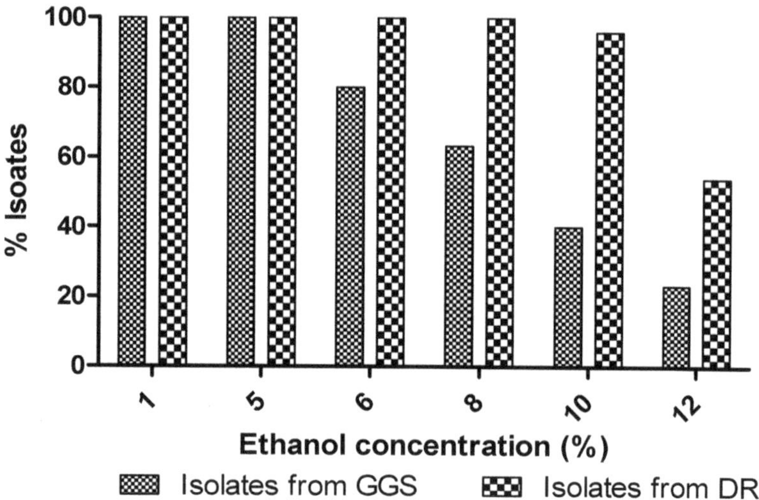

Figure 4. Effect of ethanol on the growth of yeast isolates.

3.5.3. Utilization of Carbon Sources

Figure 5 shows that the yeast isolates (GGS and DR) had a great ability to assimilate several types of carbon sources. These results corroborate with those of Gao et al. [34], who showed that isolated yeasts of marine origin could grow on several sources of carbon.

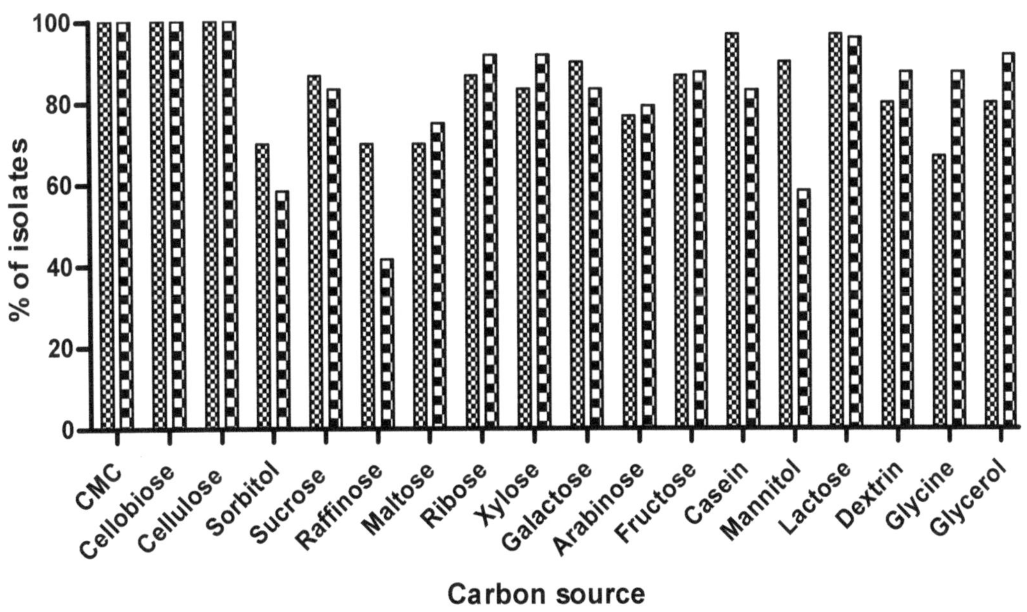

Figure 5. The assimilation of different kinds of carbon sources by yeast strains.

3.5.4. Osmotolerance

The osmotolerance test showed that all the isolates of the digestive tract and ruminant dung were osmotolerant and could survive a culture medium with 30% glucose and 10% NaCl (Table 5). In 2005, Milala et al. [58] reported that the sugar tolerance was 20% for the two strains of *S. cerevisiae* TA and C, 15% for *S. cerevisiae* MTCC 170, and 10% for the two strains *Saccharomyces rosinii* S1 and S2, as well as the strain *Rhodotorula minuta* S3.

Table 5. Osmotolerance of isolates for glucose and NaCl.

	Glucose Concentration (g/L)						NaCl Concentration (g/L)
Concentration	50	100	150	200	250	300	100
Isolates from (GGS) (%)	100	100	100	100	100	100	100
Isolates from (DR) (%)	100	100	100	100	100	100	95.83

3.5.5. Effect of pH

The effect of pH shows that all isolates were able to grow over a wide range from pH = 5 to pH = 8 (Figure 6), but from a pH equalling 9, the percentage of isolates that could endure the medium decreased. The results showed that cellulosic isolates have interesting technological properties for their application in the course. This species was well-adapted to the environmental conditions that govern fermentations, such as low pH and high concentrations of NaCl [59]. However, in other studies on volcanic yeasts, the optimum pH for growth for all isolates was between 3.5 and 5.5 [60].

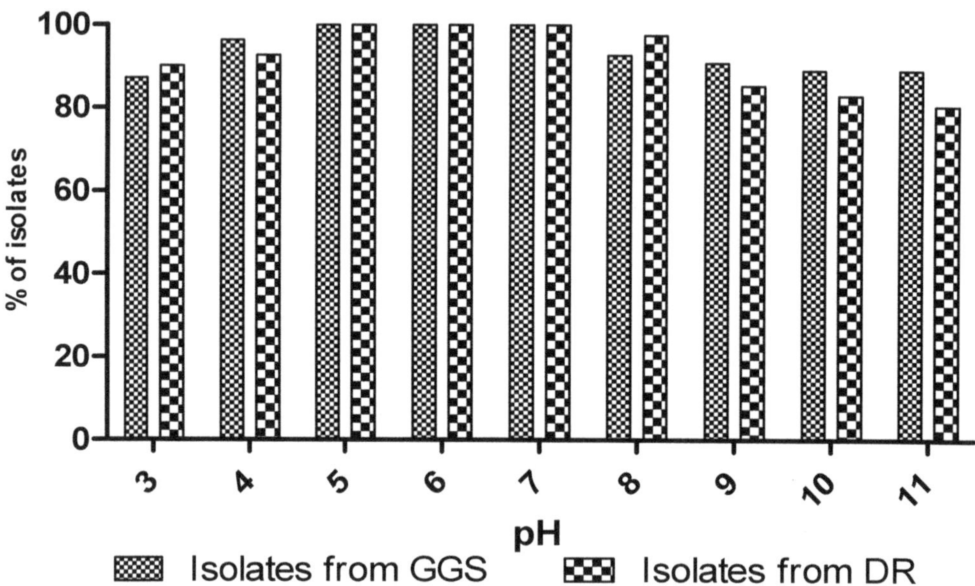

Figure 6. Effect of pH on growth of yeast isolates.

3.6. Molecular Identification

Taxonomic characterization of the 21 cellulosic isolates was carried out after morphological examination under a microscope (Figure 7). The sequence homology search was carried out using the BLASTN algorithm on the command line using ITS RefSeq Fungi database. The latter confirmed that the isolates studied belong to the genus *Trichosporon* (Table 6). All the identified sequences of strains showed high similarity with *Trichosporon insectorum*, *Trichosporon faecale*, and *Trichosporon coremiiforme*. All except L4a showed a high similarity with *Trichosporon asahii*. Many of the strains showed high similarity to *Trichosporon aquatile*. However, biotype DR also had high similarity with *Trichosporon japonicum*. The top hit of L30a was with *Trichosporon coremiiforme* with only 90.6% sequence identity. Details of the top 5 hits are shown in Figure 8, along with a phylogenetic tree. In order to understand the phylogenetic relationship of the strains, the sequences of the top hits were combined with the ITS sequences of the strains to build a phylogenetic tree. It is evident that the strains from the biotype DR are more closely related to *Trichosporon* sp. (top hits of the BLAST) as compared to biotype GGS, except B5, which shares a clade with GGS strains. All the strains were split into 4 clades, while L11C and L11a were different from other clades (Figure 9).

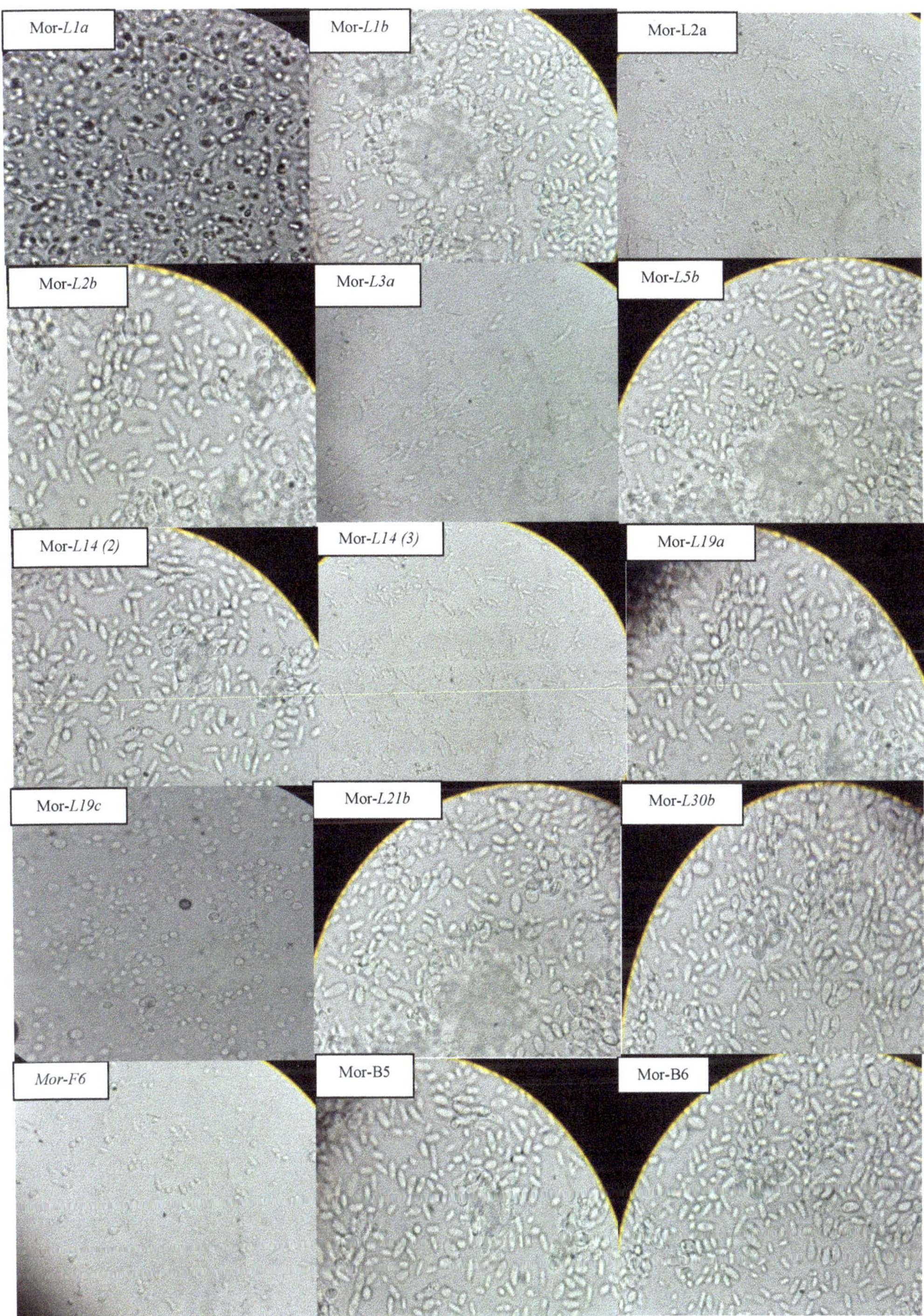

Figure 7. *Cont.*

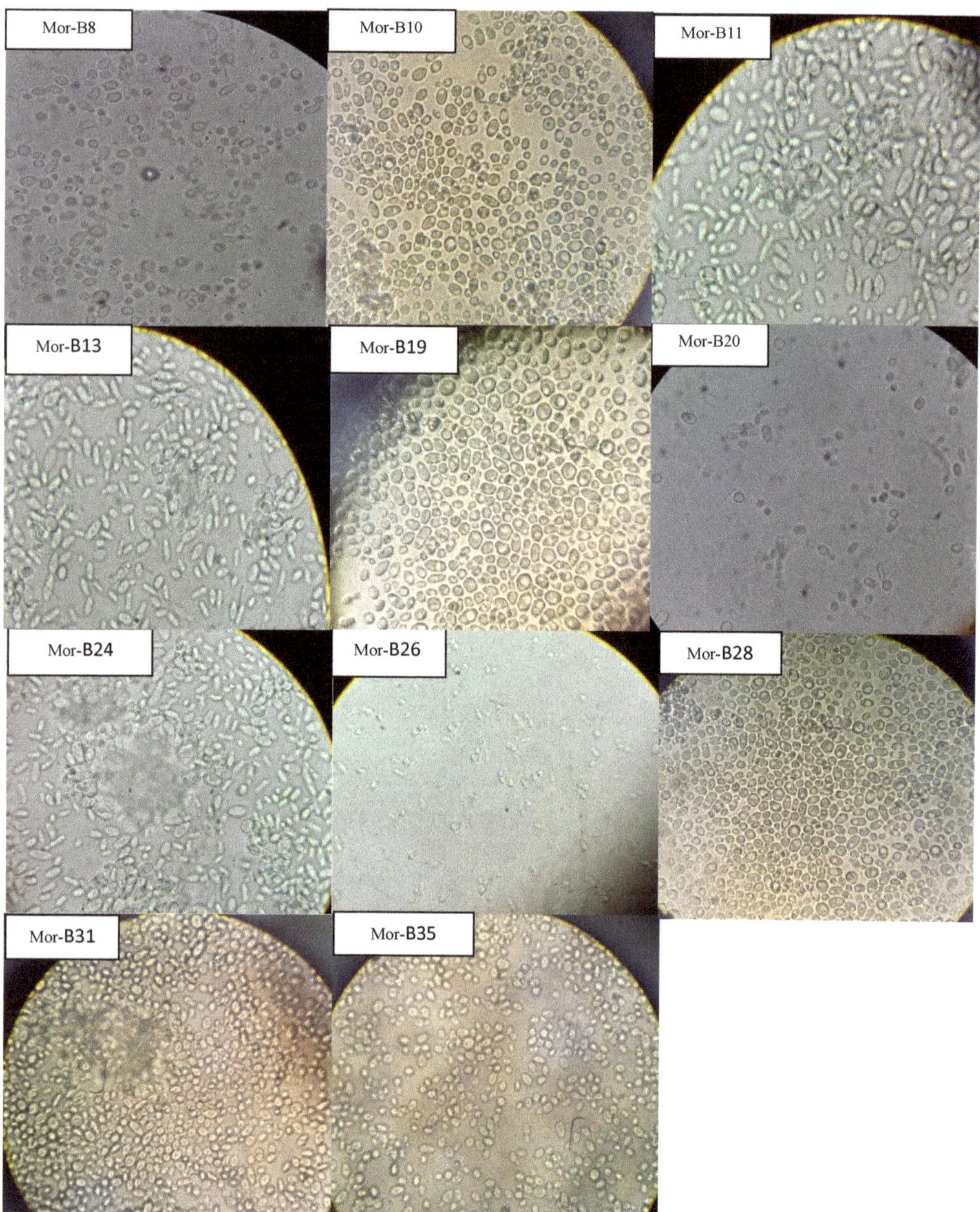

Figure 7. Morphology of different yeast isolates.

Table 6. Results of yeast identification by rDNA internal transcribed spacer (ITS).

Coded	Biotope	Homologous Strains	Accession Numbers
Mor-L1	GGS	*Trichosporon insectorum*	ON810624.1
Mor-L2a	GGS	*Trichosporon insectorum*	ON810627.1
Mor-L3b	GGS	*Trichosporon insectorum*	ON810628.1
Mor-L4a	GGS	*Trichosporon japonicum*	ON810629.1
Mor-L5	GGS	*Trichosporon insectorum*	ON810630.1
Mor-L5a	GGS	*Trichosporon insectorum*	ON862731.1
Mor-L11a	GGS	*Trichosporon insectorum*	ON810622.1
Mor-L11 (1)	GGS	*Trichosporon insectorum*	ON862728.1
Mor-L11 (2)	GGS	*Trichosporon insectorum*	ON862729.1
Mor-L11C	GGS	*Trichosporon insectorum*	ON810621.1
Mor-L14	GGS	*Trichosporon insectorum*	ON810623.1
Mor-L14 (1)	GGS	*Trichosporon insectorum*	ON862731.1
Mor-L14 (3)	GGS	*Trichosporon insectorum*	ON862732.1
Mor-L23b	GGS	*Trichosporon insectorum*	ON810625.1
Mor-L23C	GGS	*Trichosporon insectorum*	ON810626.1
Mor-L30a	GGS	*Trichosporon coremiiforme*	ON810631.1
Mor-B13	DR	*Trichosporon insectorum*	ON810633.1
Mor-B25	DR	*Trichosporon insectorum*	ON810632.1
Mor-B13a	DR	*Trichosporon insectorum*	ON862726.1
Mor-B25a	DR	*Trichosporon insectorum*	ON862727.1
Mor-B5	DR	*Trichosporon insectorum*	ON862730.1

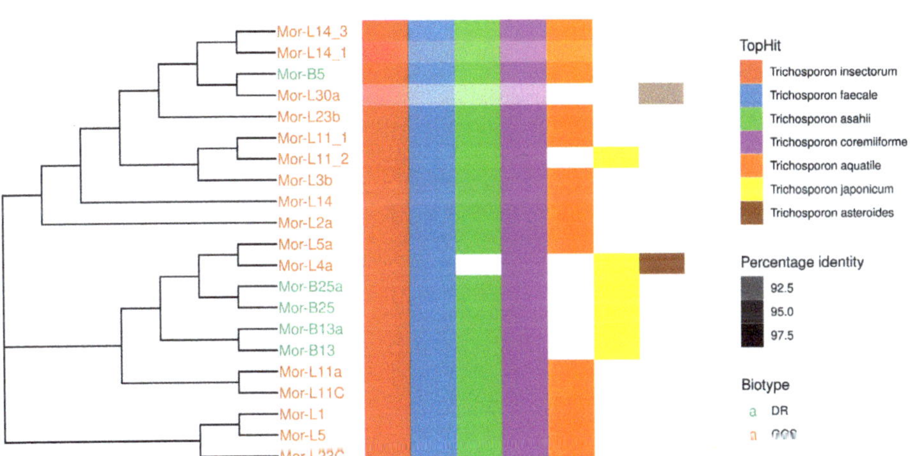

Figure 8. Phylogenetic tree and top 5 BLAST hits of all the strains. Darkness of the color increases as the percentage identity increases.

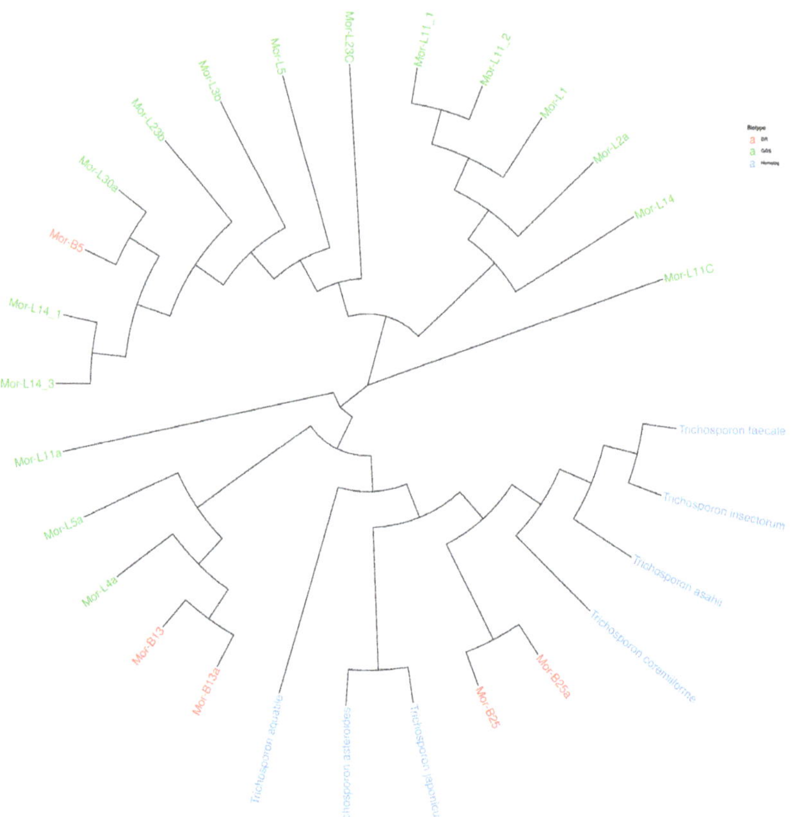

Figure 9. Phylogenetic tree of all the strains with their top BLAST hits.

4. Conclusions

Microbial resources represent natural and real assets. Their exploitation was conducted under rigorous environmental conditions, and the exploitation of their potential was oriented towards targeted biotechnological applications. The gut of coprophagous insects and the dung of ruminants are abundant and culminate an important richness of the yeasts. Purified isolates show interesting enzymatic activities. They are therefore highly desirable biomolecules in the enzyme market, and their quantification and various biotechnological applications are by far the most promising and could be explored in future studies.

Author Contributions: Conceptualization, B.H. and M.I.; methodology, T.H.; software, S.A.R.B.; validation, T.H., B.H. and M.I.; formal analysis, B.N.; investigation, T.H.; resources, H.S. and J.I.A.; data curation, B.D.; writing—original draft preparation, T.H.; writing—review and editing, T.H. and M.I.; visualization, T.H.; supervision, B.H. and M.I.; project administration, B.H.; funding acquisition, B.H., L.C. and I.A. All authors have read and agreed to the published version of the manuscript.

Funding: This work was financially supported by National Center for Scientific and Technical Research, project PPR 2015 Morocco. Scientific Research deanship at King Khalid University, Abha, Saudi Arabia for their financial support through the Large Research Group Project under grant number (RGP.02-186-43) and National Key R & D Program of China (Grant No. 2019YFD1001002) and the Discipline Construction of Professional Degree (Grant No. 880220039). The role of the funding body in the design of the study, the collection, analysis, and interpretation of data, and in the writing the manuscript is management and supervision.

Data Availability Statement: The data used to support the findings of this study are available from the corresponding author upon request.

Conflicts of Interest: The authors declare that they have no conflict of interest.

References

1. Madden, A.A.; Epps, M.J.; Fukami, T.; Irwin, R.E.; Sheppard, J.; Sorger, D.M.; Dunn, R. The ecology of insect–yeast relationships and its relevance to human industry. *Proc. R. Soc. B Biol. Sci.* **2018**, *285*, 20172733. [CrossRef]
2. Willaert, R.G. Yeast Biotechnology 4.0. *Fermentation* **2021**, *7*, 69. [CrossRef]
3. Satyanarayana, T.; Kunze, G. *Yeast Biotechnology: Diversity and Applications*; Springer: Berlin/Heidelberg, Germany, 2009.
4. Chabasse, P.; Bouchara, D.; de Gentile, J.P.; Brun, L.; Cimon, S.; Penn, B. Les Moisissures D'Intérêt Médical. *Bioforma* **2002**, 161.
5. Lee, Y.-J.; Choi, Y.-R.; Lee, S.-Y.; Park, J.-T.; Shim, J.-H.; Park, K.-H.; Kim, J.-W. Screening Wild Yeast Strains for Alcohol Fermentation from Various Fruits Screening Wild Yeast Strains for Alcohol Fermentation from Various Fruits. *Mycobiology* **2018**, *39*, 33–39. [CrossRef] [PubMed]
6. Stefanini, I. Yeast-insect associations: It takes guts Irene. *Yeast* **2018**, *35*, 315–333. [CrossRef] [PubMed]
7. Pandi, R.; Velu, G.; Devi, P.; Dananjeyan, B. Isolation and Screening of Soil Yeasts for Plant Growth Promoting Traits. *Madras Agric. J.* **2019**, *106*, 143–148. [CrossRef]
8. Quan, A.S.; Eisen, B.M. The Ecology of the Drosophila-Yeast Mutualism in Wineries. *PLoS ONE* **2018**, *13*, e0196440. [CrossRef]
9. Saberi, S.; Cliff, M.A.; Van Vuuren, H.J.J. Impact of Mixed *S. Cerevisiae* Strains on the Production of Volatiles and Estimated Sensory pro Fi Les of Chardonnay Wines. *Food Res. Int.* **2012**, *48*, 725–735. [CrossRef]
10. Mitsui, R.; Nishikawa, R.; Yamada, R.; Matsumoto, T.; Ogino, H. Construction of yeast producing patchoulol by global metabolic engineering strategy. *Biotechnol. Bioeng.* **2020**, *117*, 1348–1356. [CrossRef]
11. Romo-sánchez, S.; Alves-baf, M.; Arévalo-villena, M.; Úbeda-iranzo, J.; Briones-pérez, A. Yeast Biodiversity from Oleic Ecosystems: Study of Their Biotechnological Properties. *Food Microbiol.* **2010**, *27*, 487–492. [CrossRef]
12. Bhat, M.K. Cellulases and related enzymes in biotechnology. *Biotechnol. Adv.* **2000**, *18*, 355–383. [CrossRef]
13. Lumaret, J.-P.; Kirk, A. Ecology of dung beetles in the French Mediterranean region (Coleoptera: Scarabaeinae). *Acta Zool. Mex.* **1987**, *24*, 1–55.
14. Yun, J.-H.; Roh, S.W.; Tae, W.W.; Jung, M.-J.; Kim, M.-S.; Park, D.-S.; Yoon, C.; Nam, Y.-D.; Kim, Y.-J.; Choi, J.-H.; et al. Insect Gut Bacterial Diversity Determined by Environmental Habitat, Diet, Developmental Stage, and Phylogeny of Host. *Appl. Environ. Microbiol.* **2014**, *80*, 5254–5264. [CrossRef]
15. Darwesh, O.M.; El-Maraghy, S.H.; Abdel-Rahman, H.M.; Zaghloul, R.A. Improvement of paper wastes conversion to bioethanol using novel cellulose degrading fungal isolate. *Fuel* **2020**, *262*, 116518. [CrossRef]
16. Rosyidaa, V.T.; Indrianingsiha, W.; Maryanaa, R.; Wahono, S.K. Effect of Temperature and Fermentation Time of Crude Cellulase Production by *Trichoderma reesei* on Straw Substrate. *Energy Procedia* **2015**, *65*, 368–371. [CrossRef]
17. Rong, Y.; Zhang, L.; Chi, Z.; Wang, X. A Carboxymethyl Cellulase from a Marine Yeast (*Aureobasidium Pullulans* 98): Its Purification, Characterization, Gene Cloning and Carboxymethyl Cellulose Digestion. *J. Ocean Univ. China* **2015**, *14*, 913–921. [CrossRef]
18. Thongekkaew, J.; Kongsanthia, J. Screening and Identification of Cellulase Producing Yeast from Rongkho Forest, Ubon Ratchathani University. *Bioeng. Biosci.* **2016**, *4*, 29–33. [CrossRef]
19. Da Silva, E.G.; Borges, M.D.F.; Medina, C.; Hilsdorf Piccoli, R.; Freitas Schwan, R. Pectinolytic Enzymes Secreted by Yeasts from Tropical Fruits. *FEMS Yeast Res.* **2005**, *5*, 859–865. [CrossRef]
20. Belda, I.; Conchillo, L.B.; Ruiz, J.; Navascués, E.; Marquina, D.; Santos, A. Selection and Use of Pectinolytic Yeasts for Improving Clarification and Phenolic Extraction in Winemaking. *Int. J. Food Microbiol.* **2016**, *223*, 1–8. [CrossRef]
21. Liu, Z.; Chi, Z.; Wang, L.; Li, J. Production, Purification and Characterization of an Extracellular Lipase from *Aureobasidium Pullulans* HN2.3 with Potential Application for the Hydrolysis of Edible Oils. *Biochem. Eng. J.* **2008**, *40*, 445–451. [CrossRef]
22. Song, C.; Liu, G.L.; Xu, J.L.; Chi, Z.M. Purification and Characterization of Extracellular β-Galactosidase from the Psychrotolerant Yeast *Guehomyces Pullulans* 17-1 Isolated from Sea Sediment in Antarctica. *Process Biochem.* **2010**, *45*, 954–960. [CrossRef]
23. D'Souza, S.F.; Nadkarni, G.B. Immobilized catalase-containing yeast cells: Preparation and enzymatic properties. *Biotechnol. Bioeng.* **1980**, *22*, 2191–2205. [CrossRef] [PubMed]
24. Gong, F.; Zhang, T.; Chi, Z.; Sheng, J.; Li, J.; Wang, X. Purification and Characterization of Extracellular Inulinase from a Marine Yeast *Pichia Guilliermondii* and Inulin Hydrolysis by the Purified Inulinase. *Biotechnol. Bioprocess Eng.* **2008**, *13*, 533–539. [CrossRef]
25. Feder, V.; Kmetzsch, L.; Christian Staats, C.; Vidal-Figueiredo, N.; Ligabue-Braun, R.; Carlini, C.R.; Vainstein, M.H. *Cryptococcus gattii* urease as a virulence factor and the relevance of enzymatic activity in cryptococcosis pathogenesis. *FEBS J.* **2015**, *202*, 1406–1418. [CrossRef]
26. Loperena, L.; Soria, V.; Varela, H.; Lupo, S.; Bergalli, A.; Guigou, M.; Pellegrino, A.; Bernardo, A.; Calvino, A.; Rivas, F.; et al. Extracellular Enzymes Produced by Microorganisms Isolated from Maritime Antarctica. *World J. Microbiol. Biotechnol.* **2012**, *28*, 2249–2256. [CrossRef]
27. Lilao, J.; Mateo, J.J.; Maicas, S. Biotechnological Activities from Yeasts Isolated from Olive Oil Mills. *Eur. Food Res. Technol.* **2015**, *240*, 357–365. [CrossRef]

28. Li, H.; Chi, Z.; Wang, X.; Ma, C. Amylase Production by the Marine Yeast *Aureobasidium Pullulans* N13d. *J. Ocean Univ. China* **2007**, *6*, 60–65. [CrossRef]
29. Cripwell, R.A.; Rose, S.H.; Viljoen-Bloom, M.; Van Zyl, W.H. Improved Raw Starch Amylase Production by *Saccharomyces Cerevisiae* Using Codon Optimisation Strategies. *FEMS Yeast Res.* **2019**, *19*, foy127. [CrossRef]
30. Mahmoud, A.E.; Fathy, S.A.; Rashad, M.M.; Ezz, M.K.; Mohammed, A.T. Purification and Characterization of a Novel Tannase Produced by *Kluyveromyces Marxianus* Using Olive Pomace as Solid Support, and Its Promising Role in Gallic Acid Production. *Int. J. Biol. Macromol.* **2018**, *107*, 2342–2350. [CrossRef]
31. Patel, M.A.; Ou, M.S.; Harbrucker, R.; Aldrich, H.C.; Buszko, M.L.; Ingram, L.O.; Shanmugam, K.T. Isolation and Characterization of Acid-Tolerant, Thermophilic Bacteria for Effective Fermentation of Biomass-Derived Sugars to Lactic Acid. *Appl. Environ. Microbiol.* **2006**, *72*, 3228–3235. [CrossRef]
32. Gohel, H.R.; Contractor, C.N.; Ghosh, S.K.; Braganza, V.J.A. Comparative Study of Various Staining Techniques for Determination of Extra Cellular Cellulase Activity on Carboxy Methyl Cellulose (CMC) Agar Plates. *Int. J. Curr. Microbiol. Appl. Sci.* **2014**, *3*, 261–266.
33. Mishra, S.; Behera, N. Amylase Activity of a Starch Degrading Bacteria Isolated from Soil Receiving Kitchen Wastes. *Afr. J. Biotechnol.* **2008**, *7*, 3326–3331. [CrossRef]
34. Gao, L.; Chi, Z.; Sheng, J.; Wang, L.; Li, J.; Gong, F. Inulinase-Producing Marine Yeasts: Evaluation of Their Diversity and Inulin Hydrolysis by Their Crude Enzymes. *Microb. Ecol.* **2007**, *54*, 722–729. [CrossRef] [PubMed]
35. Villena, M.A.; Úbeda Iranzo, J.F.; Briones Pérez, A.I. β-Glucosidase Activity in Wine Yeasts: Application in Enology. *Enzym. Microb. Technol.* **2007**, *40*, 420–425. [CrossRef]
36. Hankin, L.; Zucker, M.; Sands, D.C. Improved Solid Medium for the Detection and Enumeration of Pectolytic Bacteria. *Appl. Microbiol.* **1971**, *22*, 205–209. [CrossRef]
37. Dhiman, S.; Chapadgaonkar, S. Optimization of Lipase Production Medium for a Bacterial Isolate. *Int. J. Chem. Technol. Res.* **2013**, *5*, 2837–2843.
38. Balan, S.S.; Nethaji, R.; Sankar, S.; Jayalakshmi, S. Production of Gelatinase Enzyme from *Bacillus Spp* Isolated from the Sediment Sample of Porto Novo Coastal Sites. *Asia Pacif. J. Trop. Biomed.* **2012**, *2*, S1811–S1816. [CrossRef]
39. Geweely, N.S.I. Purification and characterization of intracellular urease enzyme isolated from *Rhizopus oryzae*. *Biotechnology* **2006**, *5*, 358–364. [CrossRef]
40. Strauss, M.L.A.; Jolly, N.P.; Lambrechts, M.G.; Van Rensburg, P. Screening for the Production of Extracellular Hydrolytic Enzymes by Non-*Saccharomyces* Wine Yeasts. *J. Appl. Microb.* **2001**, *91*, 182–190. [CrossRef]
41. Hernandez, A.; Martin, A.; Aranda, E.; Perez-Nevado, F.; Cordoba, M.G. Identification and characterization of yeast isolated from the elaboration of seasoned green table olives. *Food Microbiol.* **2007**, *24*, 346–351. [CrossRef]
42. Hasanin, M.S.; Mostafa, A.M.; Mwafy, E.A.; Darwesh, O.M. Eco-friendly Cellulose Nano Fibers via First Reported Egyptian *Humicola Fuscoatra* Egyptia X4: Isolation and Characterization. *Environ. Nanotechnol. Monit. Manag.* **2018**, *10*, 409–418. [CrossRef]
43. Thongekkaew, J.; Ikeda, H.; Masaki, K.; Iefuji, H. An acidic and thermostable carboxymethyl cellulase from the yeast *Cryptococcus* sp. S-2: Purification, characterization and improvement of its recombinant enzyme production by high cell-density fermentation of Pichia pastoris. *Protein Expr. Purif.* **2008**, *60*, 140–146. [CrossRef]
44. Hasanin, M.S.; Darwesh, O.M.; Matter, I.A.; El-Saied, H. Isolation and characterization of non-cellulolytic *Aspergillus flavus* EGYPTA5 exhibiting selective ligninolytic potential. *Biocatal. Agric. Biotechnol.* **2019**, *17*, 160–167. [CrossRef]
45. Brown, J.; Scholtz, C.H.; Janeau, J.L.; Grellier, S.; Podwojewski, P. Dung beetles (Coleoptera: Scarabaeidae) can improve soil hydrological properties. *Appl. Soil Ecol.* **2010**, *46*, 9–16. [CrossRef]
46. Gonzalo, G.; Colpa, D.I.; Habib, M.H.M.; Fraaijeb, M.W. Bacterial enzymes involved in lignin degradation. *J. Biotechnol.* **2016**, *236*, 110–119. [CrossRef] [PubMed]
47. Martínkova, L.; Kotik, M.; Markova, E.; Homolka, L. Biodegradation of phenolic compounds by Basidiomycota and its phenol oxidases: A review. *Chemosphere* **2016**, *149*, 373–382. [CrossRef] [PubMed]
48. Tsioulpas, A.; Dimou, D.; Iconomou, D.; Aggelis, G. Phenolic removal in olive oil mill wastewater by strains of *Pleurotus* spp. in respect to their phenol oxidase (laccase) activity. *Bioresour. Technol.* **2002**, *84*, 251–257. [CrossRef]
49. Garrido Hoyos, S.E.; Martinez Nieto, L.; Camacho Rubio, F.; Ramos Cormenzana, A. Kinetics of aerobic treatment of olive-mill wastewater (OMW) with *Aspergillus terreus*. *Process Biochem.* **2002**, *37*, 1169–1176. [CrossRef]
50. Fadil, K.; Chahlaoui, A.; Ouahbi, A.; Zaid, A.; Borja, R. Aerobic biodegradation and detoxification of wastewaters from the olive oil industry. *Int. Biodeterior. Biodegrad.* **2003**, *51*, 37–41. [CrossRef]
51. Chi, Z.; Liu, G.-l.; Lu, Y.; Jiang, H.; Chi, Z.-M. Bio-Products Produced by Marine Yeasts and Their Potential Applications. *Bioresour. Technol.* **2016**, *202*, 244–252. [CrossRef]
52. Rai, A.K.; Kumari, R.; Sanjukta, S.; Sahoo, D. Production of Bioactive Protein Hydrolysate Using the Yeasts Isolated from Soft Chhurpi. *Bioresour. Technol.* **2016**, *219*, 239–245. [CrossRef] [PubMed]
53. Jay-Robert, P.; Errouissi, F.; Lumaret, J.P. Temporal Coexistence of Dung-Dweller and Soil-Digger Dung Beetles (*Coleoptera, Scarabaeoidea*) in Contrasting Mediterranean Habitats. *Bull. Èntomol. Res.* **2008**, *98*, 303–316. [CrossRef] [PubMed]
54. Bautista-gallego, J.; Rodríguez-gómez, F.; Barrio, E.; Querol, A. International Journal of Food Microbiology Exploring the Yeast Biodiversity of Green Table Olive Industrial Fermentations for Technological Applications. *Int. J. Food Microbiol.* **2011**, *147*, 89–96. [CrossRef]

55. Aissam, H.; Errachidi, F.; Merzouki, M.; Benlemlih, M. Identification of the yeasts isolated from the Olive Mill Wastewater and study of catalase activity. *Biotechnol. Bioeng.* **2002**, *7*, 23–30. [CrossRef]

56. Babavalian, H.; Ali Amoozegar, M.; Zahraei, S.; Rohban, R.; Shakeri, F.; Mehrdad Moosazadeh, M. Comparison of Bacterial Biodiversity and Enzyme Production in Three Hypersaline Lakes; Urmia, Howz-Soltan and Aran-Bidgol. *Indian J. Microbiol.* **2014**, *54*, 444–449. [CrossRef]

57. Ali, M.N.; Khan, M.M. Screening, Identification and Characterization of Alcohol Tolerant Potential Bioethanol Producing yeasts. *Curr. Res. Microbiol. Biotechnol.* **2016**, *2*, 316–324.

58. Milala, M.; Shugaba, A.; Gidado, A.; Ene, A.; Wafar, J. Studies on the Use of Agricultural Wastes for Cellulase Enzyme Production by *Aspegillus Niger. Res. J. Agric. Biol. Sci.* **2005**, *1*, 325–328.

59. Arroyo, F.N.; Durán-Quintana, M.C.; Garrido-Fernández, A. Evaluation of primary models to describe the growth of Pichia anomala and study by response surface methodology of the effects of temperature, NaCl and pH on its biological parameters. *J. Food Prot.* **2005**, *68*, 562–570. [CrossRef] [PubMed]

60. Russo, G.; Libkind, D.; Sampaio, J.P.; Van Broock, M.R. Yeast Diversity in the Acidic Rio Agrio-Lake Caviahue Volcanic Environment (Patagonia, Argentina). *FEMS Microb. Ecol.* **2008**, *65*, 415–424. [CrossRef]

Article

Xylitol Production from Pineapple Cores (*Ananas comosus* (L.) *Merr*) by Enzymatic and Acid Hydrolysis Using Microorganisms *Debaryomyces hansenii* and *Candida tropicalis*

Efri Mardawati [1,2,*], Agus T. Hartono [1,2], Bambang Nurhadi [2,3], Hana Nur Fitriana [1,2,4], Euis Hermiati [2,4] and Riksfardini Annisa Ermawar [2,4]

1   Department of Agroindustrial Technology, Universitas Padjadjaran, Jatinangor 45363, Indonesia
2   Research Collaboration Center for Biomass and Biorefinery between BRIN and Universitas Padjadjaran, Jatinangor 45363, Indonesia
3   Departement of Food Industrial Technology, Universitas Padjadjaran, Jatinangor 45363, Indonesia
4   Research Center for Biomass and Bioproducts, National Research and Innovation Agency, Cibinong 16911, Indonesia
*   Correspondence: efri.mardawati@unpad.ac.id

Abstract: Hydrolysis and fermentation processes are key stages in xylitol production from lignocellulosic materials. In this study, pineapple cores, one of the wastes from the canned pineapple industry, were used as raw material for xylitol production. Two methods was used for hydrolysis: enzymatically using commercial enzyme Cellic HTec2, and acid hydrolysis using 4% H_2SO_4. In contrast, the fermentation process was carried out with two selected yeasts commonly employed in xylitol fermentation, *Debaryomycess hansenii*, and *Candida tropicalis*. Before these two processes, the pineapple cores were characterized using the Van Soest method to determine their lignocellulosic content. The hemicellulose content was 36.06%, the cellulose content was 14.20%, and the lignin content was 10.05%. This result indicates that the hemicellulose content of pineapple cores has the potential to be used as a raw material in the production of xylitol. The hydrolysis efficiency of enzymatic hydrolysis was 21% higher than that of acid hydrolysis. The highest xylitol and biomass yield of 0.371 $g_{xylitol}/g_{xylose}$ and 0.225 g_{cell}/g_{xylose} were observed by *C. tropicalis* using an enzymatic hydrolysate.

Keywords: pineapple cores; *Candida tropicalis*; *Debaryomycess hansenii*; enzymatic hydrolysis; acid hydrolysis; xylitol

Citation: Mardawati, E.; Hartono, A.T.; Nurhadi, B.; Fitriana, H.N.; Hermiati, E.; Ermawar, R.A. Xylitol Production from Pineapple Cores (*Ananas comosus* (L.) *Merr*) by Enzymatic and Acid Hydrolysis Using Microorganisms *Debaryomyces hansenii* and *Candida tropicalis*. *Fermentation* **2022**, *8*, 694. https://doi.org/10.3390/fermentation8120694

Academic Editor: Timothy Tse

Received: 5 November 2022
Accepted: 20 November 2022
Published: 30 November 2022

Publisher's Note: MDPI stays neutral with regard to jurisdictional claims in published maps and institutional affiliations.

1. Introduction

Pineapple, or *Ananas comosus* (L.) *Merr* is a popular tropical fruit from South America but has been cultivated widely worldwide [1,2]. The fruit has been processed into many kinds of food, but its byproducts such as pineapple peel, core, and crown, have not been utilized properly. Indonesia is the fourth biggest pineapple producer globally, producing 2,196,456 tonnes of pineapple per year. Therefore, pineapple is a vital commodity in the Indonesian economy [1]. However, using a massive amount of pineapples results in biomass waste that causes serious environmental issues if it is not handled correctly. The utilization of agricultural biomass waste, in addition to reducing the negative impact on the environment, has the potential to produce many bio-based products. For example, pineapple cores as byproducts have several advantages, such as high bromelain enzyme content and large lignocellulose fibers. The hemicellulose concentration of pineapple cores (28.53%) is sufficient for use as raw material for xylitol production [3].

Xylitol is a non-fermentable sugar alcohol with the same level of sweetness as sucrose and a low glycemic index. Because xylitol absorption in the body does not require insulin, it is safe to use as a food sweetener for people with diabetes. Moreover, xylitol has been widely employed as a sucrose sugar alternative in processed food goods, beverage industries, and diabetic health foods. Xylitol produces a cool sensation, heat-resistant characteristics, and slow absorption by the intestine [4,5].

Xylitol is produced industrially worldwide by catalytic hydrogenation of a pure d-xylose solution at high temperatures and pressure. Biotechnological manufacture of xylitol uses cheap industrial and agricultural waste sugar and softer process conditions [4]. The first step for biotechnological xylitol production is a hydrolysis process that converts polysaccharide molecules in lignocellulosic material into sugar monomers, such as glucose and xylose. Chemical and biological hydrolysis are two types of standard hydrolysis methods. Chemical hydrolysis uses acid or base as a catalyst at high temperatures and pressure [6]. Because this is a quick process, it must be closely monitored to avoid the co-production of numerous degradative compounds. However, the process conditions require utilizing particular reactor materials and accompanying equipment. In addition, pH must be neutralized after the reaction [7,8].

Biological or enzymatic hydrolysis employing lignocellulosic enzymes can be utilized instead of chemical hydrolysis. The use of enzymes in hydrolysis requires less energy and eliminates the use of hazardous and corrosive chemicals [9]. The utilized enzymes must be compatible with the polysaccharides to be hydrolyzed. Xylan is hydrolyzed into a D-xylose monomer containing five carbon atoms using the xylanase enzyme [9,10]. Following the extraction of the xylose from the xylan hydrolysis process, the biological reduction of xylose is executed by fermentation, mainly using yeast [11]. *Candida tropicalis* and *Debaryomyces hansenii* are the most employed yeast in the xylitol production process, which is environmentally friendly and cost-effective. These yeasts generate the xylose reductase (X.R.) enzyme that converts xylose to xylitol [8,12]. Fermentation conditions such as temperature, pH, aeration conditions, substrate content, and the presence of other sugars, such as glucose, affect the bioconversion of xylose to xylitol via fermentation [8].

Recently, xylitol has been produced from corn cobs (China) and hardwoods such as birch (US). Large amounts of industrially sourced corn cobs provide an abundant substrate for xylitol production in China, whereas birch hydrolyzate, a by-product of the paper and pulp industry, provides the substrate in the United States. A chemical catalytic reaction generally is still used to convert xylose in the hydrolyzed hemicellulose fraction to xylitol [13]. The chemical catalytic reaction in xylitol production is typically carried out at high temperatures and pressures, consuming significant energy and increasing production costs. As a result, the production of xylitol via fermentation, which can be carried out under mild conditions, is expected to meet market demand for xylitol at a lower cost production [14]. However, obstacles such as the high cost of traditional feedstock, low xylitol titer due to a lack of xylose-utilizing microbial cells, and repressed xylose metabolism in the presence of glucose, have all contributed to the slow progress of biotechnological xylitol synthesis. The potency of biochemical conversion of pineapple core waste to xylitol in Indonesia has yet to be studied. The goal of the current work is to employ cheap and plentiful lignocellulosic biomass to avoid paying high substrate expenses. It is also ideal for creating a straightforward yet affordable bioprocess that increases xylitol production. As a result, major commercial manufacturing initiatives have been related to the rise in interest in xylitol bioconversion. According to our knowledge, no attempts have been made to produce xylitol from pineapple core waste. Therefore, this paper focuses on optimizing hydrolysis processes using acid and enzymatic hydrolysis, and fermentation processes using *C. tropicalis* and *D. hansenii*.

2. Materials and Methods

2.1. Tools and Materials

The pineapples were sourced from an Indonesian pineapple orchard in Subang, West Java, Indonesia. The pineapple cores were mashed in a blender (Philips, Zhuhai, China) and dried in a dry oven(B-One, Messgerate Sukses Mandiri Ltd., Tangerang, Indonesia) at 60 °C for 24 h to make a powder sieved with a 60–60 mesh size (0.25–0.177 mm).

Approximately 75 I.U./mL of the Cellic HTec2 enzyme (Novozymes, Copenhagen, Denmark) was employed for the enzymatic hydrolysis of xylan. *Debaryomyces hansenii* ITB CC R85 and *Candida tropicalis* were collected from the ITB (Institut Teknologi Bandung, Bandung, Indonesia) Department of Chemical Engineering culture collection.

All of the additional chemicals received from Sigma Aldrich (St. Louis, MO, USA) were of analytical grade and were used immediately. Each solution was produced with distilled water.

2.2. Enzymatic Hydrolysis

The pineapple core powder was sterilized by submerging 20 g of the powder in 100 mL of acetate buffer at pH 5 and then autoclaving it for 15 min at 121 °C. Then, 50 I.U./g of biomass Cellic HTec 2 was added, and the mixture was shaken at 150 rpm for 96 h at 60 °C in a shaker incubator (N-Biotek, Seoul, Republic of Korea). The xylose hydrolyzate was obtained after 20 min of centrifugation at 5000 rpm. Every 24 h, sampling was conducted. At 70 °C, the liquid hydrolysate was concentrated threefold by evaporation.

2.3. Acid Hydrolysis

A 10 g powdered pineapple core was dissolved in 250 mL of 0.72 M or 4% (v/v) of 96% H_2SO_4 stock solution and autoclaved for 20 min at 121 °C in a 500 mL Erlenmeyer flask. The hydrolyzate was filtered to remove the dregs before being neutralized with 2 M NaOH to attain a pH of 7. It was then detoxified using 15 g of activated charcoal for 1 h at 30 °C and concentrated threefold by evaporation at 70 °C.

2.4. Fermentation

D. hansenii and *C. tropicalis* seeds were cultivated on growth media containing 50 mL of xylose (20 g/L) and 50 mL of nutritional medium containing 9.438 g/L $(NH_4)_2SO_4$, 2.5 g/L KH_2PO_4, 0.05 g/L $CaCl_2 \cdot 2H_2O$, 0.5 g/L $MgSO_4 \cdot 7H_2O$, 0.5 g/L citric acid, 0.035 g/L $FeSO_4 \cdot 7H_2O$, 0.0092 g/L $MnSO_4 \cdot 7H_2O$, 0.011 g/L $ZnSO_4 \cdot 7H_2O$, 0.001 g/L $CuSO_4 \cdot 7H_2O$, 0.002 g/L $CoCl_2 \cdot 6H_2O$, 0.0013 g/L $Na_2CoO_4 \cdot 2H_2O$, 0.002 g/L H_3BO_3, 0.0035 g/L KI, 0.0005 g/L $Al_2(SO_4)_3$, 0.1 g/L myo-inositol, 0.02 g/L calcium-pantothenate, 0.005 g/L thiamine hydrochloride, 0.005 g/L pyridoxal hydrochloride, 0.005 g/L nicotine acid, 0.001 g/L aminobenzoic acid, and 0.0001 g/L D-biotin. The seeds were cultured for 48 h at 30 °C with a stirring speed of 150 rpm in a shaker incubator.

The pineapple core hydrolyzate used in fermentation was generated via enzymatic and acid hydrolysis. The microaerobic condition was produced by combining nitrogen gas and air at a volume ratio of 1:5 in the Erlenmeyer head space. The volume ratio of the inoculum (10^6 CFU/mL), hydrolysate, and fermentation medium was 2:2:3.

2.5. Purification of Fermentation Product

The fermentation product was separated from cell biomass using centrifugation at 5000 rpm for 40 min. The supernatant was filtered using filter paper before being transferred to a 250 mL Erlenmeyer flask for further purification with 15 g/L of activated charcoal. The purification process was performed at 30 °C for 1 h using a magnetic stirrer (IKA, Selangor, Malaysia) for agitation. In the final step, the purified sample was filtered with filter paper for further analysis. The total working volume in a 250 mL Erlenmeyer flask was 100 mL. The microaerobic condition was produced by combining nitrogen gas and air in the Erlenmeyer head space at a volume ratio of 1:5.

2.6. Analysis Method

Using the Van Soest method [15], the biomass, cellulose, hemicellulose, and lignin contents were measured simultaneously to estimate the composition of lignocellulose. An HPLC system with UV and refractive index (R.I.) detectors was used to evaluate the hydrolysis and fermentation samples (Waters type 1515 pump; Autosampler type 2707, Waters, Milford, MA, USA.). The concentrations of xylose, glucose, ethanol, and xylitol were measured at 65 °C, with 5 mM H_2SO_4 as the mobile phase and 0.6 mL/min as the flow rate, using an Aminex HPX-87H column with a R.I. detector. Using a calibration curve, the optical density at 650 nm, which was used to quantify cell growth, was converted to dry cell weight (D.C.W.). The OPEFB surface was examined with a scanning electron microscope (JEOL, JSM-6330F; Tokyo, Japan) before and after several processes.

2.7. Data Interpretation

The hydrolysis yield measurement was calculated based on Equations (1)–(5) Hemicellulose hydrolysis efficiency (%) is:

$$\frac{\text{Xylose (t)}}{\text{Xylose (theo)}} \times 100\%. \tag{1}$$

Maximum theoretical xylose from hydrolysis (Xylose (theo)):

$$\text{Mass of lignocellulose (g)} \times \text{hemicellulose content in lignocellulose} \times 0.88 \tag{2}$$

The yield of biomass ($Y_{X/S}$) (g/g) is:

$$Y_{X/S} = -\frac{\Delta X}{\Delta S} = \frac{X - X_o}{S_o - S} \tag{3}$$

The product yield (YP/S) (g/g) is:

$$Y_{P/S} = -\frac{\Delta P}{\Delta S} = \frac{P - P_o}{S_o - S} \tag{4}$$

where xylose(t) is the xylose concentration (g/L) produced at time t, X is the biomass concentration (g/L), S is the substrate concentration (g/L), and P is the product (xylitol or ethanol) concentration (g/L).

Xylose or glucose Utilization (%) is:

$$\frac{\Delta S}{S_0} = \frac{S - S_0}{S_0} \tag{5}$$

Xylitol fermentation efficiency (%) is:

$$\frac{\text{Xylitol (t)}}{\text{Xylitol (theo)}} \times 100\% \tag{6}$$

The theoretical yield of xylitol (xylitol (theo)) is 0.9 mol of xylitol per mol of xylose utilized [16].

The specific growth rate (μ) is:

$$\mu = \frac{1}{x}\frac{dx}{dt} \tag{7}$$

2.8. Statistical Analysis

Student's *t*-test was used to determine the statistical significance of each measurement. The information is given as a standard deviation from the mean. A *p*-value of less than 0.05 was regarded as statistically significant.

3. Results and Discussion

3.1 Lignocellulose Composition of Pineapple Core

The lignocellulose composition of the raw materials in the pineapple core was deteremined to estimate the quantity of xylose and glucose produced during hydrolysis. Table 1 summarizes the findings of the lignocellulose content analysis of the pineapple cores. The percentages of cellulose, hemicellulose, and lignin were 14.20 ± 0.31%, 36.06 ± 0.22%, and

10.05 ± 0.93%, respectively. The lignocellulose composition in this material was suitable for xylitol production by yeast fermentation. The high hemicellulose content will provides a great deal of xylose as the raw material for xylitol production [17]. The low cellulose content provides enough glucose to support microbial growth without inhibiting xylitol production [18]. The low lignin content facilitates the hydrolysis process, both acid and enzymatic, because the access of enzymes or acids to cellulose and hemicellulose is not much hindered by lignin [19,20].

Table 1. Composition of Lignocellulosic Content of Pineapple Cores.

Lignocellulose Biomass	Result (%)
NDF [1]	61.57 ± 0.65
Hemicellulose	36.06 ± 0.22
ADF [2]	25.51 ± 0.72
Cellulose	14.20 ± 0.31
Lignin	10.05 ± 0.93
Silica	1.26 ± 0.82

[1] Neutral Detergent Fiber; [2] Acid Detergent Fiber.

3.2. Impact of Hydrolysis Types on Hydrolysate Composition

In this study, two different hydrolysis methods were used to determine which resulted in the most xylose and hemicellulose hydrolysis efficiency, namely enzymatic hydrolysis and acid hydrolysis.

In the enzymatic hydrolysis of pineapple cores, auto-hydrolysis was used to break down lignocellulosic tissue and release hemicellulose from the material [9,21]. Because enzymes have specificity, their role in enzymatic hydrolysis is crucial; thus, enzyme performance is optimal if the substrate used is suitable and at the proper concentration [22]. The commercial enzyme cellic Htech2 was used in this study at a concentration of 50 IU/g biomass and a solid loading of 20%. This enzymatic hydrolysis yielded 23.792 g/L of xylose, with a hemicellulose hydrolysis efficiency of 37.550%. Furthermore, 2.73 g/L glucose was produced as a byproduct of hydrolysis (Table 2). The amount of glucose in the enzymatic hydrolysis process was low, since xylanase was more dominant than cellulase in the Cellic HTec enzyme blend composition, and the cellulose content in pineapple cores was lower than the hemicellulose content.

Table 2. Components of pineapple core hydrolysate.

Components	Type of Hydrolysis	
	Enzymatic Hydrolysis	Acid Hydrolysis
Solid Loading (%)	20.000 ± 0.800	4.00 ± 0.800
Xylose (g/L)	23.792 ± 0.163	9.844 ± 0.159
Glucose (g/L)	2.73 ± 0.080	1.22 ± 0.010
Hemicellulose hydrolysis efficiency (%)	37.550 ± 0.900	31.074 ± 0.700

For acid hydrolysis, concentrations of up to 4% or 0.72 M H_2SO_4 were utilized. Acid hydrolysis with diluted acid is frequently employed for processing lignocellulosic material due to its efficiency and cost-effectiveness in producing sugars such as d-xylose and d-glucose [23]. However, the toxicity to microbial growth (furans, aliphatic acids, and phenolic components) complicates the fermentability of hydrolyzate and even stops the fermentation process [24]. A detoxification process, such as overliming, activated charcoal, pH adjustment, or enzyme treatment, can increase biomass and xylitol production by removing 4% to 95% of these inhibitory compounds [25]. Activated charcoal was used for detoxification in this study because Winkelhausen and Kuzmanova [5] reported that this approach is a simple detoxification treatment that can be combined with additional techniques, such as alkalinization and observation of the elimination of acetic acid, lignin phenolic compounds, furan (5-HMF and furfural), and clarity. The xylose obtained from the acid hydrolysis pretreatment was 9.844 g/L, or 31.074% of the theoretical hydrolysis yield. At the same time, the amount of glucose produced was only 1.22 g/L due to the low cellulose content in the pineapple core. The factors that affected the amount of xylose produced by acid hydrolysis were temperature, acid concentration, and reaction time [26]. More reducing sugar was obtained using a higher acid concentration and a longer time. However, the disadvantage of the high acid concentration was that it caused a decomposition reaction in which the sugar components dehydrated into furfural compounds [17].

Table 2 shows the results of enzymatic hydrolysis, which are significantly better than acid hydrolysis, including the value of hemicellulose hydrolysis efficiency, which is greater than 10%. The amount of solid loading is much higher, which can be applied to enzymatic hydrolysis to produce a high concentration of xylose.

3.3. Influence of Different Hydrolysate Types on Substrate Utilization by D. hansenii and C. tropcalis

Debaryomyces hansenii consumed xylose from enzymatic hydrolysate at a rate of 47.200%, and glucose was utilized at 98.890%. On the other hand, the consumption rate for reducing sugar obtained from acid hydrolysis was 41.860% for xylose and 97.540% for glucose (Table 3). This result implies that *D. hansenii* did not thoroughly consume all xylose for metabolite product formation or cell biomass synthesis. However, compared to xylitol production using OPEFB enzymatic hydrolysate by *D. hansenii* in a previous study [27], the xylose utilization in the current study was 57.3% higher. The use of xylose in pineapple core hydrolyzate and OPEFB may differ due to the complex medium of each hydrolysate, which contains other unknown substances consumed by the microorganisms [28]. On the other hand, inhibitory compounds such as acetic acid and furfural in acid hydrolysate can cause the slow conversion of xylose to xylitol during hydrolyzate fermentation [29]; therefore, the xylitol consumption in the acid hydrolysate was lower than in enzymatic one. A graph showing the decrease in xylose and glucose concentration during fermentation by *D. hansenii* is presented in Figure 1A,B.

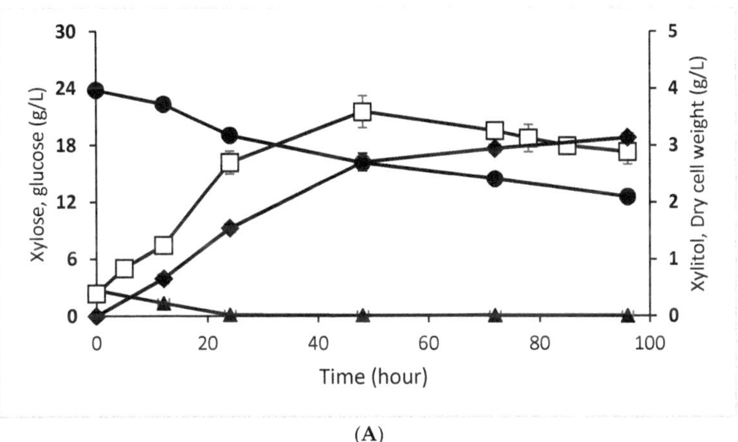

(A)

Figure 1. *Cont.*

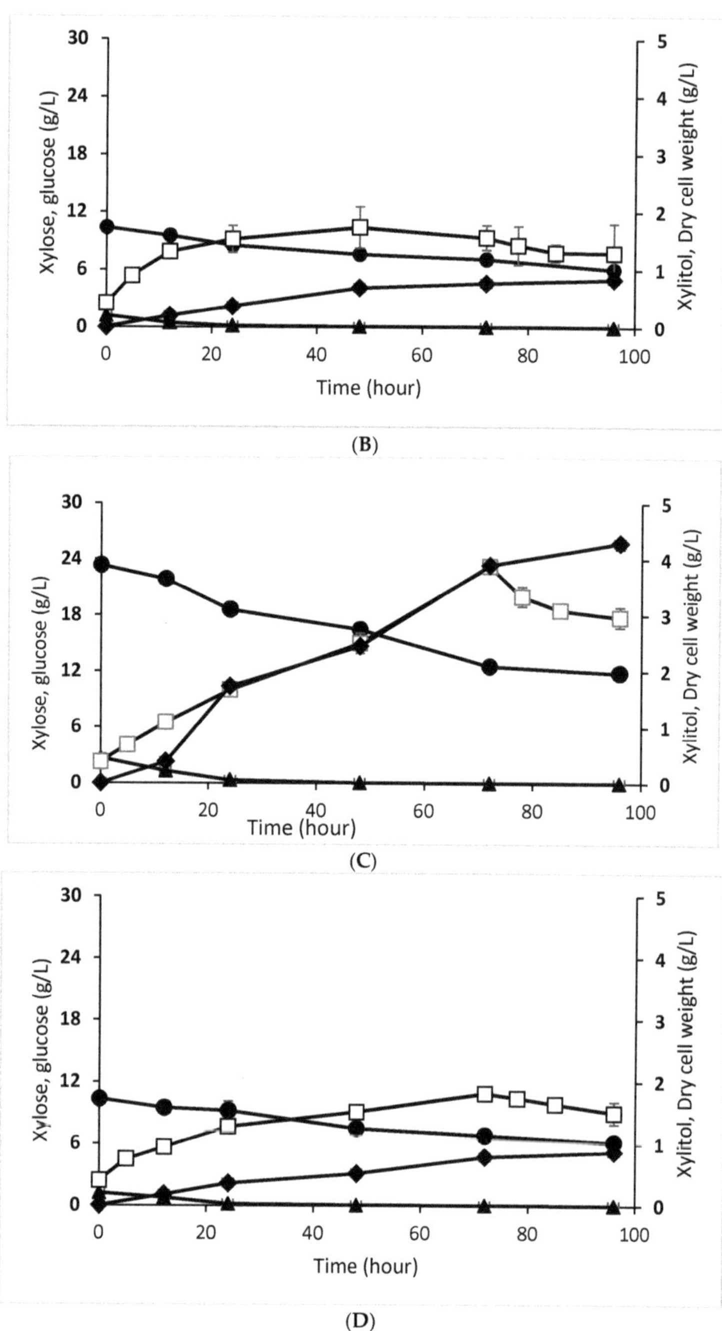

Figure 1. Xylitol fermentation profiles of *D. hansenii* in enzymatic hydrolysate (**A**), *D. hansenii* in acid hydrolysate (**B**), *C. tropicalis* in enzymatic hydrolysate (**C**), and *C. tropicalis* in acid hydrolysate (**D**). Xylose concentration (●, closed circle), glucose concentration (▲, closed triangle), dry cell weight (□, open square), xylitol concentration (♦, closed diamond). The data are the means of three replicated investigations, and the error bars are standard deviations.

Table 3. Fermentation products of enzyme hydrolysate and acid hydrolysate by *D. hansenii*.

Products	Type of Hydrolysis	
	Enzymatic Hydrolysis	Acid Hydrolysis
Initial Xylose Concentration (g/L)	23.790 ± 0.163	10.380 ± 0.050
Final xylose concentration (g/L)	12.560 ± 0.217	6.030 ± 0.030
Xylose Utilization (%)	47.200 ± 0.570	41.860 ± 0.070
Initial Glucose Concentration (g/L)	2.730 ± 0.080	1.220 ± 0.010
Final Glucose Concentration (g/L)	0.030 ± 0.010	0.030 ± 0.010
Glucose Utilization (%)	98.890 ± 0.400	97.540 ± 0.810
Initial Xylitol Concentration (g/L)	0.000 ± 0.000	0.000 ± 0.000
Final Xylitol Concentration (g/L)	3.140 ± 0.080	0.840 ± 0.040
Xylitol Yield from xylose (g/g) ($Y_{P/s}$)	0.279 ± 0.000	0.193 ± 0.000
Initial Cells Concentration (g cell/L)	0.400 ± 0.040	0.410 ± 0.080
Final Cells Concentration (g cell/L)	2.890 ± 0.022	1.290 ± 0.051
Biomass Yield of Substrate (g/g) ($Y_{X/S}$)	0.221 ± 0.000	0.203 ± 0.000
Xylitol Yield of Concentration cells (g/g) ($Y_{P/X}$)	0.279 ± 0.000	0.193 ± 0.000
Specific Growth Rate (h^{-1}) (μ)	0.108 ± 0.000	0.102 ± 0.000

C. tropicalis is an excellent yeast for converting xylose to xylitol [30]. Like the xylose consumption pattern in *D. hansenii*, *C. tropicalis* also consumed more xylose in the enzymatic hydrolysate. The percentage of *C. tropicalis* that consumed xylose from enzymatic hydrolysate was 49.600%, while that which consumed glucose was 98.370%. The consumption of xylose from acid hydrolysate was 40.490%, while for glucose, it was 99.400%. In summary, the xylose consumption in enzymatic hydrolysate by *C. tropicalis* was higher than by *D. hansenii*, while there was no significant difference in xylose utilization in acid hydrolysate between the two yeasts. A graph of the decrease in xylose and glucose concentration during *C. tropicalis* fermentation is presented in Figure 1C,D.

3.4. Effect of Enzymatic and Acid Hydrolysate on Xylitol Production Profile by D. hansenii and C. tropicalis

Xylitol production by *D. hansenii* and *C. tropicalis* was evaluated using enzymatic and acid hydrolysate as the primary substrate (Figure 1). The two strains grew using glucose and xylose as a carbon source to build up cell biomass and xylitol. Glucose was consumed up to the first 24 h, since it is the most readily consumed substrate by the yeast and its concentration was low in the fermentation medium. However, the xylose consumption, cell growth, and xylitol production in *C. tropicalis* tended to be higher than that of *D. hansenii* in both enzymatic and acid hydrolysate. Moreover, these xylitol production parameters were higher using enzymatic hydrolysate than that acid hydrolysate for both yeasts.

D. hansenii converted 49.670% of xylose from enzymatic hydrolysate to produce 3.136 g/L of xylitol with 0.279 $g_{xylitol}/g_{xylose}$ of xylitol yield and 0.221 g_{cell}/g_{xylose} of biomass yield (Table 3). In contrast, in the acid hydrolysate, the yeast converted 45.090% of xylose producing 0.837 g/L of xylitol with 0.193 $g_{xylitol}/g_{xylose}$ of xylitol yield and 0.203 g_{cell}/g_{xylose} of biomass yield. The xylitol yield obtained in the current study was low when compared to other studies using other lignocellulose material. According to Parajo et al. [31], *D. hansenii* fermentation from sawdust hydrolysate produced approximately 0.5 to 9 g/L of xylitol from 18 g/L of xylose, resulting in a maximum yield of 0.79 $g_{xylitol}/g_{xylose}$.

C. tropicalis converted 49.600% of xylose from enzymatic hydrolysate resulting in 4.296 g/L of xylitol with a xylitol yield of 0.371 $g_{xylitol}/g_{xylose}$ and a biomass yield of 0.225 g_{cell}/g_{xylose}, while in the acid hydrolysate, the yeast converted 40.490% of xylose producing 0.873 g/L of xylitol with 0.210 $g_{xylitol}/g_{xylose}$ of xylitol yield and 0.211 g_{cell}/g_{xylose} of biomass yield (Table 4). Our method produced a greater concentration of xylitol than Tran et al. study [32] in the enzymatic hydrolysate of beechwood and walnut shell which obtained 2–3.5 g/L of xylitol after 72 h fermentation using *C. tropicalis* yeast.

Table 4. Fermentation products of enzyme hydrolysate and acid hydrolysate by *C. tropicalis*.

Products	Type of Hydrolysis	
	Enzymatic Hydrolysis	**Acid Hydrolysis**
Initial Xylose Concentration (g/L)	23.340 ± 0.075	10.350 ± 0.037
Final xylose concentration (g/L)	11.480 ± 0.379	6.300 ± 0.216
Xylose Utilization (%)	49.600 ± 0.370	40.490 ± 0.410
Initial Glucose Concentration (g/L)	2.680 ± 0.050	1.220 ± 0.010
Final Glucose Concentration (g/L)	0.020 ± 0.000	0.020 ± 0.100
Glucose Utilization (%)	98.370 ± 0.280	99.400 ± 0.810
Initial Xylitol Concentration (g/L)	0.000 ± 0.000	0.000 ± 0.000
Final Xylitol Concentration (g/L)	4.290 ± 0.110	0.873 ± 0.004
Xylitol Yield from xylose (g/g) ($Y_{P/s}$)	0.371 ± 0.000	0.210 ± 0.000
Initial Cells Concentration (g cell/L)	0.380 ± 0.070	0.400 ± 0.080
Final Cells Concentration (g cell/L)	2.970 ± 0.018	1.500 ± 0.018
Biomass Yield of Substrate (g/g) ($Y_{X/S}$)	0.225 ± 0.000	0.211 ± 0.000
Xylitol Yield of Concentration cells (g/g) ($Y_{P/X}$)	0.373 ± 0.000	0.209 ± 0.000
Specific Growth Rate (h^{-1}) (μ)	0.088 ± 0.000	0.082 ± 0.000

Logarithmic growth of *D. hansenii* in both enzymatic and acid hydrolysate occurred for up to 48 h (Table 4). At this time, the cell concentration in the enzymatic hydrolysate was 3.59 g/L and 1.74 g/L in the acid hydrolysate, with a specific growth rate of 1.108 h^{-1} and 0.102 h^{-1}, respectively. On the other hand, the logarithmic growth for *C. tropicalis* occurred at 72 h in both enzymatic and acid hydrolysate resulting in the highest cell concentration of 3.880 g/L and 1.82 g/L with specific growth rates of 0.088 h^{-1} and 0.082 h^{-1}, respectively. Both yeast cells in acid hydrolysate grew slower than those in enzymatic hydrolysate due to the lower xylose concentration and, perhaps, toxic compounds remaining in the fermentation liquid. Compared to *C. tropicalis* growth with both hydrolysate, *D. hansenii* had a higher specific growth rate because this yeast reached logarithmic growth faster than *C. tropicalis* even though the xylitol yield and biomass yield were lower than *C. tropicalis* in the enzymatic hydrolysate.

3.5. Morphology Changes of Fresh and Hydrolyzed Pineapple Core

Morphological characterizations of the new and hydrolyzed pineapple cores were carried out by SEM analysis to determine the changes in the microstructure of the pineapple core during the enzymatic (xylanase) and acid hydrolysis process. The fresh pineapple core had a rigid surface and compact structure with fibers organized in bundles, as shown in Figure 2A. After hydrolysis, the system of the pineapple core was severely damaged, fractured, and roughened, and pores had formed. It was suggested that most of the lignin and hemicellulose were removed, resulting in a change of external and internal structure, and increased porosity. Because the enzyme or acid decomposed the lignin bonds, allowing it to break down the carbohydrates in the pineapple core inner structure, which contains hemicellulose and cellulose, the morphology structure was degraded after hydrolysis (Figure 2B,C), resulting in more and larger pores [33].

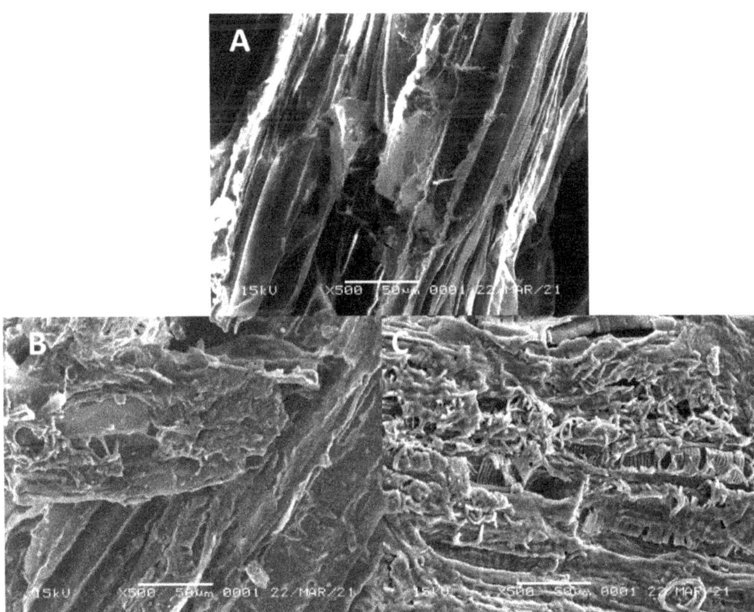

Figure 2. Scanning electron microscope images of fresh OPEFB (**A**), after enzymatic hydrolysis (**B**), after acid hydrolysis (**C**).

Comparing fresh cores and post acid hydrolysis samples, the average pore area of enzymatically hydrolyzed pineapple core was much more significant. This was caused by the 96 h hydrolysis incubation period, whereas the disconnection of fiber chains into sugar monomers dissolved in the hydrolysate liquid enlarged the pores of fiber particles [9]. Acid hydrolysis caused damage to most of the surface of the pineapple core sample.

4. Conclusions

This study showed that pineapple cores are a potential source for xylitol-based products due to their high hemicellulose content, which is 36.060%. It was found that the enzymatic hydrolysis method resulted in over 20% higher xylose concentration than acid hydrolysis. It was also found that the highest concentration of xylitol production was 4.29 g/L, which was produced by *Candida tropicalis*. The ratio between product concentration and the substrate consumed ($Y_{p/s}$) was 0.371 $g_{xylitol}/g_{xylose}$, and the ratio between product and growth cell ($Y_{p/x}$) was 0.225 g_{cell}/g_{xylose}.

Author Contributions: Conceptualization: E.M., B.N., H.N.F. and A.T.H.; data curation: A.T.H., B.N., E.H. and H.N.F.; formal analysys: H.N.F. and R.A.E.; funding acquisition: E.M. and B.N.; investigation: B.N., E.H., H.N.F. and R.A.E.; methodology: E.M., B.N., E.H. and A.T.H.; project administration: E.M. and B.N.; resources: R.A.E. and A.T.H.; writing-original draft: E.M., A.T.H. and H.N.F.; writing—review and editing: B.N., H.N.F., E.H. and R.A.E. All authors have read and agreed to the published version of the manuscript.

Funding: This research was funded by the Ministry of Education, Research and Technology of Indonesia, Universitas Padjadjaran, and the National Research and Innovation Agency (BRIN).

Institutional Review Board Statement: Not applicable.

Informed Consent Statement: Not applicable.

Data Availability Statement: The data presented in this study are available on request from the corresponding author.

Acknowledgments: We would like to thank Faculty of Agroindustrial Technology, Department of Agroindustrial Technology, Universitas Padjadjaran and the farmers in Subang, Indonesia, who provided pineapple samples for this study.

Conflicts of Interest: The authors declare no conflict of interest.

References

1. Hikal, W.M.; Mahmoud, A.A.; Said-Al Ahl, H.A.H.; Bratovcic, A.; Tkachenko, K.G.; Kačániová, M.; Rodriguez, R.M. Pineapple (*Ananas comosus* L. Merr.), Waste Streams, Characterisation and Valorisation: An Overview. *Open J. Ecol.* **2021**, *11*, 610–634. [CrossRef]
2. Zhang, Y.; Xia, X. Physicochemical Characteristics of Pineapple (Ananas Mill.) Peel Cellulose Prepared by Different Methods. *Adv. Mater. Res.* **2012**, *554–556*, 1038–1041. [CrossRef]
3. Pardo, M.E.S.; Cassellis, M.E.R.; Escobedo, R.M.; García, E.J. Chemical Characterisation of the Industrial Residues of the Pineapple (*Ananas comosus*). *J. Agric. Chem. Environ.* **2014**, *3*, 53–56. [CrossRef]
4. Dasgupta, D.; Bandhu, S.; Adhikari, D.K.; Ghosh, D. Challenges and Prospects of Xylitol Production with Whole Cell Bio-Catalysis: A Review. *Microbiol. Res.* **2017**, *197*, 9–21. [CrossRef]
5. Winkelhausen, E.; Kuzmanova, S. Microbial Conversion of D-Xylose to Xylitol. *J. Ferment. Bioeng.* **1998**, *86*, 1–14. [CrossRef]
6. Venkateswar Rao, L.; Goli, J.K.; Gentela, J.; Koti, S. Bioconversion of Lignocellulosic Biomass to Xylitol: An Overview. *Bioresour. Technol.* **2016**, *213*, 299–310. [CrossRef] [PubMed]
7. Suhartini, S.; Rohma, N.A.; Mardawati, E.; Kasbawati; Hidayat, N.; Melville, L. Biorefining of Oil Palm Empty Fruit Bunches for Bioethanol and Xylitol Production in Indonesia: A Review. *Renew. Sustain. Energy Rev.* **2022**, *154*, 111817. [CrossRef]
8. Kresnowati, M.; Mardawati, E.; Setiadi, T. Production of Xylitol from Oil Palm Empty Friuts Bunch: A Case Study on Bioefinery Concept. *Mod. Appl. Sci.* **2015**, *9*, 206. [CrossRef]
9. Mardawati, E.; Werner, A.; Bley, T.; Mtap, K.; Setiadi, T. The Enzymatic Hydrolysis of Oil Palm Empty Fruit Bunches to Xylose. *J. Japan Inst. Energy* **2014**, *93*, 973–978. [CrossRef]
10. Mardawati, E.; Purwadi, R.; Setiadi, T. Evaluation of the Enzymatic Hydrolysis Process of Oil Palm Empty Fruit Bunch Using Crude Fungal Xylanase. *ARPN J. Eng. Appl. Sci.* **2017**, *12*, 5286–5292.
11. Parajó, J.C.; Domínguez, H.; Domínguez, J.M. Biotechnological Production of Xylitol. Part 2: Operation in Culture Media Made with Commercial Sugars. *Bioresour. Technol.* **1998**, *65*, 203–212. [CrossRef]
12. Azizah, N. Biotransformation of Xylitol Production from Xylose of Lignocellulose Biomass Using Xylose Reductase Enzyme: Review. *J. Food Life Sci.* **2019**, *3*, 103–112. [CrossRef]
13. Ravella, S.R.; Gallagher, J.; Fish, S.; Prakasham, R.S. *D-Xylitol*; da Silva, S.S., Chandel, A.K., Eds.; Springer: Berlin/Heidelberg, Germany, 2012; ISBN 9783642318863.
14. Mardawati, E.; Febrianti, E.A.; Fitriana, H.N.; Yuliana, T.; Putriana, N.A.; Suhartini, S.; Kasbawati. An Integrated Process for the Xylitol and Ethanol Production from Oil Palm Empty Fruit Bunch (OPEFB) Using *Debaryomyces hansenii* and *Saccharomyces cerevisiae*. *Microorganisms* **2022**, *10*, 2036. [CrossRef] [PubMed]
15. Van Soest, P.J.; Robertson, J.B.; Lewis, B.A. Methods for Dietary Fiber, Neutral Detergent Fiber, and Nonstarch Polysaccharides in Relation to Animal Nutrition. *J. Dairy Sci.* **1991**, *74*, 3583–3597. [CrossRef] [PubMed]
16. Tamburini, E.; Costa, S.; Marchetti, M.G.; Pedrini, P. Optimized Production of Xylitol from Xylose Using a Hyper-Acidophilic *Candida tropicalis*. *Biomolecules* **2015**, *5*, 1979–1989. [CrossRef]
17. Harahap, B.M. Degradation Techniques of Hemicellulose Fraction from Biomass Feedstock for Optimum Xylose Production: A Review. *J. Keteknikan Pertan. Trop. dan Biosist.* **2020**, *8*, 107–124. [CrossRef]
18. Mardawati, E.; Wira, D.W.; Kresnowati, M.; Purwadi, R.; Setiadi, T. Microbial Production of Xylitol from Oil Palm Empty Fruit Bunches Hydrolysate: The Effect of Glucose Concentration. *J. Japan Inst. Energy* **2015**, *94*, 769–774. [CrossRef]
19. Sugiharto, Y.E.C.; Harimawan, A.; Kresnowati, M.T.A.P.; Purwadi, R.; Mariyana, R.; Andry; Fitriana, H.N.; Hosen, H.F. Enzyme Feeding Strategies for Better Fed-Batch Enzymatic Hydrolysis of Empty Fruit Bunch. *Bioresour. Technol.* **2016**, *207*, 175–179. [CrossRef]
20. Qi, B.; Chen, X.; Su, Y.; Wan, Y. Enzyme Adsorption and Recycling during Hydrolysis of Wheat Straw Lignocellulose. *Bioresour. Technol.* **2011**, *102*, 2881–2889. [CrossRef]
21. Vázquez, M.J.; Alonso, J.L.; Domínguez, H.; Parajó, J.C. Production of Xylose-Containing Fermentation Media by Enzymatic Post-Hydrolysis of Oligomers Produced by Corn Cob Autohydrolysis. *World J. Microbiol. Biotechnol.* **2001**, *17*, 817–822. [CrossRef]
22. Chen, H.Z.; Liu, Z.H. Enzymatic Hydrolysis of Lignocellulosic Biomass from Low to High Solids Loading. *Eng. Life Sci.* **2017**, *17*, 489–499. [CrossRef] [PubMed]
23. Mateo, S.; Roberto, I.C.; Sánchez, S.; Moya, A.J. Detoxification of Hemicellulosic Hydrolyzate from Olive Tree Pruning Residue. *Ind. Crops Prod.* **2013**, *49*, 196–203. [CrossRef]
24. Purwadi, R.; Niklasson, C.; Taherzadeh, M.J. Kinetic Study of Detoxification of Dilute-Acid Hydrolyzates by Ca(OH)$_2$. *J. Biotechnol.* **2004**, *114*, 187–198. [CrossRef] [PubMed]
25. Mussatto, S.I.; Roberto, I.C. Alternatives for Detoxification of Diluted-Acid Lignocellulosic Hydrolyzates for Use in Fermentative Processes: A Review. *Bioresour. Technol.* **2004**, *93*, 1–10. [CrossRef] [PubMed]

26. Ji, X.; Ma, H.; Tian, Z.; Lyu, G.; Fang, G.; Chen, J.; Saeed, H.A.M. Production of Xylose from Diluted Sulfuric Acid Hydrolysis of Wheat Straw. *BioResources* **2017**, *12*, 7084–7095. [CrossRef]
27. Mardawati, E.; Mandra Harahap, B.; Ayu Febrianti, E.; Try Hartono, A.; Putri Siahaan, N.; Wulandari, A.; Yudiastuti, S.; Suhartini, S.; Kasbawati, K. Integrated and Partial Process of Xylitol and Bioethanol Production from Oil Palm Empty Fruit Bunches. *Adv. Food Sci. Sustain. Agric. Agroind. Eng.* **2022**, *5*, 49–67. [CrossRef]
28. Mohamad, N.L.; Kamal, S.M.M.; Abdullah, N.; Ismail, I. Evaluation of Fermentation Conditions by *Candida tropicalis* for Xylitol Production from Sago Trunk Cortex. *BioResources* **2013**, *8*, 2499–2509. [CrossRef]
29. Rahman, S.H.A.; Choudhury, J.P.; Ahmad, A.L. Production of Xylose from Oil Palm Empty Fruit Bunch Fiber Using Sulfuric Acid. *Biochem. Eng. J.* **2006**, *30*, 97–103. [CrossRef]
30. Cheng, K.K.; Zhang, J.A.; Ling, H.Z.; Ping, W.X.; Huang, W.; Ge, J.P.; Xu, J.M. Optimization of PH and Acetic Acid Concentration for Bioconversion of Hemicellulose from Corncobs to Xylitol by Candida Tropicalis. *Biochem. Eng. J.* **2009**, *43*, 203–207. [CrossRef]
31. Parajó, J.C.; Domínguez, H.; Domínguez, J.M. Biotechnological Production of Xylitol. Part 1: Interest of Xylitol and Fundamentals of Its Biosynthesis. *Bioresour. Technol.* **1998**, *65*, 191–201. [CrossRef]
32. Tran, L.H.; Yogo, M.; Ojima, H.; Idota, O.; Kawai, K.; Suzuki, T.; Takamizawa, K. The Production of Xylitol by Enzymatic Hydrolysis of Agricultural Wastes. *Biotechnol. Bioprocess Eng.* **2004**, *9*, 223–228. [CrossRef]
33. Soontornchaiboon, W.; Chunhachart, O.; Pawongrat, R. Ethanol Production from Pineapple Waste by Co-Culture of *Saccharomycs cerevisiae* TISTR 5339 and *Candida shehatae* KCCM 11422. *Asia-Pacific J. Sci. Technol.* **2016**, *21*, 347–355. [CrossRef]

 *fermentation*

Article

Comparative Fatty Acid Compositional Profiles of *Rhodotorula toruloides* Haploid and Diploid Strains under Various Storage Conditions

Yue Zhang [1,2], Rasool Kamal [1], Qing Li [1], Xue Yu [1], Qian Wang [1,3,*] and Zongbao Kent Zhao [1,3]

1   Division of Biotechnology, Dalian Institute of Chemical Physics, CAS, 457 Zhongshan Road, Dalian 116023, China
2   University of Chinese Academy of Sciences, Beijing 100049, China
3   Dalian Key Laboratory of Energy Biotechnology, Dalian Institute of Chemical Physics, Chinese Academy of Sciences, Dalian 116023, China
*   Correspondence: wangqian@dicp.ac.cn; Tel.: +86-0411-8437-9066

Abstract: Microbial-based fatty acids (FAs), biofuels and oleochemicals are potential alternatives to fossil fuels and other non-renewable resources. *Rhodotorula toruloides* (formerly *Rhodosporidium toruloides*) is a basidiomycetous oleaginous yeast, and cells of the wild-type diploids can accumulate lipids to over 70 wt% on a dry cell weight basis in nutrient-limited conditions. Meanwhile, several haploid strains have been applied as hosts for producing high-value fatty acid derivatives through genetic modification and metabolic engineering. However, the differences in fatty acid compositional profiles and their stability between diploid and haploid strains remain unknown in this oleaginous yeast. Here, we grew a haploid strain *R. toruloides* NP11 and its parental diploid strain *R. toruloides* CGMCC 2.1389 (4#) under identical conditions and compared the profiles in terms of cell growth, lipid production, fatty acid compositions of lipids as well as storage stability of fatty acid methyl esters (FAMEs). It was found that lipids from *R. toruloides* composed of fatty acids in terms of chain length ranged from short-chain FAs (C6–C9) to very long-chain FAs (VLCFAs, C20–C24) and some odd-chain FAs (C15 and C17), while long-chain fatty acids (C14–C18) were the most abundant ones. In addition, NP11 produced a little more (1 wt%) VLCFAs than that of the diploid strain 4#. Moreover, no major changes were found for FAMEs being held under varied storage conditions, suggesting that FAMEs samples were stable and robust for fatty acid compositional analysis of microbial lipids. This work revealed the fatty acid profiles of lipids from *R. toruloides* haploid and diploid strains, and their stability under various storage conditions. The information is valuable for reliable assessment of fatty acid compositions of lipids from oleaginous yeasts and related microbial cell factories.

Keywords: *Rhodotorula toruloides*; haploid; diploid; fatty acid; storage conditions

check for updates

Citation: Zhang, Y.; Kamal, R.; Li, Q.; Yu, X.; Wang, Q.; Zhao, Z.K. Comparative Fatty Acid Compositional Profiles of *Rhodotorula toruloides* Haploid and Diploid Strains under Various Storage Conditions. *Fermentation* **2022**, *8*, 467. https://doi.org/10.3390/fermentation8090467

Academic Editor: Timothy Tse

Received: 8 August 2022
Accepted: 13 September 2022
Published: 18 September 2022

Publisher's Note: MDPI stays neutral with regard to jurisdictional claims in published maps and institutional affiliations.

1. Introduction

Increasing concerns over energy security, unsustainability, and climate change related to the use of fossil fuels and oleochemicals are leading to a comprehensive exploration of bio-based diesel and related chemicals which are renewable, biodegradable and clean substitutes for petroleum-based counterparts [1,2]. However, the current biodiesel and oleochemicals production from different edible and non-edible plant feedstocks puts huge stress on the food market, arable land, water consumption, and eventually on the economy [2,3]. The recent past has therefore extensively focused on microbial lipids as an alternative feedstock for biodiesel production [4,5]. Microorganisms with over 20 wt% lipids accumulating capabilities of their total cell weight are known as oleaginous microorganisms. The fuels produced from microbial lipids have higher energy density and better miscibility than other fuels [6] and are exempted from the key limitations related to plants and vegetable oil, such as the fact that they are seasonal, have a long production cycle, require huge arable land and water resources, and more importantly that they compete with edible oils affecting the community [7]. Therefore, extensive research had been conducted on oleaginous microorganisms to address the potential challenges in high titer lipid production through the integration of synthetic biology, metabolic engineering, and systems biology [8,9].

Compared with other oleaginous microorganisms, oleaginous yeasts present several leads such as higher growth frequency, adaptation to diverse substrates, and higher lipid production potentials [5,7]. The robust stress tolerance, work on a broad range of substrates and capacity to accumulate lipids to over 75 wt% per dry cell weight give *Rhodotorula toruloides* (formerly *Rhodosporidium toruloides*) an ascendant place among other oleaginous yeasts [10,11]. Previous studies have already explored the mechanism of lipid and fatty acid accumulation by *R. toruloides* by using multi-omics approaches including genomics, proteomics, metabolomics and systems analysis [8,12–14]. Most studies usually analyzed these known long-chain fatty acids, such as C14, C16, and C18 [15,16], or very-long-chain monounsaturated fatty acids produced by engineered *R. toruloides* [17]. Nevertheless, fatty acid compositional profiles of lipids produced by *R. toruloides* are still missing.

The diploid strain *R. toruloides* CGMCC 2.1389 (4#) is a wild type that had been domesticated in corn stover hydrolysates to give a strain *R. toruloides* Y4, which were used to isolate the haploid strains, NP11 (MAT A1) and NP5-2 (MAT A2) [12]. In this work, we compared fatty acid compositions of lipids from *R. toruloides* 4# and NP11 by incorporating from short chain fatty acids (SCFAs, C6–C8) to very long chain fatty acids (VLCFAs, C20–C24). Moreover, we confirmed the presence of odd chain fatty acids (OCFAs) such as C15, C17, and VLCFAs C22, C24 in lipids produced by *R. toruloides* naturally, as well as the storage stability of the fatty acid methyl esters (FAMEs) samples at different temperatures for up to 2 months. Identification of endogenous synthesis of OCFAs and VLCFAs paves the way for further engineering *R. toruloides* for high value-added fatty acid derivatives and chemicals. Moreover, this work also provides standard methods for the preservation of FAMEs being widely used for fatty acid compositional analysis of lipids produced by oleaginous microbes.

2. Materials and Methods

2.1. Strains, Media and Culture Conditions

Rhodotorula toruloides CGMCC 2.1389 (4#), which is identical to CBS 6016, was originally acquired from the China General Microbiological Culture Collection Center (CGMCC). The haploid NP11 (MAT A1, GDMCC 2.224) was stored in the Guangdong Microbial Culture Collection Center (GDMCC), and was isolated from basidiospores that were germinated from teliospores of Y4 which were derived from *R. toruloides* CGMCC 2.1389 [12].

The yeast strains were activated at 28 °C on YPD plates containing: glucose 20 g/L, peptone 20 g/L, yeast extract 10 g/L and agar 15 g/L. Precultures were prepared in a YPD broth (glucose 20 g/L, peptone 20 g/L, yeast extract 10 g/L as seed culture) at 30 °C and 200 rpm for 24 h. The nitrogen-limited lipid production medium with a C/N ratio (mol/mol) of 413 contained [10,15]: 70 g/L glucose·H_2O, 0.1 g/L $(NH_4)_2SO_4$, 0.75 g/L yeast extract, 1.5 g/L $MgSO_4$·$7H_2O$, 0.4 g/L KH_2PO_4 and supplemented with 10 mL/L trace element solution containing g/L: $CaCl_2$·$2H_2O$ 4.0, $FeSO_4$·$7H_2O$ 0.55, citric acid·H_2O 0.52, $ZnSO_4$·$7H_2O$ 0.10, $MnSO_4$·H_2O 0.076 and 100 µL 18M H_2SO_4. Moreover, the media also included 50 mM MES to maintain a pH of 6.0. All culture media were sterilized at 121 °C for 20 min before use. Shake-flask batch cultures were conducted for lipid productions. 250 mL Erlenmeyer flasks containing 45 mL nitrogen-limited medium were inoculated with 10 vol% (5 mL) 24-h-old preculture seed (around 1.0×10^7 cells/mL), incubated in an orbital shaker at 30 °C 200 r/min, and sampled every 24 h for residual glucose and OD_{600} analysis, as well as cells microscopic examination with a fluorescence microscope (Evos FL AMF4300, Thermo Fisher Scientific, Bothell. WA. USA). Each time the culture lasted until glucose exhaustion, or until it reached a glucose concentration below 10 g/L.

2.2. Cell Mass and Total Lipids

Upon completion of the culture, cells in 30 mL of culture broth were collected by centrifuging at $8000\times g$ at 4 °C for 5 min using pre-weighed 50 mL PP centrifugal tubes; the cells were then washed twice with 50 vol% ethanol to prevent bulky cell loss in the supernatant due to high lipid content. The collected cells were dried in an oven at 105 °C for 24 h to constant weight; then the dry cell weight was measured gravimetrically.

Total lipids were extracted by the classical acid heat method according to a previous report [15]. The dried cells were digested with 4 M HCl (3 mL per 0.5 g dry cell) in a shaking water bath at 78 °C, 200 r/min for 1 h. Then total intracellular lipid was extracted three times with chloroform-methanol (1:1, *v/v*). The chloroform extracts were washed with 0.1% NaCl (*w/v*) solution and passed over an anhydrous Na_2SO_4 pad; the chloroform was then eliminated by reduced pressure rotary evaporation and the pre-weighed round bottom flasks containing the lipid concentrates were dried at 105 °C to constant weight and the total lipid was measured gravimetrically.

The lipid titer was expressed in g/L (culture broth volume), and the lipid contents in weight percentage (wt%) were measured as total lipid produced per total dry cell weight. The lipid yield was calculated as gram lipid produced per gram of consumed glucose (g/g consumed glucose); the lipid productivity was calculated as lipid titer per day (g/L/d) [18].

2.3. Fatty Acid Composition Analysis

The aforesaid extracted lipids were transformed to FAMEs according to a previously established method [19], and the FAMEs samples were held at different temperatures ranging from -20 °C to room temperature for up to 2 months.

To analyze the fatty acid composition differences in lipids from different hosts as well as potential changes over time, the FAMEs samples were detected by a Gas Chromatography/Triple Quadrupole Mass Spectrometer (GC-MS, Agilent 7890B/7000D system, Agilent Technologies, Santa Clara, CA, USA) as reported [20,21] with modifications, using an HP-5MS column (30 m $\times$ 0.25 mm $\times$ 0.25 μm), FID detector. Gaseous nitrogen was used as a carrier gas (1.1 mL/min). 1 μL sample was injected and the split ratio was 1:100 (250 °C), and the oven temperature was set at 80 °C at the beginning for 1 min; increased to 230 °C with a ramp rate of 8 °C/min; then increased to 245 °C with a ramp rate of 3 °C/min. Lastly, it was increased to 280 °C with a ramp rate of 10 °C/min, held for 5 min, post-run to 320 °C, and held for 2 min. The temperatures of the MS transfer line and ion source were set at 250 °C and 230 °C, respectively. These setting conditions can detect all the short-, medium-, long-, and very-long-chain fatty acids of *R. toruloides*. Fatty acid methyl esters were qualitatively identified by standard samples (methyl lauric acid, C12:0; methyl myristic acid, C14:0; methyl palmitic acid, C16:0; methyl palmitoleate, C16:1; methyl heptadecanoate, C17:0; methyl stearic acid, C18:0; methyl oleic acid, C18:1; methyl linoleic acid, C18:2; methyl tetracosanoate, C24:0;) and mass spectrometric ion peaks, and their relative mass percentage (wt%) content was determined by the area normalization method.

2.4. Statistical Analyses

Statistical analysis was performed using the Student's *t*-test (* $p < 0.05$, ** $p < 0.01$, *** $p < 0.001$; the data is normally distributed.) by Microsoft Excel software. The Benjamini–Hochberg (BH) method was used for the correction of the *p*-value after multiple comparisons and obtaining FDR (False Discovery Rate); FDR < 0.05 was accepted as significant. All data are presented as a mean $\pm$ SD of biological triplicates.

3. Results

3.1. Difference in Growth Patterns of Haploid and Diploid Strains

Previous reports suggested the presence of discrepancies in cell growth and sugar consumption by the diploid and haploid strains [16,22]. To verify these observations, culture broths were analyzed for cell density (OD_{600}) and residual glucose every 24 h. When the initial OD_{600} was 1.0, the diploid strain *R. toruloides* 4# grew faster than the haploid NP11 (Figure 1a). Relative differences were observed in the OD_{600} from 24 h between both strains as Figure 1a showed and confirmed by the Student's *t*-test (Table S1). The μ_{max} (specific growth rate) of the strain *R. toruloides* 4# (0.14 ± 0.02 h^{-1}) and NP11 (0.11 ± 0.01 h^{-1}) also confirmed it (Figure S1). When *R. toruloides* 4# was in the stationary lipid production phase, its medium residual glucose was significantly decreased compared with that of NP11. As a result, 4# reached the end of the culture faster compared to NP11 in a nitrogen-limiting medium with the same sugar content; the glucose consumption rate in Figure 1b of 4# (10.8 g/L/d) was significantly higher than that of NP11 (7.1 g/L/d).

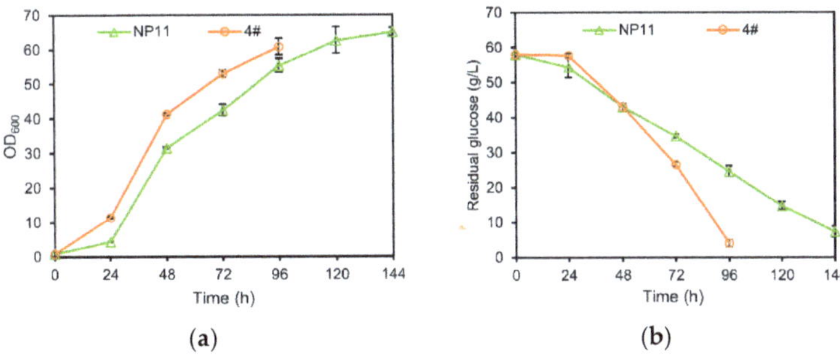

Figure 1. Results of cell density (OD_{600}) and glucose consumption time course in shaking flask culture of haploid strain *R. toruloides* NP11 and diploid strain *R. toruloides* 4#. (**a**) The growth curve of strains in nitrogen-limited culture; growth was measured as the turbidity of culture broth at 600 nm every 24 h. (**b**) Residual glucose concentration curves of strain NP11 and 4#.

Therefore, we concluded that the diploid strain has an inherent growth advantage and higher sugar utilization efficiency than the haploid strain in the culture process. In addition, its cell density increased rapidly after 24 h until the end of the culture (Figure 1a). Though the haploid strain grew a little slower, its cell density could reach the same level as the diploid strain at the end of the culture.

3.2. Lipid Accumulation Potential of Haploid and Diploid Strains

The haploid strain NP11 can accumulate over 50 wt% lipid contents of its dry cell mass [16,23], while the diploid strain could produce more than 70 wt% of lipid contents of its dry cell mass in shake flask cultures using glucose as a carbon substrate [10,15,22].

NP11 showed good lipid accumulation potential with a lipid titer of 6.9 g/L, and lipid contents of 54.4 wt% with a comparatively lower cell mass of 12.6 g/L (Figure 2a, Table S2). The strain 4# undoubtedly produced a higher cell mass of 18.5 g/L, a lipid titer of 12.0 g/L, and a lipid content of 64.9 wt%. The lipid droplet morphology differences during different culture time intervals also proved the high lipid accumulation potential of 4# following NP11, and we could observe that the cell size of the haploid strain is nearly half that of the diploid strain from microscope photos in Figure 3. The larger cell volume of the diploid strain is always the main reason for increasing storage space for intracellular production and the diploid strain also has advantages of high lipid production and robustness using different complex carbon feedstocks [24,25]. The lipid and cell mass yield as well as the productivity of 4# all were significantly two-fold higher than those of NP11 (Figure 2b). Therefore, the preparation of the diploid strain is an effective strategy to improve its production capabilities through metabolic engineering.

It is noteworthy to mention that there was an obvious color difference in the lipid extracts, e.g., the lipid extracts of 4# and NP11 were saffron yellow and deep red, respectively (Figure S2). Meanwhile, the culture broth colors of the two strains also showed differences after incubation for 24 h until the end of the culture (Figure S2). As is known, the red color of *R. toruloides* is mainly due to the presence of carotenoids, torularhodin and lycopene [26]. Therefore, we speculated that these two strains produced different pigments, and the reason for the change in the phenotype of carotenoids might be verified by analyzing the genome or metabolic flow of the mevalonate (MVA) pathway in further research. This was based on the difference between the MVA pathway flux and terpenoids compounds content in these strains that could be used as the starting material for the synthesis of different terpenoids.

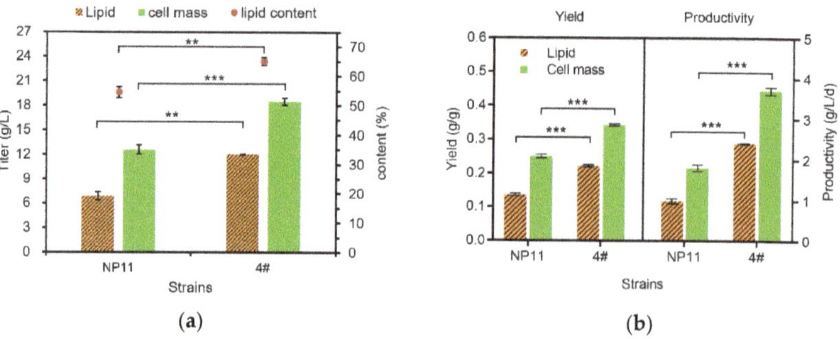

Figure 2. Lipid production results of *R. toruloides* NP11 and 4#. (**a**) Lipid and cell mass titer, and lipid content of strain NP11 and 4#. (**b**) The comparison of lipid and cell mass yield and productivity of the two strains. *** $p < 0.001$ (*p*-value); ** $p < 0.01$; of the one-tailed Student's *t*-test. All data are presented as mean $\pm$ SD of biological triplicates.

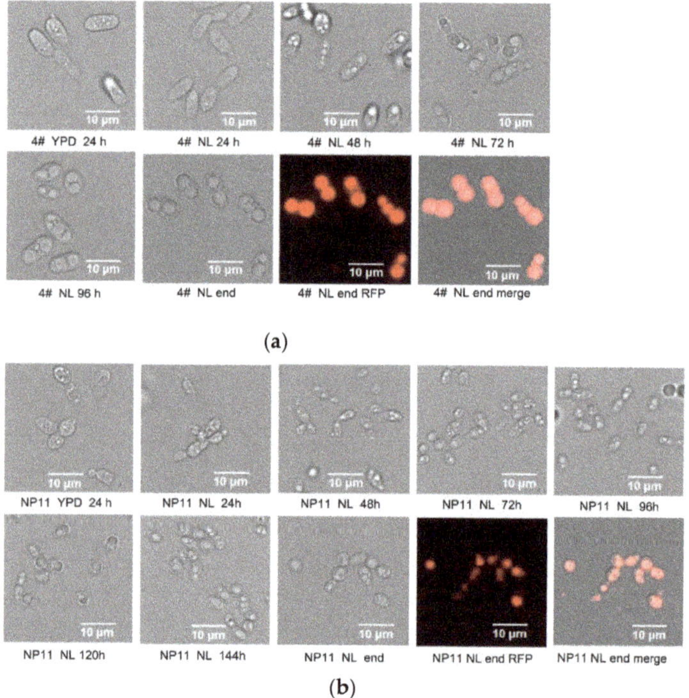

Figure 3. Microscope photos of *R. toruloides* NP11 and 4# cells during lipid production culture. (**a**) The diploid strain *R. toruloides* 4# cells during the culture and those at the end of the culture being treated by Nile Red. (**b**) The haploid strain *R. toruloides* NP11 cells during the culture and those at the end of the culture being treated by Nile Red. NL, nitrogen-limited culture. RFP, red fluorescence module of the microscope.

3.3. Analysis of Fatty Acid Compositional Difference of Haploid and Diploid Strains

As is known, the very long-chain FAs (VLCFAs) biosynthesis enzymes are evolutionarily conserved [27], and VLCFAs such as C20 and C22 can be synthesized by *R. toruloides* [17]. Although the complete FAs compositional profile of the *R. toruloides* using glucose as a carbon feedstock is still missing, and the fatty acid compositional difference between *R. toruloides* diploid and haploid strains has never been revealed. We compared the differences of FAs profile incorporated short-chain FAs (SCFAs) C6–C8, medium-chain FAs (MCFAs) C9–C12, long-chain FAs (LCFAs) C14–C18, and unexplored VLCFAs C20–C24, including the odd-chain FAs (OCFAs, C9, C15 and C17) shown in Table 1. The results showed that both strains produced oleic acid (C18:1) most abundantly followed by palmitic acid (C16:0) and stearic acid (C18:0), which was consistent with previous reports [15,16]. The content ratio of SCFAs and MCFAs was very low; the content of these two kinds of FAs in any strain was no more than 1 wt%. These two strains produced VLCFAs including C20:0, C22:0, C23:0 and C24:0, the percentage content of VLCFAs was 2.40 wt% for the haploid strain NP11 and 1.63 wt% for the diploid strain 4#. It was found that a much longer carbon chain VLCFAs such as C22:0 and C24:0 was accumulated in the haploid strain, which might take place during a longer culture time where more substrates were converted to VLCFAs by elongation enzymes. In conclusion, *R. toruloides* tend to produce long-chain fatty acids, and the proportion of LCFAs in both strains is above 96 wt%–97 wt%.

Table 1. Fatty acid compositional profiles of haploid strain *R. toruloides* NP11 and diploid strain 4#.

Fatty Acid Profiles (wt%)	Haploid-NP11	Diploid-4#	*p*-Value	FDR
C6:0	0.18 ± 0.05	0.06 ± 0.01	0.0609	0.1676
C7:0	0.07 ± 0.02	0.04 ± 0.00	0.1276	0.2160
C8:0	0.30 ± 0.08	0.16 ± 0.01	0.0963	0.1925
Short-chain FAs (SCFAs)	0.54 ± 0.16	0.26 ± 0.02	0.0861	0.1894
C9:0	0.22 ± 0.08	0.17 ± 0.02	0.3161	0.4636
C12:0	0.08 ± 0.00	0.13 ± 0.01	0.0001 ***	0.0029 *
Medium-chain FAs (MCFAs)	0.30 ± 0.09	0.30 ± 0.02	0.9984	0.9984
C14:0	1.85 ± 0.16	2.32 ± 0.02	0.0359 *	0.1315
C15:0	0.20 ± 0.03	0.21 ± 0.01	0.5810	0.7102
C16:1	0.56 ± 0.05	0.57 ± 0.01	0.8364	0.9200
C16:0	37.32 ± 2.53	38.11 ± 0.42	0.6231	0.7215
C17:1	0.25 ± 0.03	0.10 ± 0.01	0.0009 ***	0.0094 *
C17:0	0.40 ± 0.03	0.35 ± 0.00	0.1211	0.2219
C18:2	1.30 ± 0.73	0.42 ± 0.12	0.1679	0.2638
C18:1	40.99 ± 3.38	41.43 ± 0.62	0.8365	0.8763
C18:0	13.88 ± 0.93	14.32 ± 0.26	0.4770	0.6173
Long-chain FAs (LCFAs)	96.75 ± 7.87	97.81 ± 1.47	0.0526	0.1653
C20:0	0.50 ± 0.05	0.53 ± 0.00	0.4068	0.5594
C22:0	0.76 ± 0.04	0.54 ± 0.01	0.0010 **	0.0075 *
C23:0	0.09 ± 0.03	0.03 ± 0.00	0.0617	0.1509
C24:0	1.05 ± 0.09	0.53 ± 0.01	0.0094 **	0.0517
Verylong-chain FAs (VLCFAs)	2.40 ± 0.21	1.63 ± 0.02	0.0224 *	0.0987

p value, *** *p* < 0.001; ** *p* < 0.01; * *p* < 0.05 of two-tailed Student's *t*-test. All data are presented as mean ± SD of biological triplicates. FDR (False Discovery Rate) was the correction of *p*-value after multiple comparison; * FDR < 0.05.

In addition, we confirmed the OCFAs, i.e., C7:0, C9:0, C15:0, C17:0 and C17:1 with secondary ion mass spectrometry by using commercial OCFAs standards (data not shown). OCFAs hold a very small proportion of *R. toruloides* lipids and are often neglected in the detection and analysis process, which was not conducive for the related genes mining and modifying the metabolic pathway. In this work, we concluded that NP11 produced 1.14% OCFAs of total fatty acids which was 0.86 wt% in 4#. The highest content of OCFAs was C17:0 in both strains, followed by C17:1 in NP11, but the diploid strain 4# had less C17:1 than that of NP11. It was assumed that the culture time of 4# would be insufficient to accumulate some special unsaturated OCFAs such as C17:1 and elongate the fatty acid chains.

3.4. Stability of FAMEs under Different Preservation Conditions

To explore the stability of fatty acid methyl esters (FAMEs) under various preservation conditions, we selected FAMEs samples prepared from lipids by *R. toruloides* NP11 as the representative for verification. Samples in triplicates were stored in three different temperature conditions, i.e., 4 °C, −20 °C, and room temperature (RT, 25 ± 1 °C), respectively. One sample in each duplicate was vortexed well before detection to confirm the effects of low-temperature crystallization of FAMEs on compositional profiles, while the other was detected directly. There was little change in the fatty acids compositional profiles of the FAMEs samples; those were stored at 4 °C for 1 h and no significant effect was observed of the vortex treatment group (Figure 4). Although C17:1 of these two groups decreased a little compared to that of NP11, it was not significant (Table S3). This suggests that storage at 4 °C for a short time does not cause FA crystallization and change but may not be good for FAMEs with C17:1. Interestingly, all the samples showed good stability at −20 °C for 17 h with no change in FAs profile regardless of samples with vortex treatment or without vortex treatment prior analysis (Figure 4). The constant appearance suggested no crystallization effects at −20 °C on FAMEs for 17 h. Moreover, we concluded that there was no change in the FAs profile after storing the samples at −20 °C for up to 26 h and two months with vortex treatment before (data not shown). As most of the suppliers of standard FAMEs generally suggest −20 °C storage, we used to keep samples at −20 °C. We were also surprised by the fact that nearly no significant change in terms of FAs composition was found after we stored the samples at room temperature for up to two months (Figure 4 and Table S3). FAMEs were stable when stored at room temperature for a long time, and thus it is unnecessary to store them at 4 °C or −20 °C.

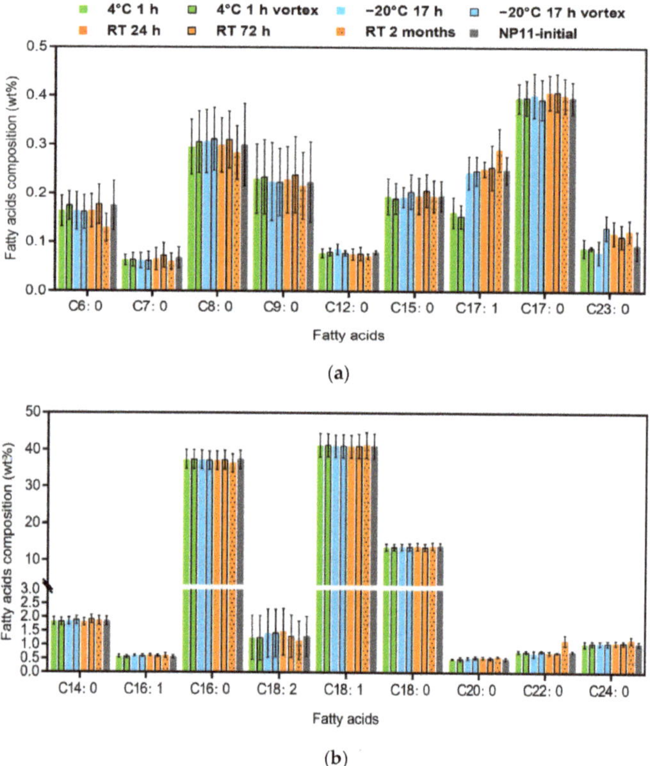

Figure 4. Effect of storage conditions on fatty acid compositions of lipids produced by *R. toruloides* NP11. (**a**) Profiles of the trace fatty acids. (**b**) Profiles of the main fatty acids. The green column for

samples being held at 4 °C for 1 h and those with a black line box for samples with vortex treatment; the blue column for samples being kept at −20 °C for 17 h and those with a black line box for samples with vortex treatment; the orange for samples being stored at RT for 24 h, the orange with a black line box for samples being held at RT for 72 h, and those with black point background for samples being held at RT for 2 months; the grey column for as-prepared samples without storage. RT, room temperature (25 ± 1 °C). All data are presented as a mean ± SD of biological triplicates. *p*-value of two-tailed Student's *t*-test and FDR were shown in Table S3.

4. Discussion

Many studies demonstrated *R. toruloides* as a promising platform for lipid production owing to its diverse substrate appetites, robust stress resistance, and other favorable features. In recent years, significant progress has been made in genome sequencing [12,13], multi-omics analysis [14], and genome-scale modeling [28]. Moreover, several molecular and genetic tools for engineering *R. toruloides* were previously developed, including gene transformation [29,30], and an RNA interference method as well as a CRISPR Cas9 genome editing system [23,31]. Meanwhile, the development of genetic components (promoters, markers, terminators, etc.) and transformation methods have paved the way towards refined genomic and metabolic engineering regulation.

In addition, it is necessary to understand the endogenous metabolic pathways and related metabolites. There have been numerous basic studies on lipid synthesis and fatty acid metabolisms, such as lipid droplets and fatty acid synthases (FASs) of *R. toruloides* [32,33]. However, we did not know exactly the fatty acid composition of *R. toruloides* as a potential chassis microorganism, especially the differences between the haploid and diploid strains, which made it difficult to choose the starting strains for engineering transformation. So, in this work, we compared the difference in fatty acid compositions between the haploid strain NP11 and the diploid strain 4#, by incorporation of the usual LCFAs C14 to C18, and unusual VLCFAs C20 to C24 and OCFAs C15 to C17, as well as SCFAs C6 to C8. Our results showed the possibility of using *R. toruloides* NP11 as the parent strain for metabolic engineering to produce special fatty acids (OCFAs and VLCFAs) and high-value fatty acids. Moreover, it is possible to hybridize it with a mating A2 strain into the diploid strain to improve production without affecting the composition of the fatty acids. Hence, the advantages of the diploid strain in growth and lipid production in large-scale culture could be utilized well, while the difficulty of genetic editing of diploid strains was also avoided.

Firstly, different cell growth and sugar consumption rates were achieved for both strains during the lipid production process (Figure 1). The diploid strain 4# grew faster with a comparatively short lag phase, and its cell density rapidly increased after 24 h during the culture. Nevertheless, the haploid strain NP11 needs more time to achieve the same level of cell density. Although the cell density of NP11 was a little higher than that of 4# in the end, its dry cell mass weight was lower than that of 4#. Similarly, significant differences were found in the growth and lipid production capacity between both strains, which could be caused by genes involved in fatty acid synthesis being formed of two copies of the diploid strain, where it has only one copy in the haploid strain. Nevertheless, NP11 is a promising oil-producing, broadly studied and well-annotated *R. toruloides* haploid strain; it showed potential applications in the genetic manipulation of biosynthesis.

Secondly, there were no significant differences in major fatty acid compositional profiles between the diploid and haploid strains. Prolonged culture time may be beneficial to the accumulation of special fatty acids such as VLCFAs (C22:0 and C24:0) and OCFAs (C17:1), or the enzymes involved in the synthesis of VLCFAs and OCFAs in *R. toruloides* were less active and thus need longer culture time; these assumptions could also be demonstrated in future work.

Lastly, we demonstrated that FAMEs are in fact very stable at room temperature. Thus, FAMEs samples can be stored normally and problems in terms of quantitative analysis associated with cryopreservation may be avoided. These findings provide sample preservation guidelines for FAs analysis as well as a method to decipher the properties of FAMEs of oleaginous yeasts. Moreover, this study showed comparison data in terms of cell growth, lipid production, and the FA composition of *R. toruloides* haploid and diploid strains. The information is valuable to explore *R. toruloides* as a host for the construction of an advanced cell factory.

5. Conclusions

In summary, the diploid strain *R. toruloides* 4# could afford higher yields of cell mass and lipid than the haploid strain NP11. There were noticeable differences in the proportion of very few fatty acids in lipids from *R. toruloides* haploid and diploid strains; however, overall FAs compositions were quite similar. Importantly, FAMEs were found to be stable under varied storage conditions to ensure the reliable estimation of fatty acid compositions of microbial lipids.

Supplementary Materials: The following supporting information can be downloaded at: https://www.mdpi.com/article/10.3390/fermentation8090467/s1, Figure S1. The growth curves fitting of *R. toruloides* NP11 and *R. toruloides* 4# in nitrogen-limited cultivation. Figure S2. The lipid extract and culture broth of the haploid strain *R. toruloides* NP11 and the diploid strain 4#. Table S1. *p*-value of OD_{600} and residual glucose comparison between the haploid strain *R. toruloides* NP11 and the diploid strain 4# at different time by statistical analysis. Table S2. *p*-value of some strain parameters comparison between the haploid strain *R. toruloides* NP11 and the diploid strain *R. toruloides* 4# by statistical analysis. Table S3. *p*-value and FDR of FAMEs composition of NP11 under different preservation conditions compared by statistical analysis. Reference [34] is cited in the supplementary materials.

Author Contributions: Z.K.Z. conceived the research; Q.W. and Y.Z. designed and conducted experiments; Q.L. contributed to the GC-MS method in analyzing fatty acids; R.K. proposed and contributed to the fatty acid samples stability part; Y.Z., Q.W. and X.Y. analyzed data; Y.Z., R.K., Q.W. and Z.K.Z. wrote and revised the manuscript. All authors have read and agreed to the published version of the manuscript.

Funding: This study was funded by the National Natural Science Foundation of China (31870042).

Institutional Review Board Statement: Not applicable.

Informed Consent Statement: Not applicable.

Data Availability Statement: The data presented in this study are available on request from the corresponding author.

Acknowledgments: We thank the National Natural Science Foundation of China (31870042) for the financial support. We thank Qi Wang for helpful discussion in writing the paper.

Conflicts of Interest: The authors declare no conflict of interest.

Nomenclature

FAs	Fatty acids
SCFAs	Short-chain fatty acids
MCFAs	Medium-chain fatty acids
LCFAs	Long-chain fatty acids
VLCFAs	Very long-chain fatty acids
OCFAs	Odd-chain fatty acids
FAMEs	Fatty acid methyl esters

References

1. Chu, S.; Majumdar, A. Opportunities and challenges for a sustainable energy future. *Nature* **2012**, *488*, 294–303. [CrossRef] [PubMed]
2. Wang, J.L.; Singer, S.D.; Souto, B.A.; Asomaning, J.; Ullah, A.; Bressler, D.C.; Chen, G.Q. Current progress in lipid-based biofuels: Feedstocks and production technologies. *Bioresour. Technol.* **2022**, *351*, 127020. [CrossRef] [PubMed]
3. Carriquiry, M.A.; Du, X.; Timilsina, G.R. Second generation biofuels: Economics and policies. *Energ. Policy* **2011**, *39*, 4222–4234. [CrossRef]
4. Madani, M.; Enshaeieh, M.; Abdoli, A. Single cell oil and its application for biodiesel production. *Process Saf. Environ. Prot.* **2017**, *111*, 747–756. [CrossRef]
5. Chintagunta, A.D.; Zuccaro, G.; Kumar, M.; Kumar, S.P.J.; Garlapati, V.K.; Postemsky, P.D.; Kumar, N.S.S.; Chandel, A.K.; Simal-Gandara, J. Biodiesel production from lignocellulosic biomass using oleaginous microbes: Prospects for integrated biofuel production. *Front. Microbiol.* **2021**, *12*, 658284. [CrossRef]
6. Nigam, P.S.; Singh, A. Production of liquid biofuels from renewable resources. *Prog. Energ. Combust.* **2011**, *37*, 52–68. [CrossRef]
7. Szczepanska, P.; Hapeta, P.; Lazar, Z. Advances in production of high-value lipids by oleaginous yeasts. *Crit. Rev. Biotechnol.* **2022**, *42*, 1–22. [CrossRef]
8. Wen, Z.; Zhang, S.; Odoh, C.K.; Jin, M.; Zhao, Z.K. *Rhodosporidium toruloides*—A potential red yeast chassis for lipids and beyond. *FEMS Yeast Res.* **2020**, *20*, foaa038. [CrossRef]
9. Uprety, B.K.; Morrison, E.N.; Emery, R.J.N.; Farrow, S.C. Customizing lipids from oleaginous microbes: Leveraging exogenous and endogenous approaches. *Trends Biotechnol.* **2022**, *40*, 482–508. [CrossRef]
10. Li, Y.; Liu, B.; Zhao, Z.K.; Bai, F. Optimization of culture conditions for lipid production by *Rhodosporidium toruloides*. *Chin. J. Biotechnol.* **2006**, *22*, 650–656. [CrossRef]
11. Lamers, D.; van Biezen, N.; Martens, D.; Peters, L.; van de Zilver, E.; Jacobs-van Dreumel, N.; Wijffels, R.H.; Lokman, C. Selection of oleaginous yeasts for fatty acid production. *BMC Biotechnol.* **2016**, *16*, 45. [CrossRef] [PubMed]
12. Zhu, Z.; Zhang, S.; Liu, H.; Shen, H.; Lin, X.; Yang, F.; Zhou, Y.J.; Jin, G.; Ye, M.; Zou, H.; et al. A multi-omic map of the lipid-producing yeast *Rhodosporidium toruloides*. *Nat. Commun.* **2012**, *3*, 1112. [CrossRef] [PubMed]
13. Coradetti, S.T.; Pinel, D.; Geiselman, G.M.; Ito, M.; Mondo, S.J.; Reilly, M.C.; Cheng, Y.F.; Bauer, S.; Grigoriev, I.V.; Gladden, J.M.; et al. Functional genomics of lipid metabolism in the oleaginous yeast *Rhodosporidium toruloides*. *eLife* **2018**, *7*, e32110. [CrossRef] [PubMed]
14. Wang, Y.; Zhang, S.; Zhu, Z.; Shen, H.; Lin, X.; Jin, X.; Jiao, X.; Zhao, Z.K. Systems analysis of phosphate-limitation-induced lipid accumulation by the oleaginous yeast *Rhodosporidium toruloides*. *Biotechnol. Biofuels* **2018**, *11*, 148. [CrossRef] [PubMed]
15. Wu, S.; Zhao, X.; Shen, H.; Wang, Q.; Zhao, Z.K. Microbial lipid production by *Rhodosporidium toruloides* under sulfate-limited conditions. *Bioresour. Technol.* **2011**, *102*, 1803–1807. [CrossRef]
16. Yang, X.B.; Sun, W.Y.; Shen, H.W.; Zhang, S.F.; Jiao, X.; Zhao, Z.K. Expression of phosphotransacetylase in *Rhodosporidium toruloides* leading to improved cell growth and lipid production. *RSC Adv.* **2018**, *8*, 24673–24678. [CrossRef]
17. Fillet, S.; Ronchel, C.; Callejo, C.; Fajardo, M.-J.; Moralejo, H.; Adrio, J.L. Engineering *Rhodosporidium toruloides* for the production of very long-chain monounsaturated fatty acid-rich oils. *Appl. Microbiol. Biotechnol.* **2017**, *101*, 7271–7280. [CrossRef]
18. Kamal, R.; Liu, Y.; Li, Q.; Huang, Q.; Wang, Q.; Yu, X.; Zhao, Z.K. Exogenous L-proline improved *Rhodosporidium toruloides* lipid production on crude glycerol. *Biotechnol. Biofuels* **2020**, *13*, 159. [CrossRef]
19. Li, Y.; Zhao, Z.K.; Bai, F. High-density cultivation of oleaginous yeast *Rhodosporidium toruloides* Y4 in fed-batch culture. *Enzym. Microb. Technol.* **2007**, *41*, 312–317. [CrossRef]
20. Zhu, Z.; Hu, Y.; Teixeira, P.G.; Pereira, R.; Chen, Y.; Siewers, V.; Nielsen, J. Multidimensional engineering of *Saccharomyces cerevisiae* for efficient synthesis of medium-chain fatty acids. *Nat. Catal.* **2020**, *3*, 64–74. [CrossRef]
21. Yu, T.; Zhou, Y.J.; Wenning, L.; Liu, Q.; Krivoruchko, A.; Siewers, V.; Nielsen, J.; David, F. Metabolic engineering of *Saccharomyces cerevisiae* for production of very long chain fatty acid-derived chemicals. *Nat. Commun.* **2017**, *8*, 15587. [CrossRef] [PubMed]
22. Wang, S.; Kamal, R.; Zhang, Y.; Zhou, R.H.; Lv, L.T.; Huang, Q.T.; Qian, S.R.G.L.; Zhang, S.F.; Zhao, Z.K. Expression of VHb improved lipid production in *Rhodosporidium toruloides*. *Energies* **2020**, *13*, 4446. [CrossRef]
23. Liu, X.; Zhang, Y.; Liu, H.; Jiao, X.; Zhang, Q.; Zhang, S.; Zhao, Z.K. RNA interference in the oleaginous yeast *Rhodosporidium toruloides*. *FEMS Yeast Res.* **2019**, *19*, foz031. [CrossRef] [PubMed]
24. de Godoy, L.M.; Olsen, J.V.; Cox, J.; Nielsen, M.L.; Hubner, N.C.; Frohlich, F.; Walther, T.C.; Mann, M. Comprehensive mass-spectrometry-based proteome quantification of haploid versus diploid yeast. *Nature* **2008**, *455*, 1251–1255. [CrossRef]
25. Li, B.Z.; Cheng, J.S.; Ding, M.Z.; Yuan, Y.J. Transcriptome analysis of differential responses of diploid and haploid yeast to ethanol stress. *J. Biotechnol.* **2010**, *148*, 194–203. [CrossRef]
26. Bao, R.Q.; Gao, N.; Lv, J.; Ji, C.F.; Liang, H.P.; Li, S.J.; Yu, C.X.; Wang, Z.Y.; Lin, X.P. Enhancement of torularhodin production in *Rhodosporidium toruloides* by agrobacterium tumefaciens-mediated transformation and culture condition optimization. *J. Agric. Food Chem.* **2019**, *67*, 1156–1164. [CrossRef]
27. Erdbrugger, P.; Frohlich, F. The role of very long chain fatty acids in yeast physiology and human diseases. *Biol. Chem.* **2020**, *402*, 25–38. [CrossRef]
28. Tiukova, I.A.; Prigent, S.; Nielsen, J.; Sandgren, M.; Kerkhoven, E.J. Genome-scale model of *Rhodotorula toruloides* metabolism. *Biotechnol. Bioeng.* **2019**, *116*, 3396–3408. [CrossRef]

29. Liu, H.; Jiao, X.; Wang, Y.; Yang, X.; Sun, W.; Wang, J.; Zhang, S.; Zhao, Z.K. Fast and efficient genetic transformation of oleaginous yeast *Rhodosporidium toruloides* by using electroporation. *FEMS Yeast Res.* **2017**, *17*, fox017. [CrossRef]

30. Lin, X.; Wang, Y.; Zhang, S.; Zhu, Z.; Zhou, Y.J.; Yang, F.; Sun, W.; Wang, X.; Zhao, Z.K. Functional integration of multiple genes into the genome of the oleaginous yeast *Rhodosporidium toruloides*. *FEMS Yeast Res.* **2014**, *14*, 547–555. [CrossRef]

31. Jiao, X.; Zhang, Y.; Liu, X.; Zhang, Q.; Zhang, S.; Zhao, Z.K. Developing a CRISPR/Cas9 system for genome editing in the basidiomycetous yeast *Rhodosporidium toruloides*. *Biotechnol. J.* **2019**, *14*, 1900036. [CrossRef] [PubMed]

32. Zhu, Z.; Ding, Y.; Gong, Z.; Yang, L.; Zhang, S.; Zhang, C.; Lin, X.; Shen, H.; Zou, H.; Xie, Z.; et al. Dynamics of the lipid droplet proteome of the oleaginous yeast *Rhodosporidium toruloides*. *Eukaryot. Cell.* **2015**, *14*, 252–264. [CrossRef] [PubMed]

33. Zhu, Z.; Zhou, Y.J.; Krivoruchko, A.; Grininger, M.; Zhao, Z.K.; Nielsen, J. Expanding the product portfolio of fungal type I fatty acid synthases. *Nat. Chem. Biol.* **2017**, *13*, 360–362. [CrossRef]

34. Zwietering, M.H.; Jongenburger, I.; Rombouts, F.M.; van't Riet, K. Modeling of the Bacterial Growth Curve. *Appl. Environ. Microbiol.* **1990**, *56*, 1875–1881. [CrossRef] [PubMed]

Article

Sugarcane Bagasse-Based Ethanol Production and Utilization of Its Vinasse for Xylitol Production as an Approach in Integrated Biorefinery

Sreyden Hor [1,2], Mallika Boonmee Kongkeitkajorn [2,3,*] and Alissara Reungsang [2,4]

1   Graduate School, Khon Kaen University, Khon Kaen 40002, Thailand; sreyden.hor@gmail.com
2   Department of Biotechnology, Faculty of Technology, Khon Kaen University, Khon Kaen 40002, Thailand; alissara@kku.ac.th
3   Research Center for Environmental and Hazardous Substance Management (EHSM), Khon Kaen University, Khon Kaen 40002, Thailand
4   Academy of Science, Royal Society of Thailand, Bangkok 10300, Thailand
*   Correspondence: mallikab@kku.ac.th

Abstract: Biorefinery of sugarcane bagasse into ethanol and xylitol was investigated in this study. Ethanol fermentation of sugarcane bagasse hydrolysate was carried out by *Saccharomyces cerevisiae*. After ethanol distillation, the vinasse containing xylose was used to produce xylitol through fermentation by *Candida guilliermondii* TISTR 5068. During the ethanol fermentation, it was not necessary to supplement a nitrogen source to the hydrolysate. Approximately 50 g/L of bioethanol was produced after 36 h of fermentation. The vinasse was successfully used to produce xylitol. Supplementing the vinasse with 1 g/L of yeast extract improved xylitol production 1.4-fold. Cultivating the yeast with 10% controlled dissolved oxygen resulted in the best xylitol production and yields of 10.2 ± 1.12 g/L and 0.74 ± 0.04 g/g after 60 h fermentation. Supplementing the vinasse with low fraction of molasses to improve xylitol production did not yield a positive result. The supplementation caused decreases of up to 34% in xylitol production rate, 24% in concentration, and 24% in yield.

Keywords: yeast; xylitol; bioethanol; sugarcane bagasse; vinasse; biorefinery; value-added products; biofuels

Citation: Hor, S.; Kongkeitkajorn, M.B.; Reungsang, A. Sugarcane Bagasse-Based Ethanol Production and Utilization of Its Vinasse for Xylitol Production as an Approach in Integrated Biorefinery. *Fermentation* **2022**, *8*, 340. https://doi.org/10.3390/fermentation8070340

Academic Editor: Timothy Tse

Received: 22 June 2022
Accepted: 17 July 2022
Published: 19 July 2022

Publisher's Note: MDPI stays neutral with regard to jurisdictional claims in published maps and institutional affiliations.

1. Introduction

Lignocellulosic biomass is a very abundant natural resource [1]. A wide range of materials can be categorized as lignocellulosic biomass, including woody materials, grasses, crops, and the associated wastes [2]. The uses of biorefinery have extended from typical heating and steam generation to the production of many high value biochemicals via various conversion processes.

Sugarcane bagasse is an agricultural excess, comprising waste from the sugar industry. Thailand is the world's third-largest sugarcane-based sugar producer [3]. In the 2019/2020 season, sugarcane production in Thailand was 74.9 million tons, from which the bagasse generated was approximately 21 million tons [4]. Although most sugar factories utilize the bagasse to produce steam and generate electricity [5], the bagasse could be used for other value-added purposes via biorefinery.

Bioethanol has been one of the target products in the biorefinery process [6,7]. It is a principal transport fuel, with 102 billion liters produced worldwide in 2021 [8]. Bioethanol from lignocellulosic biomass has immense potential as a transport fuel due to its decarbonization benefits [9]. The two principal sugars in the lignocellulosic biomass structure, glucose and xylose, can be converted to ethanol. However, our previous work has shown that converting the xylose fraction in the mixed sugar hydrolysate into ethanol with the use of wild type yeasts has its limitations. It does not allow high ethanol concentrations from the xylose fraction, and the xylose conversion was limited in a high ethanol environment [10,11]. Therefore, the xylose fraction could be more applicable for other purposes. One option for the xylose fraction of the lignocellulosic biomass is to produce xylitol.

Xylitol is a value-added chemical obtained from biomass, with no petrochemical alternative. Industrial production of xylitol is carried out through catalytic hydrogenation of xylose. Although the chemical conversion process results in high xylitol yields, certain concerns about the process exist. These included the availability of plants with high xylan content, the high energy and pressure used in the process, and the need for pure xylose in the reaction [12].

The biological approach to xylitol production is appealing, as it is more eco-friendly and less costly than the chemical path. The possibility of xylitol production in integrated biorefineries is an added benefit [13]. Xylitol production via microbial bioconversion has often focused on using hemicellulosic hydrolysate prepared from various biomass types, including grasses, agricultural residues, and woods [13,14]. Hydrolysate can be detoxified and concentrated to obtain a higher concentration of xylose [15–17].

In addition to xylitol production using hemicellulosic hydrolysate, integrated xylitol production with ethanol has been used in various operations. A well-studied scheme involved the separate fermentation of ethanol and xylitol from cellulosic and hemicellulosic fractions of biomass [15,17–20]. Two-stage hydrolysis was required for this operation, to obtain hemicellulosic hydrolysate. The remaining cellulignin, when hydrolyzed, would obtain glucose-rich hydrolysate, which could be a substrate in microbial fermentation. Alternatively, cellulignin fraction could be used directly in simultaneous saccharification and fermentation (SSF) for ethanol production. Fermentation using a single strain of yeast, and co-culture of two yeast strains to co-produce ethanol and xylitol were also reported [16,21].

Other proposed operating configurations have included model-based platforms involving separate hydrolysis, fermentation, and SSF for ethanol production, followed by xylitol fermentation. These configurations could be applied as an extension of a lignocellulosic bioethanol plant process, involving single-stage hydrolysis of biomass [22].

In this study, integrated biorefinery of sugarcane bagasse was considered, with enzymatic hydrolysate of the bagasse as the substrate for ethanol production. The vinasse containing unused xylose obtained from simple distillation was investigated for xylitol production as a value-added step in the overall process. Nitrogen supplementation and aeration level in xylitol production were evaluated as prime parameters in xylitol production from vinasse. A study of the addition of molasses to the vinasse to improve the xylitol production was also carried out.

2. Materials & Methods

2.1. Microorganisms

This study used a commercial strain of *Saccharomyces cerevisiae* (Thermal Resistance Yeast, Angel Yeast Co., Ltd., Hubei, China) for ethanol production. The yeast was supplied in powder form and stored in a refrigerator at 4 °C. Xylitol fermentation employed *Candida guilliermondii* TISTR 5068. The stock of *C. guilliermondii* was maintained at −20 °C as cell suspension in 30% (*v/v*) glycerol. In preparing pre-inoculation cells, the −20 °C yeast stock was streaked on YMX agar plates and incubated at 30 °C for 48 h or until the colonies appeared. Fresh colonies were used to prepare the inoculum. Yeasts on agar plates were refreshed every two weeks to maintain good cell activity.

2.2. Culture Media for Yeast Growth and Fermentation

YM media were used as the inoculum media. One liter of YM-based solution contained 3 g of yeast extract, 3 g of malt extract, and 5 g of peptone. The carbon source was glucose or xylose (designated as YMG or YMX medium) at final concentrations of 20 g/L for *S. cerevisiae* or *C. guilliermondii* cultivation.

Hydrolysate medium, prepared from sugarcane bagasse, was used in all ethanol fermentation. The bagasse properties and the hydrolysate preparation have been described previously [23]. In brief, 15% w/v of sugarcane bagasse was heat-pretreated with 2 M NaOH. After washing and drying, it was then hydrolyzed at a solid loading of 15% w/v with 40 FPU/g of a mixed enzyme, Cellic CTec2 (Novozymes, Bagsvaerd, Denmark), at pH 4.8 and 50 °C for 3 days. The hydrolysate obtained consisted of 120.5 ± 1.8 g/L of reducing sugar, with 96.52 ± 9.13 g/L of glucose and 19 ± 5 g/L of xylose. The hydrolysate and nitrogen source solutions were mixed at the volumetric mixing ratio of 95:5 (hydrolysate: nitrogen source) to make up the medium. Hydrolysate and nitrogen sources were autoclaved separately before they were mixed after cooling to obtain the final concentration of the nitrogen source as planned.

Vinasse medium was used as the main medium for xylitol fermentation. In vinasse preparation, the sugarcane bagasse hydrolysate was prepared and used in ethanol fermentation as described below. After fermentation, the liquid broth was separated from the cells by centrifugation at 11,200× g for 10 min (Sorvall LYNX 4000/6000, Thermo Scientific, Massachusetts, USA). The broth was subjected to simple distillation to remove ethanol. The vinasse contained 20.00 ± 1.19 g/L of xylose, 1.22 ± 0.08 g/L of xylitol, 10.00 ± 1.51 g/L of glycerol, and 2.40 ± 0.77 g/L of ethanol. In medium preparation, the nitrogen source solution was autoclaved and mixed with the vinasse at a volumetric ratio of 95:5.

2.3. Ethanol Fermentation

Ethanol fermentation employed a commercial strain of *Saccharomyces cerevisiae* (Thermal Resistance Yeast, Angel Yeast Co., Ltd., Hubei, China). The inoculum of ethanol fermentation was prepared by adding 0.01 g/L of the yeast powder to the YMG medium and incubated at 30 °C with 200 rpm orbital shaking for 24 h. The inoculum volume was increased to suit the fermentation scale by transferring 10% (v/v) of the seed inoculum into a fresh YMG medium and further incubating under the same condition. All fermentation used 10% inoculum.

Fermentation at flask scale was done in a 250-mL Erlenmeyer flask using 100 mL of hydrolysate media with yeast extract, diammonium phosphate, or YM-based solution as nitrogen sources. The cultures were incubated at 30 °C with 100 rpm orbital shaking.

Fermenter-scale fermentation employed a 5-L fermenter with 3.5-L of hydrolysate medium supplemented with the selected nitrogen source. The cultivation was controlled at 30 °C and constantly stirred at 100 rpm. The fermentation ended when there was no change in glucose concentration. The analysis of the final sample was necessary to control the properties of the fermentation broth used in preparing the vinasse. The fermentation broth obtained from the hydrolysate without nitrogen source supplementation contained 43.70 ± 5.27 g/L of ethanol, 0.65 ± 0.04 g/L of xylitol, and 17.17 ± 0.04 g/L of xylose.

2.4. Xylitol Fermentation

Xylitol fermentation employed *C. guilliermondii* TISTR 5068 with cultivation in 250-mL baffled Erlenmeyer flasks or a 2-L fermenter. The inoculum was prepared by transferring a few single colonies from a YMX agar plate to a YMX medium. It was cultivated at 30 °C, 200 rpm for 24 h before inoculation. All cultivations involving xylitol production employed vinasse medium.

Flask-scale fermentation was carried out using 100 mL of the vinasse supplemented with nitrogen sources or mixed with molasses. The cultivation was carried out at 30 °C with 200 rpm orbital shaking for 72 h with interval samplings.

Fermenter-scale cultivation was conducted in a 2-L bioreactor with 1 L of working volume. The temperature was controlled at 30 °C. Air was supplied through a ring sparger. The 100% dissolved oxygen (DO) was calibrated using 2 vvm of airstream at a stirring speed of 900 rpm. Under these calibrating conditions, dissolved oxygen was 7.7 mg/L (SensoDirect 150, Lovibond, Dortmund, Germany). The fermentation was carried out for 72 h with interval samplings.

2.5. Analytical Methods

Cell optical density was measured using a spectrophotometer at 600 nm. Cell dry weight (CDW) analysis involved drying the known volume of cell suspension at 105 °C for at least 16 h or until constant weight. The weight of the dried cell was calculated using the known volume of the cell. Reducing sugar was analyzed using 3,5-dinitrosalicylic acid [24]. Total sugar was analyzed using the phenol-sulfuric method.

A high-performance liquid chromatography system (LC-20A, Shimadzu, Kyoto, Japan) equipped with an Aminex HPX-87H column (Bio-Rad, California, USA) was employed to analyze xylose, xylitol, and ethanol. The compounds were detected by a refractive index detector (RID-6A, Shimadzu, Kyoto, Japan). The column's temperature was 40 °C. The injection volume of the sample was 20 µL. The mobile phase was 5 mM sulfuric acid with a 0.75 mL/min flow rate.

2.6. Statistical Analysis

The SPSS program version 20 was employed for general statistical analysis. One-way ANOVA was used to compare differences between groups and factors. Duncan's test was the post hoc test for mean comparison within the dataset. All tests were carried out at a 95% confidence level.

3. Results and Discussions

3.1. Nitrogen Source Supplementation for Ethanol Fermentation Using Sugarcane Bagasse Hydrolysate

This part of the study aimed to use the vinasse from ethanol fermentation using sugarcane bagasse as the raw material for xylitol production. Because ethanol fermentation from bagasse has not been commercially practiced, we first evaluated the necessity of supplementing nitrogen sources from enzymatic hydrolysate of sugarcane bagasse. The results from this part of the study were applied in the ethanol fermentation. Vinasse was prepared from the ethanol broth and used for xylitol production.

As representatives of organic and inorganic nitrogen sources, yeast extract (YE) and diammonium phosphate (DAP) were compared as potential nitrogen sources for ethanol production. Commercial or industrial grades of these nitrogen sources are available for use on a larger scale if necessary. In the experiments, the amount of each nitrogen source used varied from 1 g/L to 10 g/L. Controls included no nitrogen source supplementation, and supplementation with a YM-based solution.

Nitrogen source supplementation did not have a significant effect on sugar utilization during ethanol fermentation (Figure 1a). Glucose was fully consumed by the yeast, even without any nitrogen source supplementation to the hydrolysate. This result indicated that the sugarcane bagasse hydrolysate prepared in this study contained sufficient nutrients to support the cell activities, which resulted in complete glucose consumption. As well as glucose, xylose consumption was evident during fermentation. A similar amount of xylose was consumed regardless of the amount of nitrogen source supplemented, which resulted in final xylose in the ethanol broth of 14.76 ± 1.96 g/L.

Ethanol production in all the test conditions was similar (Figure 1b). Although there were some fluctuations in the results, these were not significantly different. The ethanol concentrations varied from 43.14 ± 5.46 g/L to 50.80 ± 1.93 g/L (85–96% conversion efficiency). For the diammonium phosphate supplementation, the range of ethanol produced was narrower, from 45.81 ± 2.95 g/L to 48.48 ± 0.89 g/L (84–99% conversion efficiency). Nonetheless, these values were not significantly different, even when compared with the concentration obtained from non-supplemented hydrolysate, which produced 51.15 ± 5.41 g/L of ethanol, or 91% conversion efficiency.

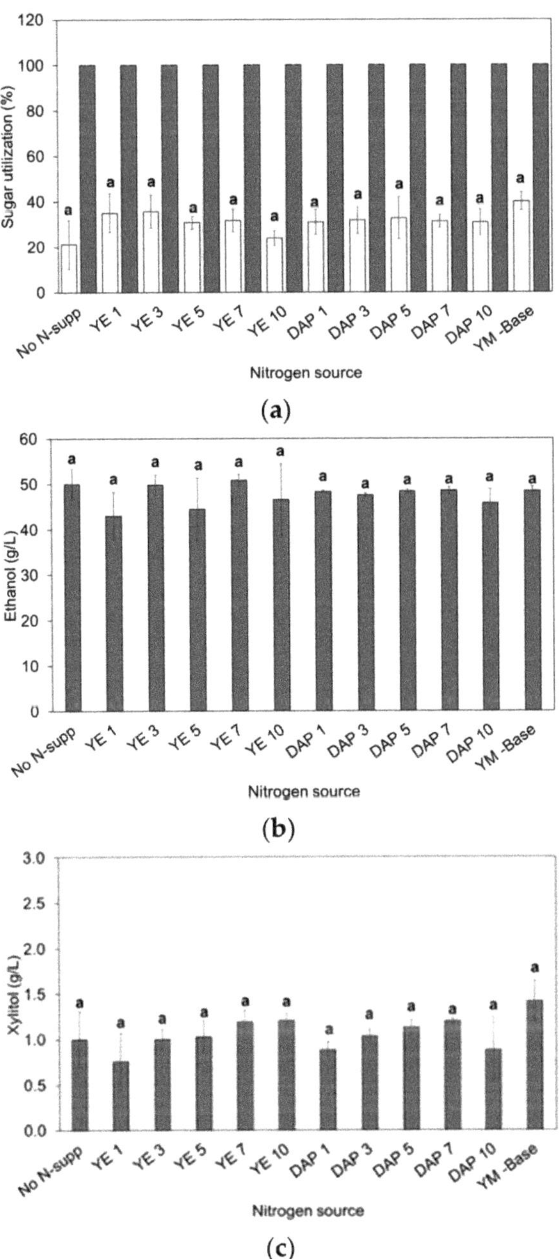

Figure 1. Comparisons of (**a**) sugar utilization (dark bars—glucose, white bars—xylose), (**b**) ethanol, and (**c**) xylitol from ethanol fermentation by *S. cerevisiae* using sugarcane bagasse hydrolysate, supplemented with yeast extract (YE) or diammonium phosphate (DAP), cultivated at 30 °C, 100 rpm for 36 h. Numbers following YE and DAP are concentrations of the respective nitrogen sources in g/L. The same letters above the bars indicated non-significant difference in mean values.

Generally, nitrogen sources had a positive effect on the growth and activity of microorganisms [25]. In our study, the addition of a nitrogen source did not influence cell activity, as the sugar consumption and ethanol production remained at similar levels without an added nitrogen source. The results implied that the hydrolysate contained sufficient nitrogen levels to support yeast growth and hence ethanol production. The nitrogen content in the hydrolysate was 0.13 g/100 g or approximately 1.3 g/L. The presence of nitrogen in the hydrolysate was possibly due to a mixed commercial cellulolytic enzyme used in hydrolysate preparation. As the crude hydrolysate was used for ethanol fermentation, the enzyme proteins remained in the hydrolysate and could serve as the nitrogen source in the fermentation. However, supplementation of nitrogen sources has been reported to increase the fermentation rate, thus reducing fermentation time [26]. We did not verify that report, as it was beyond the scope of our current work to consider the necessity of nitrogen source addition when producing ethanol broth. The broth would be used further to prepare the vinasse, which would be the feedstock for the xylitol production.

In addition to glucose consumption for ethanol production, there was an unexpected utilization of xylose, with a small amount of xylitol detected (Figure 1c). *S. cerevisiae* has no specialized xylose transporter; xylose uptake occurs through native hexose transporters [27,28]. This occurrence agreed with a study claiming that enhanced sugar consumption occurred with the addition of nitrogen sources [29]. The uptake was superior as the condition was more aerobic [30], which was not the case in our study as the ethanol fermentation was carried out at 100 rpm, which was not an aerobic condition. Xylitol conversion from xylose in *S. cerevisiae* was due to a nonspecific aldose reductase [31].

Because ethanol production in sugarcane bagasse hydrolysate is not significantly different with or without nitrogen source supplementation, we chose to carry out the ethanol fermentation using sugarcane bagasse hydrolysate without an additional nitrogen source. The ethanol broth was used to prepare the vinasse for xylitol fermentation.

3.2. Evaluating the Supplementation of Nitrogen Sources for Xylitol Production Using Vinasse from Ethanol Fermentation

Following the same investigating principle as in ethanol fermentation, supplementation of yeast extract (YE) and diammonium phosphate (DAP) were applied in xylitol fermentation using vinasse prepared from distillation of ethanol produced from sugarcane bagasse hydrolysate. Nitrogen source supplementation may be necessary as available nitrogen in the hydrolysate could be depleted in the ethanol fermentation stage. The vinasse was used without any treatment or detoxification, although *Candida mogii* fermentation in concentrated stillage (vinasse) after ethanol fermentation from sweet sorghum juice and bagasse showed the necessity of detoxifying the stillage [32].

Supplementation of nitrogen sources into the vinasse improved xylose utilization in the media (Figure 2a). Adding as little as 1 g/L of YE significantly improved xylose utilization in the vinasse media to 95.2 ± 1.93%, compared with 79.8 ± 0.5% without added nitrogen source. Adding YE at higher concentrations did not help improve the overall xylose consumption nor the consumption rate, according to the monitoring of reducing sugar during fermentation (profiles not shown). The addition of YE increased the reducing sugar consumption rates from 0.377 g/L.h (no YE addition) to the range of 0.514–0.582 g/L.h (1–10 g/L of YE).

A different response was observed after the addition of DAP. Low concentrations of DAP resulted in a similar xylose consumption of 96.0 ± 2.6% to 95.6 ± 4.4%. However, at higher DAP of 7 g/L and 10 g/L, xylose consumption slightly decreased. Regarding the reducing sugar consumption rate (profiles not shown), the addition of DAP at 1 g/L increased the sugar consumption rate to 0.511 g/L.h. The rates continuously decreased as DAP concentration increased.

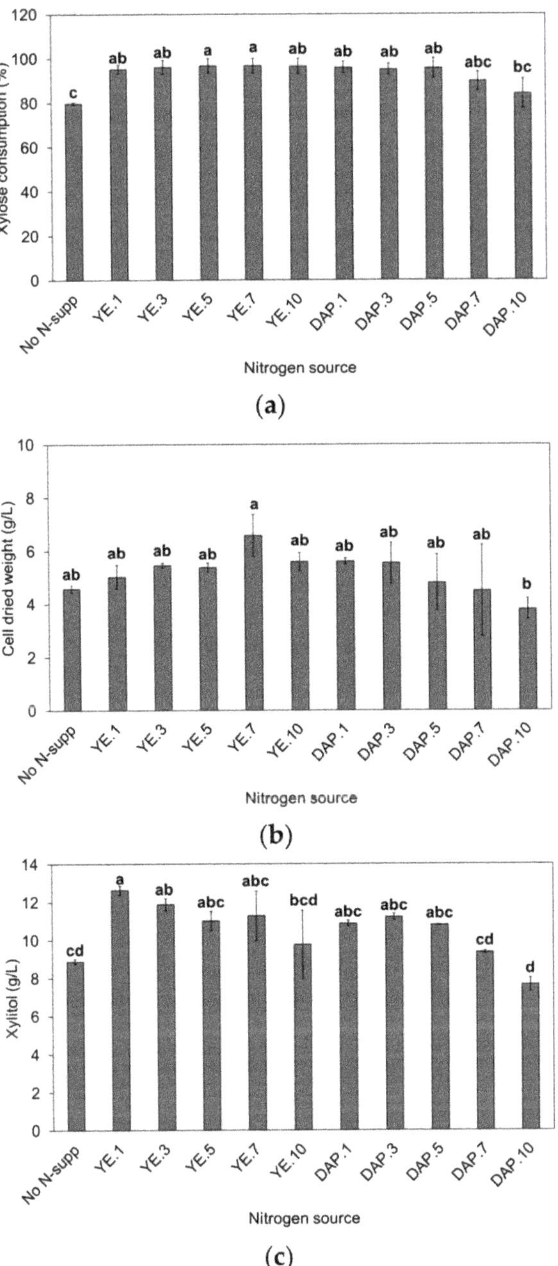

Figure 2. Comparisons of (**a**) xylose consumption, (**b**) cell dry weight, and (**c**) xylitol from fermentation of *C. guilliermondii* TISTR5068 in vinasse from sugarcane hydrolysate-based ethanol broth supplemented with yeast extract (YE) or diammonium phosphate (DAP), cultivated at 30 °C, 200 rpm for 60 h. Numbers following the abbreviations indicate concentration of the respective nitrogen source in g/L. The same letters above the bars indicated non-significant difference in mean values.

Cell growth in terms of dried weight benefited from nitrogen source addition, especially with yeast extract (Figure 2b). Cell dry weight showed an increasing trend with nitrogen source addition. However, a high concentration of DAP showed an adverse effect on cell growth. A decreasing biomass trend was evident when the DAP concentration was 5 g/L and above. This reduction in cell growth followed decreased xylose consumption at high DAP levels. Regardless of the concentrations, cell yields did not show any significant differences according to the type of nitrogen sources or their concentration (Table 1).

Table 1. Xylitol yields and cell yields obtained from 60-h xylitol fermentation by *C. guilliermondii* TISTR5068 using vinasse supplemented with yeast extract (YE) or diammonium phosphate (DAP).

N Source	Xylitol Yield (g/g)		Cell Yield (g/g)	
(g/L)	YE	DAP	YE	DAP
No N-supp	0.671 ± 0.012 [ab]		0.339 ± 0.006 [a]	
1	0.796 ± 0.039 [a]	0.645 ± 0.005 [ab]	0.315 ± 0.018 [a]	0.332 ± 0.004 [a]
3	0.732 ± 0.065 [a]	0.690 ± 0.025 [ab]	0.334 ± 0.015 [a]	0.341 ± 0.055 [a]
5	0.692 ± 0.056 [ab]	0.658 ± 0.008 [ab]	0.336 ± 0.002 [a]	0.291 ± 0.061 [a]
7	0.704 ± 0.098 [ab]	0.612 ± 0.011 [ab]	0.411 ± 0.060 [a]	0.295 ± 0.116 [a]
10	0.622 ± 0.143 [ab]	0.526 ± 0.007 [b]	0.354 ± 0.038 [a]	0.261 ± 0.018 [a]

Note: Comparisons were made between the same results i.e., within the xylitol yields and the cell yields, regardless of type of nitrogen sources. The same letters indicated non-significant difference in mean values. Theoretical xylitol yield from xylose is 1.013 g/g.

When supplementing the vinasse with nitrogen sources at different concentrations, xylitol production showed an inversing trend with cell dry weight, especially in the case of yeast extract (Figure 2c). Increasing the concentration of YE resulted in a decreasing trend in xylitol production, while the cell dry weights were statistically unchanged. The decrease in xylitol production was evident with increasing DAP concentrations from 5 g/L and above, following drops in xylose consumption and growth. The highest xylitol of 12.6 ± 0.2 g/L was obtained with the addition of 1 g/L of yeast extract. The corresponding yield was also the highest, at 0.796 ± 0.039 g/g_{xylose} (Table 1).

Organic and inorganic nitrogen sources have shown a positive effect on xylitol production by yeasts when used in suitable amounts. Yeast extract has been a popular organic nitrogen source for use in xylitol production. A few studies have reported that excessive levels of yeast extract could be detrimental to xylitol production. Xylitol production from D-xylose by *C. tropicalis* DMS 7524 degenerated when 15 g/L of yeast extract were added to the media [33]; the study recommended adding no more than 1 g/L of yeast extract to the medium for xylitol production. A similar result was also obtained in *C. guilliermondii* FTI 20037, where adding yeast extract of more than 1 g/L resulted in decreasing xylitol concentration, although the cell growth improved [34]. A decreasing trend in xylitol production with increasing yeast extract was reported for *C. tropicalis* NRRLY-12968 growing in the acid hydrolysate of mung bean hull [35].

Among inorganic nitrogen sources, urea and ammonium salts are the most commonly used in xylitol production studies. Any form of ammonium salts, including the dihydrogen phosphate form, resulted in similar xylitol production, except for the chloride form, which showed an inferior result when used at the same concentration [35,36]. In our results, high concentrations of diammonium phosphate reduced the fermentation performance, especially the xylitol production. A study of xylitol production from corncob hemicellulose hydrolysate indicated the significant influence of $(NH_4)_2SO_4$, KH_2PO_4, and yeast extract [37]. Upon optimizing those components, the optimal concentration of $(NH_4)_2SO_4$ was in the higher level of the varied range, whereas the optimal concentrations of KH_2PO_4 and yeast extract were at the lower ends of the ranges. Those results implied the negative effect of high phosphate concentration on xylitol production, which could explain the lower xylitol production observed in this study when supplementing the vinasse with high concentrations of diammonium phosphate.

In a study employing a high concentration of phosphate buffer, it was indicated that reduced xylose consumption and xylitol yield could be caused by high osmotic pressure due to high ionic concentrations in the media [38]. In addition, high salt concentrations could reduce the oxygen solubility in the media. Both of these reasons could apply in our study. Although the phosphate concentrations we used were much lower than that in that previously reported study, the vinasse used as the media for xylitol production could fit the category of high salt concentration, even at the maximum of 10 g/L (75.7 mM) DAP added into the media.

The results from this section indicate that nitrogen sources, especially yeast extract (as an organic nitrogen source), improved sugar consumption rates and hence fermentation rates. High xylose consumption, accompanied by high cell growth due to excess nitrogen source supply, did not favor xylitol production. Based on the above observations, the vinasse was supplemented with 1 g/L of yeast extract.

3.3. Effect of Dissolved Oxygen (DO) on Xylitol Production Using Vinasse from Ethanol Fermentation of Sugarcane Bagasse Hydrolysate

This part of the study involved supplying different levels of dissolved oxygen (DO) to the fermentation medium. The fermentation was carried out in a fermenter system. The vinasse supplemented with 1 g/L of yeast extract was used for xylitol production by *C. guilliermondii* TISTR5068. Two patterns of aeration were investigated in the study. Controlled dissolved oxygen levels at 5%, 10%, and 15% were achieved through cascade control (DO/stirrer speed) with a constant airflow of 2 vvm. An uncontrolled dissolved oxygen pattern employed a constant stirring speed at 100 rpm with 1 vvm airflow. A negative control experiment with 100 rpm stirring speed and no aeration was also carried out for comparison.

The results in Figure 3 and Table 2 demonstrate the importance of aeration on xylitol fermentation. The fermentation profiles indicate that the dissolved oxygen level affected fermentation rates, such that the higher DO resulted in faster xylose consumption (Figure 3a). The xylose profiles demonstrated that the xylose consumption rate at 5% DO was slower than those at 10% and 15%, at which similar rates were evident for each. In the uncontrolled aeration using a constant stirring speed at 100 rpm with 1 vvm airflow, the dissolved oxygen was high at the beginning of the fermentation, dropped to 0% and remained at that level throughout the fermentation. Note that in this case the DO of 0% implied that the yeast used the dissolved oxygen as fast as it was supplied to the medium, not that the medium did not have enough oxygen supply as in the negative control run (100 rpm with no aeration). Xylose consumption was similar at approximately 90%. Regardless of the incomplete xylose consumption, the residual concentrations were low, ranging from 1–2 g/L.

Table 2. Parameters from xylitol fermentation by *C. guilliermondii* using vinasse supplement by 1 g/L of yeast extract cultivated at different dissolved oxygen levels for 72 h.

Aeration Conditions	Xylose Used (%)	Xylitol * (g/L)	Q_p * (g/L.h)	$Y_{p/s}$ * (g/g$_{xylose}$)	CDW (g/L)	$Y_{x/s}$ (g/g$_{xylose}$)
100 rpm	28.1 ± 0.0 [b]	0.04 ± 0.00 [b]	0.001 ± 0.000 [d]	0.010 ± 0.001 [c]	1.3 ± 0.8 [c]	0.308 ± 0.202 [c]
1 vvm & 100 rpm	90.9 ± 1.5 [a]	9.67 ± 0.10 [a]	0.134 ± 0.001 [c]	0.610 ± 0.014 [b]	5.4 ± 0.8 [b]	0.342 ± 0.064 [c]
5% DO	88.3 ± 4.7 [a]	9.54 ± 0.53 [a]	0.159 ± 0.009 [b]	0.656 ± 0.022 [b]	6.0 ± 0.8 [b]	0.408 ± 0.048 [bc]
10% DO	91.9 ± 2.1 [a]	9.96 ± 0.55 [a]	0.186 ± 0.010 [a]	0.685 ± 0.029 [a]	11.0 ± 0.5 [a]	0.765 ± 0.106 [ab]
15% DO	90.3 ± 1.9 [a]	10.1 ± 0.3 [a]	0.177 ± 0.010 [a]	0.697 ± 0.027 [a]	12.3 ± 0.8 [a]	0.826 ± 0.044 [a]

Note: $Y_{p/s}$ = xylitol yield, Q_p * = xylitol productivity, CDW = cell dry weight and $Y_{x/s}$ = biomass yield; * Values are reported for the times of maximum concentration. Comparisons of mean values were made within the same column. The same letters indicated non significant difference in mean values.

Cell growth followed the xylose consumption rates (Figure 3b). Faster growth was evident with higher dissolved oxygen levels. In high DO conditions, yeast growth continued even when xylose consumption almost stopped, as observed after 48 h and 60 h of fermentation. A significant increase in the cell dry weight resulted from aeration at any level (Table 2). Increasing the dissolved oxygen level to 10% significantly increased the cell biomass. Further increase in DO to 15% did not significantly promote further cell growth. Nonetheless, cell yield had an increasing trend with higher dissolved oxygen levels in the range investigated.

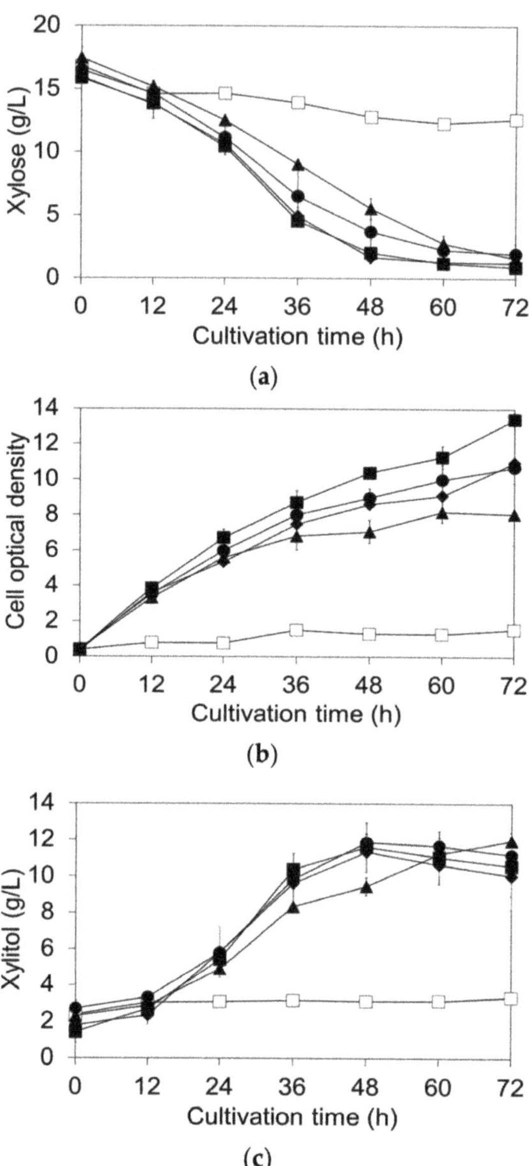

Figure 3. Profiles of (**a**) xylose, (**b**) cell optical density, and (**c**) xylitol for cultivated *C. guilliermondii* TISTR5068 in vinasse medium at 30 °C and various aeration patterns for 72 h. □—no aeration, 100 rpm; ▲—1 vvm, 100 rpm, •—5% DO; v—10% DO; ■—15% DO.

Xylose consumption and hence fermentation rate had been reported to increase with an increase in oxygen transfer coefficient in *Candida boidinii* NRRL Y-17213 [39]. This report explains our results, as higher dissolved oxygen levels allowed higher oxygen transfer. Our results also suggested a limit on the xylose consumption rate, as increasing the DO level from 10% to 15% did not further increase the xylose consumption rate.

The xylitol profiles showed dependency on the level of dissolved oxygen. Notwithstanding the distinct effect of DO level on xylose consumption and cell growth, the xylitol production rate was similar for controlled DO of 10% and 15% (Figure 3c and Table 2). A slower production rate was observable in the uncontrolled DO conditions. However, the final xylitol concentrations produced at all the aerated conditions were insignificantly different (Table 2). The dissolved oxygen level showed a slight but significant effect on xylitol yield, where the yield increased with an increase in DO level, up to DO of 10%. A further increase to 15% DO did not significantly improve the xylitol yield.

In the controlled DO conditions, xylitol profiles showed a drop in concentration after 48 h for fermentation at 10% and 15% DO and after 60 h at 5% DO. These decreases in xylitol occurred after the xylose consumption had ended. The reductions shared a similar rate, regardless of the dissolved oxygen levels.

Controlling dissolved oxygen to a certain level, as we applied in this study, can help with continuous and complete consumption of xylose. In studies of *Candida* sp. strains under constant agitation, incomplete utilization of xylose was evident at low aeration rates or low k_La [39,40]. A similar result was observed in another study of *C. guilliermondii* at a constant aeration rate, where a low agitation rate resulted in incomplete xylose consumption [41]. A common characteristic of incomplete xylose consumption was slow or no cell growth following faster growth early in the fermentation. Controlling the dissolved oxygen promoted complete xylose consumption, even at a low level of 0.5%, although with a slower fermentation rate [42].

The metabolisms of xylose and xylitol depend on NAD(P)+/NAD(P)H redox balance [43]. Oxygen-limited conditions result in the ineffective regeneration of NAD(P)+, causing intracellular NAD(P)H levels to increase. High NAD(P)H represses the activity of xylitol dehydrogenase, causing xylitol to accumulate. Our results fit this scenario. The range of dissolved oxygen covered in this study gave rise to oxygen-limited conditions, where the imbalance of the cofactors occurred with a fast fermentation rate, as demonstrated in Figure 3. We also investigated xylitol production at a higher dissolved oxygen level of 20%, and found that the yeast grew to a much higher cell density with a fast xylose consumption rate, but with no xylitol production (results not shown). Proper aeration led to high cell growth and no xylitol production [42].

In our study, a drop in xylitol concentration occurred after 48 h when xylose levels were low. Xylitol consumption or reassimilation have been reported in xylitol-producing yeasts such as *Starmerella meliponinorum* and a strain of *C. guilliermondii* [44,45]. More aerobic conditions caused faster xylitol consumption in a recombinant *Saccharomyces cerevisiae* [43]. In our study, small but noticeably faster xylitol assimilation could be observed when the dissolved oxygen level increased from 5% to 15%.

From the results, we selected controlled dissolved oxygen at 10% as the condition for xylitol production from the vinasse medium. A higher DO level was not required as it did not improve the production in terms of xylitol concentration, yield, or productivity. The results also suggested that xylitol production performance was not particularly sensitive to changes in dissolved oxygen ranging from 5% to 15%. In addition, close attention should be paid to the harvesting time, as the yeast could reassimilate the produced xylitol during prolonged fermentation.

3.4. Effect of Molasses Addition to the Vinasse Medium on Xylitol Production

Investigation of molasses addition to the xylitol fermentation followed the claims made by several studies that the small addition of carbon sources such as glucose, maltose, or sucrose could improve xylitol production [16,46–48]. We were interested in supplementation with molasses, a by-product of the sugar crystallization process that contains sucrose and other invert sugars that could support the yeast function. Molasses was added to the vinasse medium on a weight basis, from 0.5% w/v to 3.5% w/v of molasses. At these amounts, the total sugar corresponding to the molasses in the medium would range from 11 g/L to 16 g/L, which accounts for 0.59:1 to 0.94:1 of the xylose contained in the vinasse medium. Two control runs were also carried out, using only the vinasse (V-only) and only the molasses at 3.5 g/L (M-only).

For sugar consumption performance, only the xylose profiles are shown (Figure 4a). However, the yeast was able to utilize the sugar in molasses, which was evident in the M-only control run (results not shown). The results demonstrate that supplementing the molasses to the vinasse did not promote xylose utilization rates (Figure 4a and Table 3). It prolonged the fermentation time such that supplementing molasses from 2 g/L and above increased the fermentation time to 72 h, instead of 60 h when supplementing at lower amounts.

Table 3. Parameters from xylitol fermentation by *C. guilliermondii* supplemented by various molasse concentrations into the vinasse medium.

Molasses Supp.	Q_{xylose} (g/L.h)	Xylitol ** (g/L)	$Y_{p/s}$ (g/g_{xylose})	Q_p (g/L.h)	CDW (g/L)	$Y_{x/s}$ (g/$g_{total\ sugar}$)
V-only	0.357 ± 0.029 [a]	9.27 ± 0.05 [a]	0.499 ± 0.004 [a]	0.150 ± 0.001 [a]	5.82 ± 0.05 [c]	0.193 ± 0.002 [b]
M-only	-	-	-	-	6.57 ± 0.27 [a]	0.349 ± 0.013 [a]
0.5 g/L	0.239 ± 0.038 [b]	8.19 ± 0.09 [b]	0.443 ± 0.005 [b]	0.136 ± 0.002 [b]	5.25 ± 0.08 [d]	0.176 ± 0.020 [b]
1 g/L	0.299 ± 0.000 [ab]	8.33 ± 0.04 [b]	0.471 ± 0.002 [b]	0.139 ± 0.001 [b]	5.80 ± 0.03 [c]	0.174 ± 0.001 [b]
2 g/L	0.247 ± 0.003 [b]	7.70 ± 0.05 [c]	0.454 ± 0.003 [b]	0.107 ± 0.001 [c]	6.03 ± 0.03 [bc]	0.168 ± 0.002 [b]
3 g/L	0.263 ± 0.006 [b]	6.87 ± 0.14 [e]	0.360 ± 0.006 [d]	0.096 ± 0.003 [d]	6.42 ± 0.02 [ab]	0.176 ± 0.001 [b]
3.5 g/L	0.242 ± 0.009 [b]	7.26 ± 0.01 [d]	0.394 ± 0.020 [d]	0.101 ± 0.000 [cd]	6.47 ± 0.20 [ab]	0.185 ± 0.006 [b]

Q_{xylose} = xylose utilization rates calculated from the slope of the xylose profiles, Q_p = xylitol productivity calculated when xylose was depleted, $Y_{p/s}$ = xylitol yield based on xylose, $Y_{x/s}$ = biomass yield based on total sugar; ** Xylitol was reported as the difference between the value for depleted xylose and the initial value. Comparisons of mean values were made within the same column. The same letters indicated non-significant difference in mean values.

Yeast growth profiles, according to cell optical density, showed that molasses at 3.5 g/L (M-only control) resulted in the fastest growth rate during the first 12 h of fermentation (Figure 4b) and the highest cell dry weight (Table 3). The addition of molasses into the vinasse medium slightly increased cell growth with increased molasses supplementation. However, the cell yields ($Y_{x/s}$) did not change. It should be emphasized that the control with molasses medium contained the lowest total sugar concentration at approximately 21 g/L, while vinasse media with and without molasses supplementation contained total sugar in the range of 44.7 to 61.3 g/L. Low sugar concentration would benefit cell growth, as the effect of osmotic pressure was less than that in the higher sugar concentrations.

Supplementation with molasses did not improve the xylitol production rate or the concentrations. The xylitol profiles in Figure 4c indicate that by using the vinasse medium without molasses supplementation, the xylitol production rate was faster and resulted in the highest xylitol concentration. The more molasses, the lower the xylitol obtained (Table 3). This trend was also true for xylitol yields and productivities.

In addition to xylitol, ethanol was another product of fermentation. In the M-only control, ethanol concentration increased until 24 h, reaching a concentration of 5.17 g/L. It consistently dropped after prolonged fermentation (Figure 4d). The drop in ethanol matched the increase in cell optical density, implying that the yeast cells utilized ethanol as a carbon source to support their growth. In the media supplemented with molasses, the yeast produced ethanol early in the fermentation process. The amount of ethanol corresponded to the amount of molasses supplemented. Supplementing 0.5 g/L of molasses resulted in 2.40 ± 0.01 g/L of ethanol, while supplementing 3 g/L and 3.5 g/L of molasses resulted in 6.13 ± 0.02 g/L and 5.89 ± 0.07 g/L of ethanol, respectively.

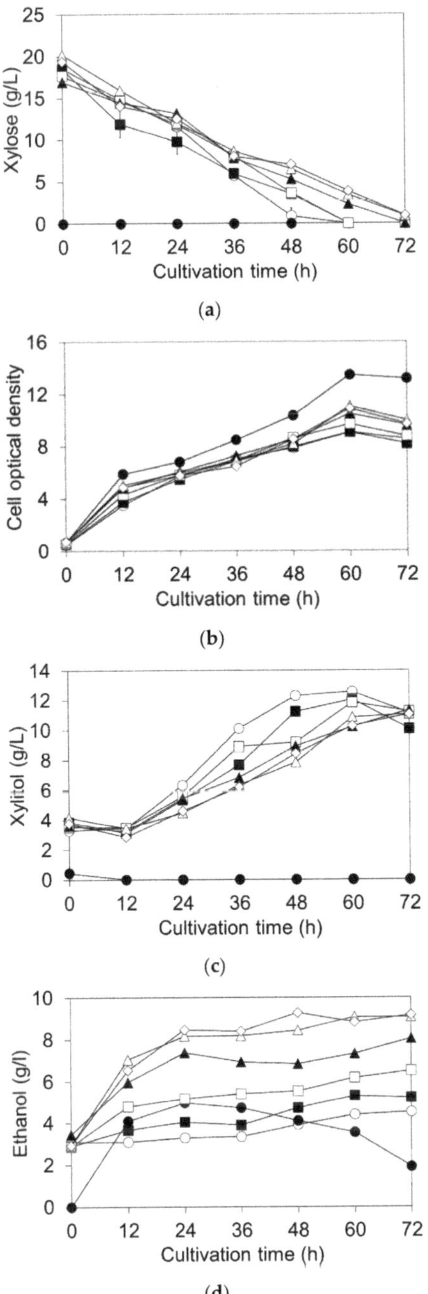

Figure 4. Profiles of (**a**) xylose, (**b**) cell optical density, (**c**) xylitol, and (**d**) ethanol when cultivating *C. guilliermondii* TISTR5068 in vinasse medium supplemented with various molasses concentrations at 30 °C, 200 rpm for 72 h. ○—vinasse medium, ●—only molasses at 3.5 g/L, ■—0.5 g/L molasses supplemented, □—1 g/L molasses supplemented, ▲—2 g/L molasses supplemented, △—3 g/L molasses supplemented, and ◇—3.5 g/L molasses supplemented.

The results obtained in our study did not suggest a positive effect of molasses supplementation on xylitol production by C. guillermondii TISTR5068 in a vinasse medium. The results contradicted claims that supplementation of other carbon sources helps improve various aspects of xylitol production including xylose consumption rate, xylitol production rate, xylitol concentration, and yield [16,46–48]. Supplementation with molasses appeared to deteriorate xylitol production while improving yeast growth.

Several studies into carbon source supplementation to improve xylitol production involved supplementing carbon sources in their pure forms. These studies did not report any ethanol production by yeast, even with distinctively high concentrations of the supplementary glucose and sucrose [47,48]. However, other studies used complex carbon sources, such as sugarcane syrup and molasses, as supplements to xylitol production using sugarcane bagasse and straw hydrolysates, and reported the production of ethanol [16,47]. The amount of ethanol depended on the amount of sugar supplemented; over 20 g/L was reported. Our results showed that maximum ethanol of ~5 g/L was produced from the supplementation of sugar via molasses addition. Ethanol has been reported to harm the cell viability of xylose-utilizing yeast, with viability dropping to 50% at high ethanol concentrations [49]. Ethanol tolerance in xylose-utilizing yeast was also strain-dependent [50]. Exposure to ethanol during the first 12 h to 24 h could affect the yeast performance, especially in conversion of xylose to xylitol, causing inferior xylitol production when supplementing with molasses.

Although slight improvement in xylitol production, especially in xylose uptake rate, when supplemented with sugar syrup or molasses was reported in other studies [16,47], our results did not follow those earlier findings. A significant decrease in the xylose utilization rate was evident in our case. Nonetheless, the reduction in xylitol yield that we observed concurred with a study that used molasses as a supplement [16].

Based on our results and those reported, we inferred that the differences in various aspects of xylitol production performance would depend on substrate type and characteristics. The amount and type of supplementary carbon did not necessarily guarantee improved xylitol production. In our study, we employed vinasse of ethanol broth obtained from the fermentation of sugarcane bagasse hydrolysate. It contained mainly xylose with other sugars and oligosaccharides, and a low level of ethanol remaining after the distillation.

3.5. Proposed Process for Ethanol and Xylitol Production from Sugarcane Bagasse

By integrating the results obtained in this study, each step in the proposed process was combined, and the flow chart of the process is demonstrated in Figure 5. In this process, separate streams of ethanol and xylitol resulted. Ethanol obtained in this study was within the same range as a previous study with a similar design using sweet sorghum juice and bagasse [32]. Increased quantities of xylitol were obtained in this study, compared with less than 5 g/L obtained from the stillage (vinasse) of the study employing *Candida mogii.*

Other studies have reported co-production of ethanol and xylitol with separate streams of hydrolysate. A similar range of ethanol concentrations resulted, with values between 48–56 g/L [16–18]. These values were comparable to results from this study at approximately 50 g/L. The higher xylitol obtained in those studies, ranging from 24.0 g/L to 34.5 g/L, was the result of high xylose concentration after concentrating the hemicellulosic hydrolysate, or from high solid loading in hydrolysis by xylanase. Regardless of the lower xylitol concentration in this study (~10 g/L) due to the use of non-concentrated vinasse, the conversion yield was comparable.

From the diagram (Figure 5), it should be noted that some xylose was converted to xylitol during the ethanol fermentation, due to the activities of native hexose transporters and a nonspecific aldose reductase, as discussed earlier. Regarding the separation of ethanol by simple distillation, ethanol cannot be fully separated from the aqueous solution so a small amount of ethanol was present in the vinasse. The mass balance showed a 15% loss from distillation. This loss could be due to evaporation and residual liquid holdup in the instrument, because a simple laboratory glass distillation set was used. In addition, a noticeably large loss of bagasse after pretreatment was due to lignin loss and small residue loss during the washing steps after pretreatment, when mesh cloth was used for draining the liquid.

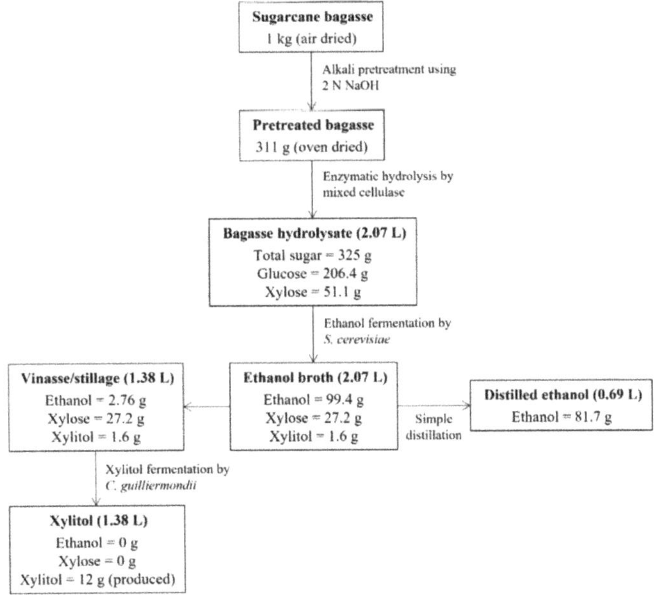

Figure 5. Diagram of the proposed ethanol and xylitol production process, illustrating important substrates and products at each stage of the operation.

4. Conclusions

This study demonstrated the integrated biorefinery of sugarcane bagasse for ethanol and xylitol production. The results showed the possibility of integrating xylitol production into the lignocellulosic-based ethanol process by utilizing vinasse from the distillation stage containing remaining xylitol. A small supplementation with nitrogen sources was necessary to improve xylitol production from the vinasse. Limited oxygen conditions promoted xylitol production, whereas excessive aeration was observed to result in xylitol consumption in the later stages of fermentation, resulting in a low xylitol yield. It was not necessary to supplement a carbon source to improve the xylitol production when using vinasse as the substrate. Using a waste stream from cellulosic ethanol production, as demonstrated in this study, represents waste utilization through biological processes and serves as a part of the circular bioeconomy.

Author Contributions: Conceptualization, M.B.K. and A.R.; methodology, S.H. and M.B.K.; formal analysis, S.H. and M.B.K.; investigation, S.H.; resources, A.R.; data curation, S.H. and M.B.K.; writing—original draft preparation, S.H.; writing—review and editing, M.B.K.; supervision, M.B.K.; funding acquisition, M.B.K. and A.R. All authors have read and agreed to the published version of the manuscript.

Funding: This research was financially supported by the Research Center for Environmental and Hazardous Substance Management (EHSM), Khon Kaen University, Thailand (grant number EHSM64-014); Thailand Science Research and Innovation (TSRI) Senior Research Scholar (grant number RTA6280001).

Institutional Review Board Statement: Not applicable.

Informed Consent Statement: Not applicable.

Data Availability Statement: The data presented in this study are available on request from the corresponding author.

Acknowledgments: The authors appreciated the support for S.H. from the Royal Scholarship under Her Royal Highness Princess Maha Chakri Sirindhorn Education Project to the Kingdom of Cambodia year 2019–2021.

Conflicts of Interest: The authors declare no conflict of interest.

References

1. Zoghlami, A.; Paës, G. Lignocellulosic Biomass: Understanding Recalcitrance and Predicting Hydrolysis. *Front. Chem.* **2019**, *7*, 874. [CrossRef] [PubMed]
2. Ullah, K.; Kumar Sharma, V.; Dhingra, S.; Braccio, G.; Ahmad, M.; Sofia, S. Assessing the Lignocellulosic Biomass Resources Potential in Developing Countries: A Critical Review. *Renew. Sustain. Energy Rev.* **2015**, *51*, 682–698. [CrossRef]
3. OECD/FAO. Sugar Production Classified by Traditional Crops. In *OECD-FAO Agricultural Outlook 2021–2030*; OECD Publishing: Paris, France, 2021. [CrossRef]
4. Office of the Cane and Sugar Board. Report on Sugarcane Plantation, Production Year 2019/20. Available online: http://www.ocsb.go.th/upload/journal/fileupload/923-1854.pdf (accessed on 6 September 2021).
5. Office of the Cane and Sugarcane Board. Electricity and Steam Production of the Sugar Cane and Sugar Industry. Available online: http://www.ocsb.go.th/upload/bioindustry/fileupload/10169-3442.pdf (accessed on 17 May 2021).
6. Cheng, H.H.; Whang, L.M. Resource Recovery from Lignocellulosic Wastes via Biological Technologies: Advancements and Prospects. *Bioresour. Technol.* **2022**, *343*, 126097. [CrossRef]
7. Velvizhi, G.; Balakumar, K.; Shetti, N.P.; Ahmad, E.; Kishore Pant, K.; Aminabhavi, T.M. Integrated Biorefinery Processes for Conversion of Lignocellulosic Biomass to Value Added Materials: Paving a Path towards Circular Economy. *Bioresour. Technol.* **2022**, *343*, 126151. [CrossRef]
8. Renewable Fuels Association. 2022 Ethanol Industry Outlook. Available online: https://ethanolrfa.org/media-and-news/category/news-releases/article/2022/02/rfa-s-2022-ethanol-industry-outlook-zeroes-in-on-new-opportunities (accessed on 13 May 2022).
9. Raj, T.; Chandrasekhar, K.; Naresh Kumar, A.; Rajesh Banu, J.; Yoon, J.J.; Kant Bhatia, S.; Yang, Y.H.; Varjani, S.; Kim, S.H. Recent Advances in Commercial Biorefineries for Lignocellulosic Ethanol Production: Current Status, Challenges and Future Perspectives. *Bioresour. Technol.* **2022**, *344*, 126292. [CrossRef] [PubMed]
10. Yuvadetkun, P.; Boonmee, M. Ethanol Production Capability of *Candida shehatae* in Mixed Sugars and Rice Straw Hydrolysate. *Sains Malays.* **2016**, *45*, 581–587.
11. Kongkeitkajorn, M.B.; Sae-Kuay, C.; Reungsang, A. Evaluation of Napier Grass for Bioethanol Production through a Fermentation Process. *Processes* **2020**, *8*, 567. [CrossRef]
12. Delgado Arcaño, Y.; Valmaña García, O.D.; Mandelli, D.; Carvalho, W.A.; Magalhães Pontes, L.A. Xylitol: A Review on the Progress and Challenges of Its Production by Chemical Route. *Catal. Today* **2020**, *344*, 2–14. [CrossRef]
13. Felipe Hernández-Pérez, A.; de Arruda, P.V.; Sene, L.; da Silva, S.S.; Kumar Chandel, A.; de Almeida Felipe, M.D.G. Xylitol Bioproduction: State-of-the-Art, Industrial Paradigm Shift, and Opportunities for Integrated Biorefineries. *Crit. Rev. Biotechnol.* **2019**, *39*, 924–943. [CrossRef]
14. West, T.P. Xylitol Production by *Candida* Species from Hydrolysates of Agricultural Residues and Grasses. *Fermentation* **2021**, *7*, 243. [CrossRef]
15. Antunes, F.A.F.; Thomé, L.C.; Santos, J.C.; Ingle, A.P.; Costa, C.B.; Dos Anjos, V.; Bell, M.J.V.; Rosa, C.A.; Silva, S.S.D. Multi-Scale Study of the Integrated Use of the Carbohydrate Fractions of Sugarcane Bagasse for Ethanol and Xylitol Production. *Renew. Energy* **2021**, *163*, 1343–1355. [CrossRef]
16. De Souza Queiroz, S.; Jofre, F.M.; dos Santos, H.A.; Hernández-Pérez, A.F.; das Graças de Almeida Felipe, M. Xylitol and Ethanol Co-Production from Sugarcane Bagasse and Straw Hemicellulosic Hydrolysate Supplemented with Molasses. *Biomass Convers. Biorefinery* **2021**, 1–10. [CrossRef]
17. Unrean, P.; Ketsub, N. Integrated Lignocellulosic Bioprocess for Co-Production of Ethanol and Xylitol from Sugarcane Bagasse. *Ind. Crops Prod.* **2018**, *123*, 238–246. [CrossRef]
18. Raj, K.; Krishnan, C. Improved Co-Production of Ethanol and Xylitol from Low-Temperature Aqueous Ammonia Pretreated Sugarcane Bagasse Using Two-Stage High Solids Enzymatic Hydrolysis and *Candida tropicalis*. *Renew. Energy* **2020**, *153*, 392–403. [CrossRef]
19. Shankar, K.; Kulkarni, N.S.; Sajjanshetty, R.; Jayalakshmi, S.K.; Sreeramulu, K. Co-Production of Xylitol and Ethanol by the Fermentation of the Lignocellulosic Hydrolysates of Banana and Water Hyacinth Leaves by Individual Yeast Strains. *Ind. Crops Prod.* **2020**, *155*, 112809. [CrossRef]
20. Dasgupta, D.; Ghosh, D.; Bandhu, S.; Adhikari, D.K. Lignocellulosic Sugar Management for Xylitol and Ethanol Fermentation with Multiple Cell Recycling by *Kluyveromyces marxianus* IIPE453. *Microbiol. Res.* **2017**, *200*, 64–72. [CrossRef]
21. Zahed, O.; Jouzani, G.S.; Abbasalizadeh, S.; Khodaiyan, F.; Tabatabaei, M. Continuous Co-Production of Ethanol and Xylitol from Rice Straw Hydrolysate in a Membrane Bioreactor. *Folia Microbiol.* **2016**, *61*, 179–189. [CrossRef]

22. Morales-Rodriguez, R.; Perez-Cisneros, E.S.; de Los Reyes-Heredia, J.A.; Rodriguez-Gomez, D. Evaluation of Biorefinery Configurations through a Dynamic Model-Based Platform: Integrated Operation for Bioethanol and Xylitol Co-Production from Lignocellulose. *Renew. Energy* **2016**, *89*, 135–143. [CrossRef]

23. Ngamsirisomsakul, M.; Reungsang, A.; Kongkeitkajorn, M.B. Assessing Oleaginous Yeasts for Their Potentials on Microbial Lipid Production from Sugarcane Bagasse and the Effects of Physical Changes on Lipid Production. *Bioresour. Technol. Rep.* **2021**, *14*, 100650. [CrossRef]

24. Miller, G.L. Use of Dinitrosalicylic Acid Reagent for Determination of Reducing Sugar. *Anal. Chem.* **1959**, *31*, 426–428. [CrossRef]

25. Martínez-Moreno, R.; Morales, P.; Gonzalez, R.; Mas, A.; Beltran, G. Biomass Production and Alcoholic Fermentation Performance of *Saccharomyces cerevisiae* as a Function of Nitrogen Source. *FEMS Yeast Res.* **2012**, *12*, 477–485. [CrossRef] [PubMed]

26. Li, Z.; Wang, D.; Shi, Y.-C. Effects of Nitrogen Source on Ethanol Production in Very High Gravity Fermentation of Corn Starch. *J. Taiwan Inst. Chem. Eng.* **2017**, *70*, 229–235. [CrossRef]

27. Zhao, Z.; Xian, M.; Liu, M.; Zhao, G. Biochemical Routes for Uptake and Conversion of Xylose by Microorganisms. *Biotechnol. Biofuels* **2020**, *13*, 21. [CrossRef] [PubMed]

28. Subtil, T.; Boles, E. Competition between Pentoses and Glucose during Uptake and Catabolism in Recombinant *Saccharomyces cerevisiae*. *Biotechnol. Biofuels* **2012**, *5*, 14. [CrossRef]

29. Salmon, J.M.; Vincent, O.; Mauricio, J.C.; Bely, M.; Barre, P. Sugar Transport Inhibition and Apparent Loss of Activity in *Saccharomyces cerevisiae* as a Major Limiting Factor of Enological Fermentations. *Am. J. Enol. Vitic.* **1993**, *44*, 56–64.

30. Batt, C.A.; Caryallo, S.; Easson, D.D.; Akedo, M.; Sinskey, A.J. Direct Evidence for a Xylose Metabolic Pathway In *Saccharomyces cerevisiae*. *Biotechnol. Bioeng.* **1986**, *28*, 549–553. [CrossRef]

31. Toivari, M.H.; Salusjärvi, L.; Ruohonen, L.; Penttilä, M. Endogenous Xylose Pathway in *Saccharomyces cerevisiae*. *Appl. Environ. Microbiol.* **2004**, *70*, 3681–3686. [CrossRef]

32. Stoklosa, R.J.; Nghiem, N.P.; Latona, R.J. Xylose-Enriched Ethanol Fermentation Stillage from Sweet Sorghum for Xylitol and Astaxanthin Production. *Fermentation* **2019**, *5*, 84. [CrossRef]

33. Silva, S.S.; Afschar, A.S. Microbial Production of Xylitol from D-Xylose Using Candida tropicalis. *Bioprocess Eng.* **1994**, *11*, 129–134. [CrossRef]

34. Silva, S.S.; Quesada-Chanto, A.; Vitolo, M. Upstream Parameters Affecting the Cell Growth and Xylitol Production by *Candida guilliermondii* FTI 20037. *Zeitschrift Für Naturforsch. C* **1997**, *52*, 359–363. [CrossRef]

35. Mushtaq, Z.; Imran, M.; Zahoor, T.; Ahmad, R.S.; Arshad, M.U. Biochemical Perspectives of Xylitol Extracted from Indigenous Agricultural By-Product Mung Bean (Vigna Radiata) Hulls in a Rat Model. *J. Sci. Food Agric.* **2014**, *94*, 969–974. [CrossRef] [PubMed]

36. Vongsuvanlert, V.; Tani, Y. Xylitol Production by a Methanol Yeast, *Candida boidinii* (*Kloeckera* Sp.) No. 2201. *J. Ferment. Bioeng.* **1989**, *67*, 35–39. [CrossRef]

37. Ling, H.; Cheng, K.; Ge, J.; Ping, W. Statistical Optimization of Xylitol Production from Corncob Hemicellulose Hydrolysate by Candida tropicalis HDY-02. *New Biotechnol.* **2011**, *28*, 673–678. [CrossRef] [PubMed]

38. Cortez, D.V.; Roberto, I.C. Effect of Phosphate Buffer Concentration on the Batch Xylitol Production by *Candida guilliermondii*. *Lett. Appl. Microbiol.* **2006**, *42*, 321–325. [CrossRef] [PubMed]

39. Winkelhausen, E.; Amartey, S.A.; Kuzmanova, S. Xylitol Production from D-Xylose at Different Oxygen Transfer Coefficients in a Batch Bioreactor. *Eng. Life Sci.* **2004**, *4*, 150–154. [CrossRef]

40. Ding, X.; Xia, L. Effect of Aeration Rate on Production of Xylitol from Corncob Hemicellulose Hydrolysate. *Appl. Biochem. Biotechnol.* **2006**, *133*, 263–270. [CrossRef]

41. Mussatto, S.I.; Roberto, I.C. Xylitol Production from High Xylose Concentration: Evaluation of the Fermentation in Bioreactor under Different Stirring Rates. *J. Appl. Microbiol.* **2003**, *95*, 331–337. [CrossRef]

42. Faria, L.F.F.; Gimenes, M.A.P.; Nobrega, R.; Pereira, N. Influence of Oxygen Availability on Cell Growth and Xylitol Production by *Candida guilliermondii*. In *Biotechnology for Fuels and Chemicals*; Humana Press: Totowa, NJ, USA, 2002; pp. 449–458. [CrossRef]

43. Tani, T.; Taguchi, H.; Akamatsu, T. Analysis of Metabolisms and Transports of Xylitol Using Xylose- and Xylitol-Assimilating *Saccharomyces cerevisiae*. *J. Biosci. Bioeng.* **2017**, *123*, 613–620. [CrossRef]

44. Schirmer-Michel, Â.C.; Flôres, S.H.; Hertz, P.F.; Matos, G.S.; Ayub, M.A.Z. Production of Ethanol from Soybean Hull Hydrolysate by Osmotolerant *Candida guilliermondii* NRRL Y-2075. *Bioresour. Technol.* **2008**, *99*, 2898–2904. [CrossRef]

45. Da Silva, R.O.; Serpa, M.D.N.; Brod, F.C.A. Influence of Agitation and Aeration on Xylitol Production by the Yeast *Starmerella meliponinorum*. *Quim. Nova* **2020**, *43*, 705–710. [CrossRef]

46. Wannawilai, S.; Lee, W.C.; Chisti, Y.; Sirisansaneeyakul, S. Furfural and Glucose Can Enhance Conversion of Xylose to Xylitol by *Candida magnoliae* TISTR 5663. *J. Biotechnol.* **2017**, *241*, 147–157. [CrossRef] [PubMed]

47. Hernández-Pérez, A.F.; de Arruda, P.V.; Felipe, M.D.G.D.A. Sugarcane Straw as a Feedstock for Xylitol Production by *Candida guilliermondii* FTI 20037. *Braz. J. Microbiol.* **2016**, *47*, 489–496. [CrossRef] [PubMed]

48. Tamburini, E.; Bianchini, E.; Bruni, A.; Forlani, G. Cosubstrate Effect on Xylose Reductase and Xylitol Dehydrogenase Activity Levels, and Its Consequence on Xylitol Production by *Candida tropicalis*. *Enzyme Microb. Technol.* **2010**, *46*, 352–359. [CrossRef]

49. Castañón-Rodríguez, J.F.; Domínguez-González, J.M.; Ortíz-Muñiz, B.; Torrestiana-Sanchez, B.; de León, J.A.R.; Aguilar-Uscanga, M.G. Continuous Multistep versus Fed-Batch Production of Ethanol and Xylitol in a Simulated Medium of Sugarcane Bagasse Hydrolyzates. *Eng. Life Sci.* **2015**, *15*, 96–107. [CrossRef]
50. Snoek, T.; Verstrepen, K.J.; Voordeckers, K. How Do Yeast Cells Become Tolerant to High Ethanol Concentrations? *Curr. Genet.* **2016**, *62*, 475–480. [CrossRef]

fermentation

Article

Isolation of Novel Yeast from Coconut (*Cocos nucifera* L.) Water and Phenotypic Examination as the Potential Parameters in Bioethanol Production

Getari Kasmiarti [1], Dwita Oktiarni [2,3], Poedji Loekitowati Hariani [4], Novia Novia [5] and Hermansyah Hermansyah [4,*]

1   Doctoral Program of Environmental Science, Graduate Program Universitas Sriwijaya, Jalan Padang Selasa No.524, Bukit Besar, Palembang 30139, South Sumatra, Indonesia; getariikasmiarti@gmail.com
2   Mathematics and Natural Sciences Study Program, Graduate School of Universitas Sriwijaya, Jalan Raya Palembang Prabumulih KM32, Indralaya 30662, South Sumatra, Indonesia; dwita.oktiarni@gmail.com
3   Chemistry Department, Faculty of Mathematics and Natural Sciences, Jl. WR. Supratman Kandang Limun, Universitas Bengkulu, Muara Bangka Hulu, Bengkulu 38371, Indonesia
4   Department of Chemistry, Faculty of Mathematics and Natural Sciences, Universitas Sriwijaya, Jalan Raya Palembang Prabumulih KM32, Indralaya 30662, South Sumatra, Indonesia; puji_lukitowati@mipa.unsri.ac.id
5   Chemical Engineering Department, Faculty of Engineering, Universitas Sriwijaya, Jalan Raya Palembang Prabumulih KM32, Indralaya 30662, South Sumatra, Indonesia; novia@ft.unsri.ac.id
*   Correspondence: hermansyah@unsri.ac.id

Abstract: Yeast is a fermentation agent for producing bioethanol as an environmentally friendly alternative energy. Therefore, this study aims to find novel yeasts with the capability to persevere under acidic, high temperature, and high sugar content conditions, which are required in the bioethanol industry. The yeasts were isolated and identified from coconut (*Cocos nucifera* L.) water by a DNA sequencing method and phenotypic test. Yeast isolation has been completed with a serial dilution procedure and purification was conducted with HiPurA Genomic DNA Purification Spin Kits, which were analyzed by DNA Sequencing. The phenotypic test was carried out with thermotolerant (30 °C and 41 °C), high acidity (lactic acid), and sugar content (molasses 35 °brix) parameters in the media as the initial step of yeast ability screening. Based on the results, the three species of *Candida tropicalis* K5 (*Candida tropicalis* strain L2), K15 (*Candida tropicalis* strain MYA-3404), and K20 (*Candida tropicalis* strain Y277) obtained met the phenotypic standards. This showed that the yeasts have the potential to produce molasses-based bioethanol.

Keywords: candida; coconut water (*Cocos nucifera* L.); phenotypic; DNA sequencing

Citation: Kasmiarti, G.; Oktiarni, D.; Hariani, P.L.; Novia, N.; Hermansyah, H. Isolation of Novel Yeast from Coconut (*Cocos nucifera* L.) Water and Phenotypic Examination as the Potential Parameters in Bioethanol Production. *Fermentation* **2022**, *8*, 283. https://doi.org/10.3390/fermentation8060283

Academic Editor: Timothy J. Tse

Received: 26 May 2022
Accepted: 9 June 2022
Published: 16 June 2022

1. Introduction

Yeast is a unicellular microorganism in the biotechnology process that plays an important role in bioconversion activity for the production of alcoholic products such as wine, pickles, beer, and bread. The wide applications of yeast make it possible for it to be used in bioethanol fermentation. There are various kinds of raw materials for bioethanol production including lignocellulosic biomass [1], molasses [2], and agriculture waste [3]. Meanwhile, molasses is an advantageous source of fermentable sugar because it contains 48–55% sucrose. During its fermentation, yeast is usually added to convert the sucrose content into bioethanol through its enzymatic metabolism [4].

The effective and efficient exploration of yeast in the production of bioethanol from molasses is always challenging. The novelty of this study is discovering novel yeast in coconut (*Cocos nucifera* L.) water that can get through stress tolerance. Several parameters have shown successful bioethanol processing such as thermotolerance, acid resistance, and sugar tolerance. This stress tolerance is important in yeast screening, and it is commonly referred to for phenotypic identification.

The discovery of yeast that meets phenotypic standards can be carried out through isolation. Previous studies showed that indigenous yeast isolation from a certain source is a reliable method to discover a novel product with a specific ability and a significant effect on the yield of bioethanol [5]. Coconut water was selected as the source of yeast because it contains many nutrients needed by cells, which are sugar in the form of sucrose, as well as amino and organic acids [6]. Saraswati (2014) tested the effect of coconut water on the growth of *Saccharomyces cerevisiae*, with the hypothesis being accepted [7]. The presence of nutrients and compounds possessed by coconut water can be the first step in characterizing and identifying microorganisms from coconut water. The identification of microorganisms in coconut water is expected to uncover better novel yeast compared to *Saccharomyces cerevisiae* in biotechnology exploration.

Saccharomyces cerevisiae is not resistant to the high concentrations of ethanol produced [8]. Its optimum activity occurs at 28–35 °C and 3.5–6.0 pH [9]. Meanwhile, the bioethanol industry runs the production in a reactor that uses a high temperature inlet with a huge amount of molasses. In conclusion, the purpose of this study is to examine the phenotypic responses of yeast isolated from coconut water by observing the feedback of non-saccharomyces yeast on high temperature, acidity, and sugar concentration (°brix) in the growing medium. Yeast with a higher probability of stress tolerance was selected as the novel yeast.

2. Materials and Methods

2.1. Sample Preparation

Coconut (*Cocos nucifera* L.) was obtained from a traditional market in Ogan Ilir Regency, South Sumatra, Indonesia. The coconut was peeled, and the obtained water was transferred to a sterile Erlenmeyer for further processing.

2.2. Yeast Isolation from Coconut Water (Cocos Nucifera L.)

A total of 10 mL of the sample was diluted using a serial dilution procedure (10^{-1}–10^{-5}). The yeast population was carried out through the modification of Maciel et al.'s [10] method by taking aliquots (0.1 mL) of the serial dilution and spreading it on Yeast Malt Agar (YMA) with 2 g glucose, 2 g peptone, 1 g malt extract, 1 g yeast extract, 2 g agar, plus chloramphenicol at 100 mg per 100 mL of distilled water. The samples were incubated at 30 °C for 3 days, and the obtained yeast cultures were stored in 20% glycerol stock for further identification.

2.3. Extraction of Yeast DNA Genome

DNA purification was conducted using HiPurA Genomic DNA Purification Spin Kits, which provide a quick and easy method for application in PCR (Polymerase Chain Reaction) based on the manufacturer's procedures.

2.4. DNA Sequencing

DNA sequences were amplified using two common primers, namely ITS1 (5′-TCCGTAGGTGAACCTGCGG-3′) and ITS4 (5′-TCCTCCGCTTATTGATATGC-3′). The amplification was carried out through a reaction mixture containing 1×25 µL (9.5 sterile water, 12.5 MyTaq Red Mix, 10 M ITS1, 10 M ITS4, and 1 DNA template). Amplicons were amplified according to a previous method under PCR conditions of 95 °C for 3 min (initial denaturation), continued (95 °C, denatured 10 s at 95 °C, annealing 30 s at 52 °C, extension at 72 °C for 45 s) at 35 cycles, with a final extension at 72 °C for 5 min. Subsequently, the PCR product was electrophoresed using 1% agarose gel. Sequencing data were taken from the National Center for Biotechnology Information NCBI/BLAST (blast.ncbi.nlm.nih.gov/BLAST.cgi) to create a phylogenetic tree based on the neighbor-joining algorithm.

2.5. Phenotypic Test

The isolated yeast was grown in YPDA (Yeast Peptone Dextrose Agar) at a pH of 3.5 and in YPA-Molasse 3 °brix media at temperatures of 30 °C and 41 °C. The pH of 3.5 was made by adding lactic acid to YPDA media until the pH value was obtained. Meanwhile, YPA-molasses 35 °brix was made by adding pretreated 35 °brix molasses into YPA media. The growth of yeast was observed and compared to control Saccharomyces cerevisiae on each representative medium after 24 h of incubation time.

3. Results

Nucleic acid (Genomic DNA) quantification was completed between 3.80–79.90 ng/µL (Table 1), while sequencing produced a specific ratio of $A_{260/280}$ and $A_{260/230}$ through PCR amplification and bi-directionality. The ratio resulted from the uv-vis spectrum ranging from 1.63–5.35 and 0.48–2.01, respectively. Therefore, the absorbance scale indicated the DNA purity, and ≥ 1.8 signified a pure DNA sample.

Table 1. The genomic DNA concentration of the yeast isolates.

Isolate	Concentration (ng/µL)	$A_{260/280}$	$A_{260/230}$
K1	8.20	2.09	0.77
K2	18.60	2.08	1.20
K3	45.10	2.28	2.01
K4	8.00	2.01	0.81
K5	10.30	2.16	1.01
K6	3.80	1.63	0.48
K7	36.50	2.25	1.52
K8	79.90	1.84	0.60
K9	33.50	1.94	0.60
K10	59.80	2.34	2.03
K11	43.40	1.85	0.49
K12	7.00	2.28	0.97
K13	17.70	2.12	1.11
K14	4.80	1.94	0.62
K15	11.10	1.86	0.73
K16	4.30	5.35	0.48
K17	5.00	1.64	0.52
K18	5.10	1.77	0.57
K19	7.10	2.98	0.92
K20	7.00	1.64	0.48

Figure 1 shows the PCR result, which was amplified and assessed by electrophoresis, and the band on each DNA fragment describes the purity of the gene. Based on the photo, the 20 samples (K1–K20) read around 500–900 base pairs. All the isolates have a single band, with the exception of isolates K4 and K12, which have two bands. The band below describes the total impurities.

Figure 2 shows the phylogenetic tree of isolates K1, K2, K3, K4, and K5. Meanwhile, K1 was identified as *Henseniaspora opuntiae* strain NS02 (KT226114.1) with a 99.86% similarity of identity. *Henseniaspora meyeri* CBS:8775 (KY103531.1) was identical with isolates K2 and K3 as well as having a 99.81% identity percentage. Furthermore, isolate K4 was analogous with *Meyerozima carpophila* strain CBS5256 (MK394110.1) with a 99.83% resemblance, while K5 was *Candida tropicalis* strain L2 (MK752673.1) because it was comparable with a 99.81% identity percentage. The phylogenetic tree of isolates K6, K7, K8, K9, and K10 is shown in Figure 3. It was discovered that isolate K6 had a 100% similarity of identity with *Henseniaspora opuntiae* strain F173 (KY497945.1). Meanwhile, isolates K7 and K8 were both equally interpreted as *Henseniaspora meyeri* culture CBS:8775 (KY103531.1) with a 100% identity percentage. An isolate was classified as *Saccharomyces cerevisiae* strain KSD-Yc (CP024006.1), which was 100% identical to isolate K9. *Henseniaspora meyeri* culture CBS:8775 (KY103531.1) was described as isolate K10 with a 100% resemblance.

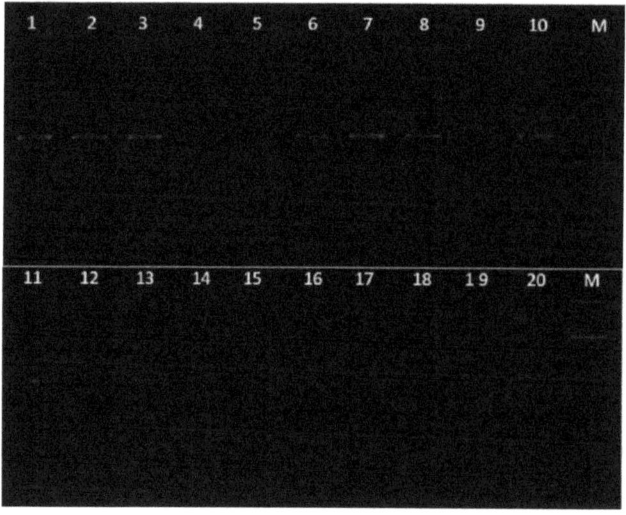
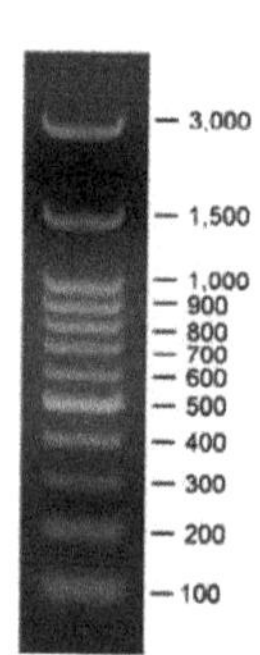

Figure 1. Gel photo of the isolate as a PCR product in an electrophoresis gel.

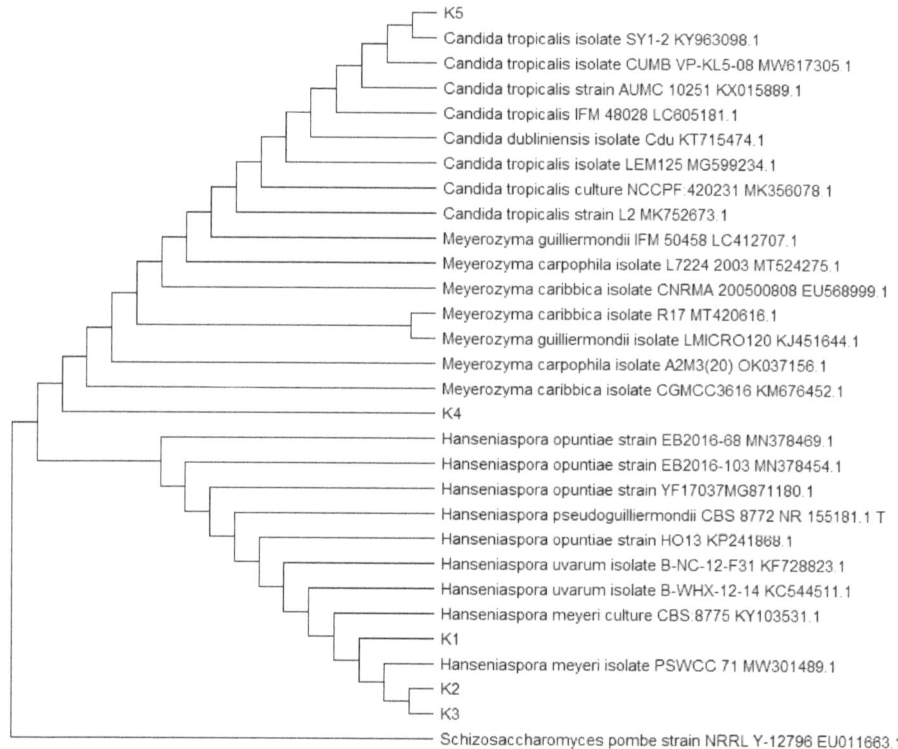

Figure 2. Dendrogram of yeast isolates K1, K2, K3, K4, and K5.

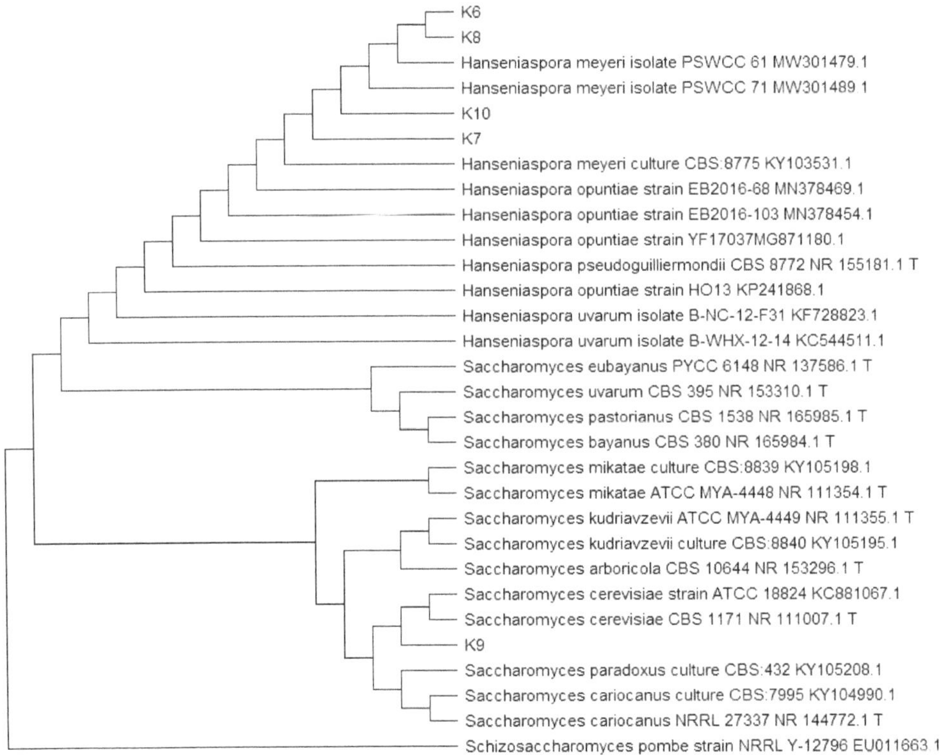

Figure 3. Dendrogram of yeast isolates K6, K7, K8, K9, and K10.

Figures 4 and 5 show the phylogenetic tree of isolates K11, K12, K13, K14, K15, K16, K17, K18, K19, and K20. *Henseniaspora thailandica* (AB501145.1) was discovered as isolate K11 with a 100% identity percentage, while isolate K12 was 99.52% identical with *Meyerozima carribica* Strain UFLA CWFY11 (KM402049.1). *Henseniaspora meyeri* culture CBS:8775 (KY103531.1) was transcribed as K13 with a 99.81% similarity of identity. Isolate K14 was also considered to be *Saccharomyces cerevisiae* strain KSD-Yc (CP024006.1), like K9, with a 100% identity percentage. Furthermore, K15 was transcribed as *Candida tropicalis* strain MYA-3404 (CP047875.1) with a 99.61% identity percentage, while *Lachance fermentati* strain CNRMA8.216 (KP132361.1) was isolate K16 with a 99.70% identity similarity. The isolate K17 was found to be *Meyerozima carribica* strain CBS 5256 (MK394110.1) at 100%, while K18 resembled *Candida othopsilosis* (FM178396.1), which had a 100% identity percentage. *Henseniaspora ovarum* culture CBS:2580 (KY103573.1) was similar to isolate K19, at 99.91%. Another *Candida tropicalis* strain Y277 (KT459476.1) was also discovered in isolate K20, which was 100% identical.

The phenotypic identification was carried out at two different temperatures for the comparison, namely 30 °C and 41 °C. The 20 isolates were compared to *Saccharomyces cerevisiae* (X) in various growing media, which included YPDA, pH 3.5 (by adding lactic acid), and YPA-Molasses 35 °brix. Figures 6–8 (a) are the visualization of the growth after 24 h of incubation at 30 °C and Figures 6–8 (b) at 41 °C. The result showed that isolates K5, K15, and K20 had constant growth during the observations.

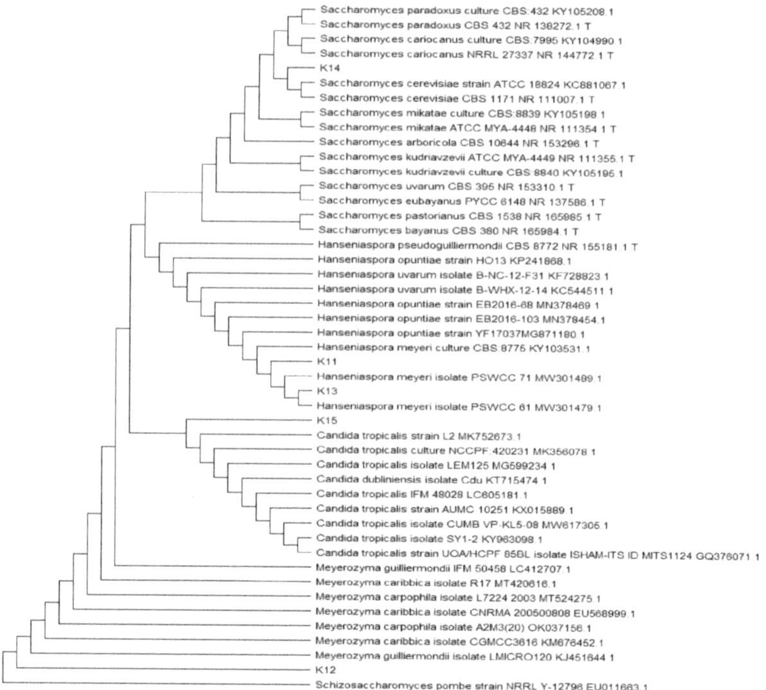

Figure 4. Dendrogram of yeast isolates K11, K12, K13, K14, and K15.

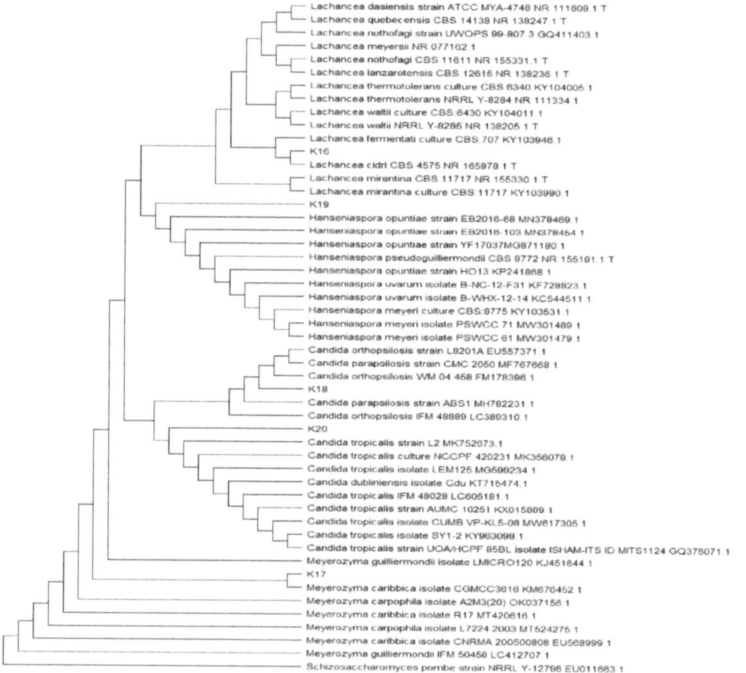

Figure 5. Dendrogram of yeast isolates K16, K17, K18, K19, and K20.

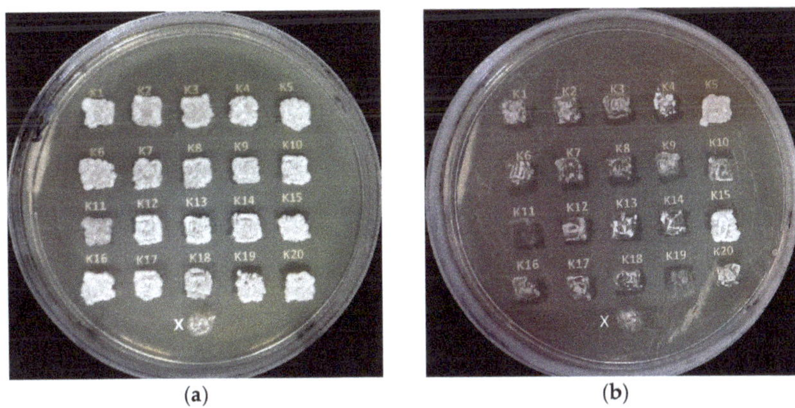

(a)

(b)

Figure 6. The growth of 20 isolates and X (Saccharomyces cerevisiae) in YPDA at (**a**) 30 °C and (**b**) 41 °C.

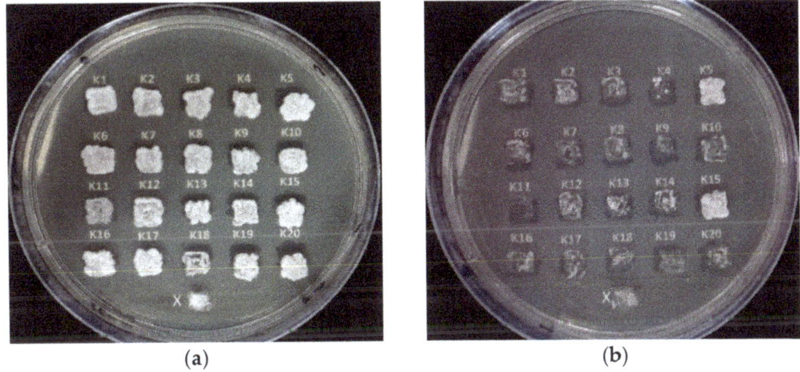

(a)

(b)

Figure 7. The growth of 20 isolates and X (Saccharomyces cerevisiae) in YPDA pH 3.5 at (**a**) 30 °C and (**b**) 41 °C.

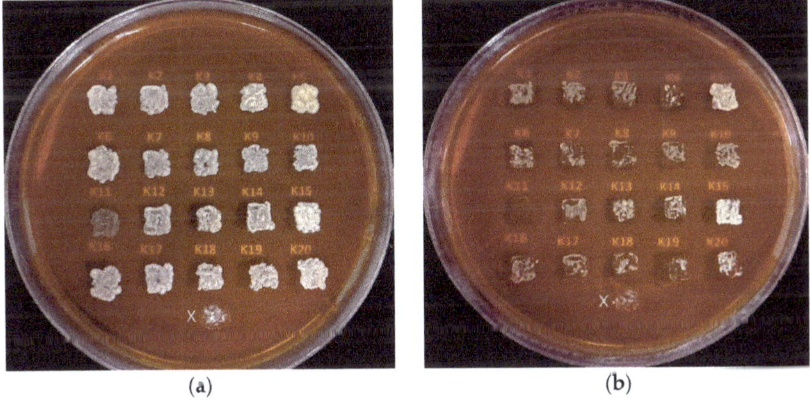

(a)

(b)

Figure 8. The growth of 20 isolates and X (Saccharomyces cerevisiae) in YPA-Molasses 35 °brix at (**a**) 30 °C and (**b**) 41 °C.

4. Discussion

Microorganisms play an important role in fermentation. The fermentation process depends on the isolation/development of yeast, which can ferment various type of sugars. In order to achieve this goal, investigations were carried out to obtain indigenous yeast from various sources such as fermented natural ingredients [11], fruit skins [12], fruits [13], and coconut water. The identification of yeast in coconut (*Cocos nucifera* L.) water was expected to contribute to the genetic diversity, with a significant potential for bioethanol production. There are limited investigations of coconut water that focus on this because the majority of the research focuses only on the specific use of biochemical composition and preservation techniques.

Coconut water contains minerals, a high sugar content, acidity, protein, crude fat, and total phenolic compounds [14]. Furthermore, the sugar content is 80% sucrose, 3% glucose, and 7% fructose, with a low glycemic index (35–54) [15]. The nutrients and other compounds in coconut water can be the initial step in characterizing and identifying the existence of indigenous yeast that is needed for living.

In this study, 20 isolates were obtained from coconut water. Based on the identification result in Table 2, a total of 14 different indigenous yeast strains were discovered among the isolates. These include K1 (*Henseniaspora opuntiae* strain NS02), K2-K3-K7-K8-K10-K13 (*Henseniaspora meyeri* culture CBS:8775), K4 (*Meyerozima carpophila* strain CBS5256), K5 (*Candida tropicalis* strain L2), K6 (*Henseniaspora opuntiae* strain F173), K9-K14 (*Saccharomyces cerevisiae* strain KSD-Yc), K11 (*Henseniaspora thailandica*), K12 (*Meyerozima carribica* Strain AUMC 7262), K15 (*Candida tropicalis* strain MYA-3404), K16 (*Lachance fermentati* strain CNRMA8.216), K17 (*Meyerozima carribica* strain CBS 5256), K18 (*Candida othopsilosis*), K19 (*Henseniaspora ovarum* culture CBS:2580), and K20 (*Candida tropicalis* strain Y277). It was necessary to test the obtained group of yeasts to examine their potential for surviving under stress control by a phenotypic test as the first testing step through selective media and temperatures.

Table 2. The summary of BLAST-N 16 rRNA genes of yeast isolates.

Isolate	Accession Number	Species	Max Score	Total Score	Query Coverage	E Value	Percent Identity
K1	KT226114.1	*Henseniaspora opuntiae* strain NS02	1351	1351	99%	0.0	99.86%
K2	KY103531.1	*Henseniaspora meyeri* culture CBS:8775	959	959	100%	0.0	99.81%
K3	KY103531.1	*Henseniaspora meyeri* culture CBS:8775	952	952	100%	0.0	99.81%
K4	MK394110.1	*Meyerozima carpophila* strain CBS5256	1083	1083	99%	0.0	99.83%
K5	MK752673.1	*Candida tropicalis* strain L2	898	898	100%	0.0	99.81%
K6	KY497945.1	*Henseniaspora opuntiae* strain F173	935	935	100%	0.0	100%
K7	KY103531.1	*Henseniaspora meyeri* culture CBS:8775	957	957	100%	0.0	100%
K8	KY103531.1	*Henseniaspora meyeri* culture CBS:8775	953	953	99%	0.0	100%
K9	CP024006.1	*Saccharomyces cerevisiae* strain KSD-Yc	1520	1520	99%	0.0	100%
K10	KY103531.1	*Henseniaspora meyeri* culture CBS:8775	957	957	100%	0.0	100%
K11	AB501145.1	*Henseniaspora thailandica*	1347	1347	97%	0.0	100%
K12	KM402049.1	*Meyerozima carribica* Strain UFLA CWFY11	1105	2215	100%	0.0	99.51%
K13	KY103531.1	*Henseniaspora meyeri* culture CBS:8775	952	952	99%	0.0	99.81%
K14	CP024006.1	*Saccharomyces cerevisiae* strain KSD-Yc	1515	10,502	100%	0.0	100%
K15	CP047875.1	*Candida tropicalis* strain MYA-3404	948	2845	99%	0.0	99.61%
K16	KP132361.1	*Lanchance fermentati* strain CNRMA8.216	1221	1221	99%	0.0	99.70%
K17	MK394110.1	*Meyerozima carribica* strain CBS 5256	1123	1123	99%	0.0	100%
K18	FM178396.1	*Candida othopsilosis*	948	948	99%	0.0	100%
K19	KY103573.1	*Henseniaspora ovarum* culture CBS:2580	1312	1312	99%	0.0	99.91%
K20	KT459476.1	*Candida tropicalis* strain Y277	974	974	100%	0.0	100%

A phenotypic test was carried out under some stressful conditions to discover yeast with a superior strain. The conditions were thermotolerance, acidity tolerance and a high molasses content. The phenotypic scheme was relevant for bioethanol production based on the result in Table 3 and Figures 6–8. On an industrial scale, bioethanol production usually takes place at 41 °C [16] because the fermentation reactor sets a high energy intake [17]. Due to this reason, the temperature setting was 30 °C and 40 °C. Bioethanol was produced in an anaerobic and acidity setup, while pH 3.5 was expected as the maximum acid environment for yeast to convert the sugar into ethanol [18,19]. It was discovered that yeast worked sensitively at a certain maximum sugar content concentration based on the raw material [20, 21]. The analysis of the physicochemical characteristics of molasses with 48–54% sucrose shows the potential for producing alcohol products [22] because it is fermentable without modification. Based on this fact, the investigation of high sugar content was completed in yeast agar molasses media with a 35 °brix concentration. Therefore, the phenotypic landscape was completed by comparing the superiority between *Saccharomyces cerevisiae* and indigenous yeast from coconut water.

Table 3. The phenotypic results at 30 °C and 41 °C.

Isolate	Growth					
	30 °C			41 °C		
	YPDA Media	YPDA pH 3.5 Media	YPA-Molase 35°brix Media	YPDA Media	YPDA pH 3.5 Media	YPA-Molase 35 °brix Media
K1	+++	+++	+++	++	++	++
K2	+++	+++	+++	++	++	++
K3	+++	+++	+++	++	++	++
K4	+++	+++	+++	+++	++	++
K5	+++	+++	+++	+++	+++	+++
K6	+++	+++	+++	++	++	++
K7	+++	+++	+++	++	++	++
K8	+++	+++	+++	++	++	++
K8	+++	+++	+++	++	++	++
K9	+++	+++	+++	++	+	++
K10	+++	+++	+++	++	++	++
K11	+++	+++	+	+	+	+
K12	+++	+++	+++	++	++	++
K13	+++	+++	+++	++	++	++
K14	+++	+++	+++	++	++	++
K15	+++	+++	+++	++	+++	+++
K16	+++	+++	+++	++	++	++
K17	+++	+++	+++	++	++	++
K18	+++	+++	+++	++	++	++
K19	+++	+++	+++	+	++	+
K20	+++	+++	+++	+++	++	+++
X	++	+++	+++	+	+	+

Where (+++): extremely good, (++): very good, and (+): moderate.

Isolates K5 (*Candida tropicalis* strain L2), K15 (*Candida tropicalis* strain MYA-3404), and K20 (*Candida tropicalis* strain Y277) were considered to be the distinguished yeasts in Table 3. Based on the results, *Candida tropicalis* resisted a high temperature (40 °C and above) [23] and acidic pH, which indicated its potential for use in the biotechnological process. *C. tropicalis* isolates showed an ability to produce 6.55% (v/v) and 4.58% (v/v) ethanol at 30 °C and 42 °C [24]. It also resisted the presence of inhibitors such as furfural, which hinder cell growth and decrease bioethanol productivity [25]. Furthermore, *C. tropicalis* is also capable of releasing α-amylase for fermentation [26].

In conclusion, *Saccharomyces cerevisiae* is generally used in fermentation. On the other hand, the application of novel yeast from coconut water in the fermentation process is promising due to its high temperature tolerance, acidity, and sugar content, as these can lead to a quicker fermentation by reducing the cooling cost, inhibitors, and acidity treatment with lower production expenses. In addition, the phenotypic test can be designed based on the conditions that are needed in the production. It has a flexibility setting in terms of regulating the temperature, pH, and molasses concentration. Isolates K5 (*Candida tropicalis* strain L2), K15 (*Candida tropicalis* strain MYA-3404), and K20 (*Candida tropicalis* strain Y277) have a high capacity to achieve this condition.

5. Conclusions

The three discovered isolates were K5 (*Candida tropicalis* strain L2), K15 (*Candida tropicalis* strain MYA-3404), and K20 (*Candida tropicalis* strain Y277), which were *Candida tropicalis* species from the isolation and identification of yeast from coconut water. These isolates probably contribute to fermenting molasses because they survived the phenotypic test with a high temperature, high acidity, and high molasses concentration. These novel yeasts have the potential to effectively and efficiently ferment molasses for further studies in bioethanol production.

Author Contributions: Conceptualization, G.K. and H.H.; methodology, G.K. and H.H; validation, H.H.; formal analysis, G.K.; investigation, G.K.; resources, G.K.; data curation, G.K.; writing—original draft preparation, G.K.; writing—review and editing, G.K., H.H., D.O., N.N. and P.L.H.; visualization, G.K.; supervision, H.H., N.N. and P.L.H. All authors have read and agreed to the published version of the manuscript.

Funding: This study was funded by the Directorate of research, technology, and community services, Directorate general of higher education, research, and technology through PMDSU research grant with contract number 057/ES/PG.02.00.PT/2022.

Institutional Review Board Statement: Not applicable.

Informed Consent Statement: Not applicable.

Data Availability Statement: Not applicable.

Acknowledgments: We acknowledge The Ministry of Research, Technology, and Higher Education of the Republic of Indonesia, the Directorate of Research, Technology, and Community Services, Directorate general of higher education, research, and technology on Master's Education toward a Superior Doctor of Bachelor (PMDSU) program.

Conflicts of Interest: The authors declare no conflict of interest.

References

1. Rezania, R.; Oryani, B.; Cho, J.; Talaiekhozani, A.; Sabbagh, F.; Hashemi, B.; Rupani, F.P.; Mohammadi, A.A. Different pretreatment technologies of lignocellulosic biomass for bioethanol production: An overview. *Energy* **2020**, *199*, 117457. [CrossRef]
2. Zentou, H.; Abidin, Z.Z.; Zouwanti, M.; Greetham, D. Effect of operating conditions on molasses fermentation for bioethanol production. *Int. J. Appl. Eng. Res.* **2017**, *12*, 5202–5506.
3. Carillo-Nieves, D.; Alanis, J.R.M.; Quiroz, C.R.; Ruiz, A.H.; Iqbal, M.N.H.; Parra-Saldivar, R. Current status and future trends of bioethanol production from agroindustrial wastes in Mexico. *Renew. Sustain. Energy Rev.* **2019**, *102*, 63–74. [CrossRef]
4. Vilela, A. The Importance of yeasts on fermentation quality and human health-promoting compounds. *Fermentation* **2019**, *5*, 46. [CrossRef]
5. Balia, L.R.; Kurniani, A.B.; Utama, L.G. The Combination of Mozzarella Cheese Whey and Sugarcane Molasses in the Production of Bioethanol with the Inoculation of Indigenous Yeasts. *J. Jpn. Inst. Energy* **2018**, *97*, 266–269. [CrossRef]
6. Kailaku, I.S.; Setiawan, B.; Sulaeman, A. The effects of ultrafiltration and ultraviolet process on nutritional composition, physicochemical and organoleptic properties of coconut water drink. *J. Littri* **2016**, *22*, 43–51. [CrossRef]
7. Saraswati, D. Pengaruh Konsentrasi Air Kelapa Muda Terhadap Pertumbuhan *Saccharomyces cereviceae*. Master's Thesis, Universitas negeri Gorontalo, Gorontalo, Indonesia, 2014.
8. Bagaskara, A.; Wuijaya, M.M.I.; Antara, S.N. Isolation and characterization of bioethanol-producing bacteria from the arak industry of environment in the Tri Eka Buana village, Sidemen district, Karangasem Bali. *J. Rekayasa Dan Manaj. Agroind.* **2020**, *8*, 290–300. [CrossRef]
9. Anita, W.N.; Admadi, H.B.; Arnata, W.I. Optimasi Konsentrasi Enzim Amiloglukosidase dan Saccharomyces cerevisiae dalam Pembuatan Bioetanol dari ubi jalar (*Ipomoea batatas* L.) varietas daya dengan proses sakarifikasi fermentasi simultan (SFS). *J. Rekayasa Dan Manaj. Agroind.* **2015**, *3*, 30–39.
10. Maciel, N.O.P.; Piló, B.F.; Freitas, F.D.L.; Gomes, C.O.F.; Johann, S.; Nardi, M.D.R.; Lachance, M.; Rosa, A.C. The diversity and antifungal susceptibility of the yeasts isolated from coconut water and reconstituted fruit juices in Brazil. *Int. J. Food Microbiol.* **2013**, *160*, 201–205. [CrossRef]
11. Arachchige, A.M.; Yoshida, S.; Toyama, H. Thermo-and salt-tolerant Saccharomyces cerevisiae strains isolated from fermenting coconut toddy from Sri Lanka. *Biotechnol. Biotechnol. Equip.* **2019**, *33*, 937–944. [CrossRef]

12. Talukdera, A.A.; Easmina, F.; Mahmudb, A.S.; Yamada, M. Thermotolerant yeasts capable of producing bioethanol: Isolation from natural fermented sources, identification and characterization. *Biotechnol. Biotechnol. Equip.* **2016**, *30*, 1106–1114. [CrossRef]

13. Demiray, E.; Karatay, E.S.; Dönmez, G. Efficient bioethanol production from pomegranate peels by newly isolated Kluyveromyces marxianus. *Energy Sources Part A Recovery Util. Environ. Eff.* **2019**, *42*, 709–718. [CrossRef]

14. Kenga, S.E.; Easaa, M.A.; Muhamedb, M.A.; Ooib, C.; Chewa, T.T. Composition and physicochemical properties of fresh and freeze-concentrated coconut (*Cocos nucifera*) water. *J. Agrobiotechnol.* **2017**, *8*, 13–24.

15. Alves, V.; Scapini, T.; Camargo, F.A.; Bonatto, C.; Stefanski, S.F.; Jesus, P.E.; Diniz, T.G.L.; Bertan, C.L.; Maldonado, R.R.; Treichel, H. Development of fermented beverage with water kefir in water-soluble coconut extract (*Cocos nucifera* L.) with inulin addition. *LWT-Food Sci. Technol.* **2021**, *145*, 111364. [CrossRef]

16. Nouri, H.; Azin, M.; Mousafi, L.M. Xylan-hydrolyzing thermotolerant *Candida tropicalis* HNMA-1 for bioethanol production from sugarcane bagasse hydrolysate. *Ann. Microbiol.* **2017**, *67*, 633–641. [CrossRef]

17. Balat, M.; Balat, H.; Oz, C. Progress in bioethanol processing. *Prog. Energyand Combust. Sci.* **2008**, *34*, 551–571. [CrossRef]

18. Oliveira, A.C.; Fuess, T.L.; Soares, A.L.; Damianovic, Z.R.H.M. Thermophilic biohydrogen production from sugarcane molasses under low pH: Metabolic and microbial aspects. *Int. Hydrog. Energy* **2020**, *45*, 4182–4192. [CrossRef]

19. Anggraini, A.S.; Yuniningsih, S.; Sota, M.M. Pengaruh pH terjadap kualitas produk etanol dari molase melalui proses fermentasi. *J. Reka Buana* **2017**, *2*, 99–105.

20. Raharja, R.; Murdiyatmo, U.; Sutrisno, A.; Wardani, A.K. Bioethanol production from sugarcane molasses by instant dry yeast. *IOP Conf. Ser. Earth Environ. Sci.* **2019**, *230*, 012076. [CrossRef]

21. Arshad, M.; Hussain, T.; Iqbal, M.; Abbas, M. Enhanced ethanol production at commercial scale from molasses using high gravity technology by mutant S. cerevisiae. *Braz. J. Microbiol.* **2017**, *48*, 403–409. [CrossRef]

22. Kusmiyati, K.; Shitophyta, L.M. Produksi Bioetanol dari Bahan Baku Singkong, Jagung dan Iles-iles: Pengaruh Suhu Fermentasi dan Berat Yeast Saccharomyces cerevisiae. *Reaktor* **2014**, *15*, 97–103. [CrossRef]

23. Shariq, M.; Sohail, M. Application of Candida tropicalis MK-160 for the production of xylanase and ethanol. *J. King Saud Univ. -Sci.* **2019**, *31*, 1189–1194. [CrossRef]

24. Hermansyah, H.; Novia, N.; Minetaka, S.; Satoshi, H. *Candida tropicalis* isolated from tuak, a North Sumatera-Indonesian traditional beverage, for bioethanol production. *Microbiol. Biotechnol. Lett.* **2015**, *43*, 241–248. [CrossRef]

25. Wang, S.; Cheng, G.; Joshua, C.; He, Z.; Sun, X.; Li, R.; Liu, L.; Yuan, Q. Furfural tolerance and detoxification mechanism in Candida tropicalis. *Biotechnol. Biofuels* **2016**, *9*, 250. [CrossRef] [PubMed]

26. Azoulay, E.; Jouanneau, F.; Bertrand, J.; Raphael, A.; Janssens, J.; Lebeault, M. Fermentation methods for protein enrichment of cassava and corn with *Candida Tropicalis*. *Appl. Environ. Microbiol.* **1980**, *39*, 41–47. [CrossRef] [PubMed]

Article

Trials of Commercial- and Wild-Type *Saccharomyces cerevisiae* Strains under Aerobic and Microaerophilic/Anaerobic Conditions: Ethanol Production and Must Fermentation from Grapes of Santorini (Greece) Native Varieties

Kalliopi Basa, Seraphim Papanikolaou *, Maria Dimopoulou, Antonia Terpou, Stamatina Kallithraka and George-John E. Nychas

Department of Food Science and Human Nutrition, Agricultural University of Athens, 75 Iera Odos, 11855 Athens, Greece; kbas1995@gmail.com (K.B.); mdimopoulou@uniwa.gr (M.D.); aterpou@agro.uoa.gr (A.T.); stamatina@aua.gr (S.K.); gjn@aua.gr (G.-J.E.N.)
* Correspondence: spapanik@aua.gr; Tel./Fax: +30-210-5294700

Abstract: In modern wine-making technology, there is an increasing concern in relation to the preservation of the biodiversity, and the employment of "new", "novel" and wild-type *Saccharomyces cerevisiae* strains as cell factories amenable for the production of wines that are not "homogenous", expressing their terroir and presenting interesting and "local" sensory characteristics. Under this approach, in the current study, several wild-type *Saccharomyces cerevisiae* yeast strains (LMBF Y-10, Y-25, Y-35 and Y-54), priorly isolated from wine and grape origin, selected from the private culture collection of the Agricultural University of Athens, were tested regarding their biochemical behavior on glucose-based (initial concentrations *ca* 100 and 200 g/L) shake-flask experiments. The wild yeast strains were compared with commercial yeast strains (*viz.* Symphony, Cross X and Passion Fruit) in the same conditions. All selected strains rapidly assimilated glucose from the medium converting it into ethanol in good rates, despite the imposed aerobic conditions. Concerning the wild strains, the best results were achieved for the strain LMBF Y-54 in which maximum ethanol production (EtOH$_{max}$) up to 68 g/L, with simultaneous ethanol yield on sugar consumed = 0.38 g/g were recorded. Other wild strains tested (LMBF Y-10, Y-25 and Y-35) achieved lower ethanol production (up to ≈47 g/L). Regarding the commercial strains, the highest ethanol concentration was achieved by *S. cerevisiae* Passion Fruit (EtOH$_{max}$ = 91.1 g/L, yield = 0.45 g/g). Subsequently, the "novel" strain that presented the best technological characteristics regards its sugar consumption and alcohol production properties (*viz.* LMBF Y-54) and the commercial strain that equally presented the best previously mentioned technological characteristics (*viz.* Passion Fruit) were further selected for the wine-making process. The selected must originated from red and white grapes (Assyrtiko and Mavrotragano, Santorini Island; Greece) and fermentation was performed under wine-making conditions showing high yields for both strains (EtOH$_{max}$ = 98–106 g/L, ethanol yield = 0.47–0.50 g/g), demonstrating the production efficiency under microaerophilic/anaerobic conditions. Molecular identification by rep-PCR carried out throughout fermentations verified that each inoculated yeast was the one that dominated during the whole bioprocess. The aromatic compounds of the produced wines were qualitatively analyzed at the end of the processes. The results highlight the optimum technological characteristics of the selected "new" wild strain (*S. cerevisiae* LMBF Y-54), verifying its suitability for wine production while posing great potential for future industrial applications.

Keywords: wild-type yeast strains; high-gravity fermentation; *Saccharomyces cerevisiae*; Assyrtiko; Mavrotragano; molecular identification; wine production

---

Citation: Basa, K.; Papanikolaou, S.; Dimopoulou, M.; Terpou, A.; Kallithraka, S.; Nychas, G.-J.E. Trials of Commercial- and Wild-Type *Saccharomyces cerevisiae* Strains under Aerobic and Microaerophilic/Anaerobic Conditions: Ethanol Production and Must Fermentation from Grapes of Santorini (Greece) Native Varieties. *Fermentation* **2022**, *8*, 249. https://doi.org/10.3390/fermentation8060249

Academic Editor: Timothy J. Tse

Received: 6 April 2022
Accepted: 22 May 2022
Published: 26 May 2022

Publisher's Note: MDPI stays neutral with regard to jurisdictional claims in published maps and institutional affiliations.

---

1. Introduction

The beginning of Greek viticulture dates back to the early Neolithic Age as favorable soil and climate of Greece have allowed the wide spread of wine-making [1]. The technologies of viticulture and wine-making were widely developed along the Mediterranean region since grapes are an excellent raw material for wine-making [2]. Nowadays, wine is an important component of the Mediterranean dietary tradition, while recent studies have moderated that its consumption reduces the incidences of coronary heart disease, atherosclerosis and protects against oxidative damage [3–5].

In the last decade, a significant number of studies targeted to improve the fermentation performance and productivity of wines and alcoholic beverages have been carried out [6–8]. An important factor in wine-making is the grape microbial ecosystem which is highly composed of diverse microorganisms, including yeasts, bacteria and fungi. The grape microbial ecosystem can result in a spontaneous wine fermentation which is a complicated process where many indigenous microorganisms may occur, including both *Saccharomyces cerevisiae* and non-*Saccharomyces* yeasts, lactic acid or acetic acid bacteria and fungi [9]. The final quality of wine is highly dependent on wine yeasts' metabolic activities; therefore, many wineries use selected yeast strains targeting to control fermentation and achieve high-quality wines [10].

Considering ethanol industrial production, yeasts represent the most significant microbial group providing high yields of productivity and also high tolerance and resistance to antagonistic microflora [11]. Biotechnological applications upon ethanol production mainly target bioprocess optimization technologies and application of selected/"optimized" strains. Regarding bioprocess optimization, research mainly focuses on the technological optimization of bioprocesses, such as the effect of the agitation/aeration upon the process, the "very high-gravity" application of the fermentation (*viz.* "the preparation and fermentation to completion of mashes containing 27 g or more dissolved solids per 100 g mash"), the utilization of "new" fermentation feedstocks (e.g., algae, crude glycerol, food-processing wastes, hemicellulose hydrolysates) and the development of innovative bioreactor designs [12–15]. On the other hand, biotechnological applications on bioethanol production relay on the yeast's "system biology" studies, including mostly screening and biochemical characterization of "novel" yeast strains and genetic engineering/random mutation/adaptive laboratory evolution strategies in order to "construct" "new" robust high-performing yeast strains with specified and desired fermentation characteristics [12,13,16], with a major topic currently developed being linked towards the isolation of novel robust high-performing microorganisms presenting technological interest [12,13,17]. Likewise, wine-making science is highly interested in the discovery of novel high-performing microorganisms aiming to be applied for wine-making providing quality wines with unique sensory characters [18,19].

It is widely known that the ordinary practice in wine-making facilities regarding the utilization of yeasts performing alcoholic fermentation refers to the employment of commercial "optimized" *Saccharomyces cerevisiae* strains, which results in the production of highly "homogenous" types of wines [20,21]. On the other hand, the use of "indigenous" ("local", *viz.* isolated from various "wine-type" products or various origins) strains amenable to be used as starters represents a potentially very useful tool to safeguard several types of sensory characteristics from specific regions, while it also demonstrates the potential of the biodiversity on the mentioned process [21–23]. Another approach demonstrates the use of *S. cerevisiae* yeast strains for aerobic fermentations targeting to obtain reduced ethanol yields in wines with acceptable volatile composition [24]. Moreover, in several cases, newly isolated (*viz.* non-commercial) *Saccharomyces cerevisiae* strains can present even better biotechnological properties compared to the commercial strains (i.e., higher vitality, killer factor, resistance to high concentrations of sugar, ethanol and SO_2, increased trehalose and glycogen cellular content), which could enhance their performance during the grape must fermentation process [21,22,25]. Therefore, according to the previously mentioned analysis, and in regard of the need in high-performing novel yeast strains, while avoiding the long term and strain-stressful techniques of genetic engineering or the controversial techniques of mutation, the present study was conducted with the aim to assess the biochemical and technological properties of three commercial and four wild-type yeasts concerning comparison of strains' performance and productivity. Subsequently, two *S. cerevisiae* strains that were chosen based on their ethanol production capabilities (one commercial and one wild-type "novel" strain not previously systematically studied regarding its fermentation potential), were applied in a wine-making process, and applied for the fermentation of white and red musts from indigenous *Vitis vinifera* Greek grape varieties of the Santorini region. The strains were monitored via molecular analysis throughout fermentation targeting the evaluation of strains' adaptation, performance and productivity. Technological considerations regarding the performances of the yeast strains were critically considered and assessed.

2. Materials and Methods

2.1. Yeast Strains and Culture Conditions

Commercial- and wild-type yeast strains of the species *Saccharomyces cerevisiae* were selected and studied in synthetic media targeting application of selected yeast strains as starter cultures for wine-making of *Vitis vinifera* Greek grape varieties from the Santorini region. Specifically, the new isolated strains LMBF Y-10, Y-25, Y-35 and Y-54, originated from the culture collection of the Laboratory of Microbiology and Biotechnology of Foods (Department of Food Science and Human Nutrition, Agricultural University of Athens, Greece), originated from various types of food-stuffs (*viz.* commercial wine, grape musts and grapes) and not having been previously systematically studied regarding their fermentation potential, were used as cell factories in the present study. The commercial strains: Symphony, Cross X and Passion Fruit, were also used for comparison reasons. Strains were regenerated in YPDA slants (20 g/L glucose, 10 g/L yeast extract, 10 g/L peptone and 25 g/L agar) every 4 months to maintain the yeast viability [26]. Pre-cultures were performed in 250 mL non-baffled conical flasks filled with 50 mL of medium (YPD medium: 20 g/L glucose, 10 g/L yeast extract, 10 g/L peptone, pH $\approx$ 3.5) previously autoclaved at $T = 115\,^{\circ}C/1.5$ atm for 15 min.

Aerobic experiments were performed in 250 mL (filled with 50 ± 1 mL medium) agitated flasks (use of a ZHWY-211B Rocking Incubator) at 180 ± 5 rpm in which commercial glucose (Hellenic Industry of Sugar SA, Orestiada, Greece) was used as carbon source. Glucose-based media presented the following salt composition in g/L: KH_2PO_4 7.0; Na_2HPO_4 2.5; $MgSO_4 \cdot 7H_2O$ 1.5; $CaCl_2 \cdot 2H_2O$ 0.15; $FeCl_3 \cdot 6H_2O$ 0.15; $ZnSO_4 \cdot 7H_2O$ 0.02; $MnSO_4 \cdot H_2O$ 0.06 [26]. The nitrogen sources used were peptone and yeast extract (at concentrations 3.0 and 3.0 g/L respectively), while initial glucose (Glc_0) was added to *ca* 100 g/L and *ca* 200 g/L. The pH value was adjusted to 3.5 ± 0.2, while incubation temperature $T = 30 \pm 1\,^{\circ}C$ was employed. Flasks were aseptically inoculated with 1 mL of yeast preculture (thus, a 2% *v/v* inoculation occurred). In the elaborated kinetics, each flask constituted the experimental point for the relevant experiments.

Microaerophilic/anaerobic trials were performed during wine-making experiments in static cultures [14]; grape must from two different *Vitis vinifera* Greek grape varieties of Santorini indigenous, Assyrtiko and Mavrotragano, were used for white and red wine production, respectively. The initial concentration of total reducing sugars (TS) for Assyrtiko must was 221.5 g/L, pH was =3.3 and total acidity was =6.0 g/L (expressed as tartaric acid) while for Mavrotragano must the respective values were 213.5 g/L, 3.4 and 5.8 g/L. In addition, SpringFerm™ (Fermentis, France) nutrients (0.2 g/L) were applied in each fermentation batch targeting enhanced fermentation rates [27]. Subsequently, alcoholic fermentations were carried out by applying the strains LMBF Y-54 and Passion Fruit under static cultures, ensuring microaerophilic (initially) and nearly anaerobic (after the initial steps of the fermentation and the CO_2 accumulation in the must and the bottle) conditions [12,14] in 1.0 L Duran bottles containing 500 mL of grape must. Alcoholic fermentations were performed at $T = 18\,^{\circ}C$ constant temperature incubating an initial yeast population of 5×10^6 cfu per 500 mL of must (*viz.* 5 mL of exponential pre-culture in 500 mL of must). During fermentation, samples were obtained aseptically at 24-h intervals for further analysis, while sugar depletion signified the end of the wine-making process [7]. At the end of fermentation, each wine sample was treated with 0.08 g/L potassium metabisulfite (Fluka, Switzerland) and placed for 5 days in refrigerator storage ($T = 4\,^\circ$C) targeting clarification. Finally, wine samples were transferred in sterile clean bottles and evaluated regarding their sensory attributes.

2.2. Analytical Methods

The whole content of the 250 mL flasks (*viz.* 50 ± 1 mL) or 5–10 mL of the content of static Duran bottles was collected at predetermined intervals to correctly assess the kinetic studies. Yeast biomass (for the shake-flask trials) was harvested by centrifugation at 9055 g, for 10 min at $T = 21 \pm 1\,^\circ$C (Suprafuge, Heraeus Sepatech), washed with distilled water and re-centrifuged again. Yeast cell concentration was determined gravimetrically by placement of wet biomass at $T = 95 \pm 5\,^\circ$C until constant weight (usually within 24 ± 2 h) and was expressed as dry cell weight (DCW) (X, g/L). Intra-cellular polysaccharides (IPS, expressed as % w/w in DCW) were measured by collecting 0.05 g of dry biomass weighted in a precision scale of four decimal digits (Ker new 420-3NM) and placed in McCartney glass containers. The dried yeast cell mass was hydrolyzed using 10 mL of 2.5 M HCl at $T = 80\,^\circ$C for 30 min. The whole was neutralized to pH = 7.0 with 2.5 M NaOH and the volume was adjusted to 100 mL. Samples containing total sugars were then filtered (through No. 2 Whatman filters) and subjected to the DNS assay [28,29].

Ethanol, glucose, fructose and glycerol were quantified through high performance liquid chromatography (HPLC) analysis carried out in a Waters Association 600E apparatus equipped with a RI detector (Waters 410). A Rezex ROA-Organic Acid H^+ column (300 mm $\times$ 7.8 mm) (Phenomenex, Torrance, California, USA) was used for the separation of the compounds. The mobile phase was H_2SO_4 at 0.005 M. The column temperature was set at $T = 40\,^\circ$C with a flow rate of 0.5 mL/min. The injection volume was 20 μL [26]. For quantitative analysis, standard solutions of the compounds were prepared in pure water (Milli-Q, Merk) at various concentrations. The range of concentration used to build the calibration curve for each compound analyzed (*viz.* glucose, fructose, glycerol, ethanol) was between 0.0 and 20.0 g/L.

In the shake-flask experiments, dissolved oxygen concentration (DOC, in % v/v) was off-line determined using a selective electrode (OXI 96, B-SET, Germany) according to previously published procedure [29]. Before harvesting, the shaker was stopped and the probe was placed into the flask, after which the shaker was again switched on and the measurement was taken after DOC equilibration (within *ca* 10 min). In all experiments, and irrespective of the initial concentration of glucose set in the medium or the used strain, DOC values were for all culture phases $\geq 20\%\ v/v$, indicating that in the shake-flask experiments performed, full aerobic conditions were maintained [26,29].

At the end of the microaerophilic/anaerobic trials, the volatiles of produced wines were determined using Gas Chromatography/Mass Spectrometry with Headspace Solid-Phase Micro-Extraction sampling (HS-SPME/GC–MS) [30]. Specifically, 2 mL of each wine sample was mixed with 7.5 mL of deionized water, 2 g of sodium chloride (dried at >100 $^\circ$C prior to weighing), and 500 μL of 1,4-dioxane solution (1000 mg/L) as IS and placed in 20 mL glass vials. Each amber headspace vial was sealed with a cap (Teflon-lined septum) and placed in a water bath (40 $^\circ$C) under constant stirring. Each vial was sealed with a silicone septum and placed in the water bath for 5 min reaching 40 $^\circ$C under constant stirring (250 rpm). Subsequently, the SPME fibre (SPME; fibre DVB/CAR/PDMS, 2 cm; Sigma–Aldrich, Germany) was exposed to the headspace for 30 min at 40 $^\circ$C under constant stirring. Then the fibre containing the absorbed volatiles was exposed in the injection port of the chromatograph (GC), in a split mode (split ratio 1/10), at 240 $^\circ$C for 5 min. The chromatograph (GCMS-QP2010 Ultra, Shimadzu Inc., Kyoto, Japan) was equipped with a DB-Wax capillary column (30 m, 0.25 μm film thickness, Agilent, Santa Clara, California,

USA) and was applied as carrier gas. The mass spectrometer operated in an electron ionization mode, at an ionization energy of 70 eV and 4 at 0–300 m/z mass scan range. The source and interface temperatures were set at 200 °C and 240 °C, respectively.

For the identification of volatiles, the following were compared: (i) retention index (RI) based on the homologous series of C8-C24 n-alkanes with those of available authentic compounds and those available in the NIST14 library (NIST, Gaithersburg, Maryland, USA), and (ii) AMDIS software (v. 2.65 build 116.66) was employed, based on retention time and mass spectra, with a parallel use of NIST library as confirmation. The analysis of volatile profile of wines was based on the absolute values of the peak area of each compound, and they were expressed as a percentage of the total peak area [(compound peak area/sum of peak areas) × 100].

2.3. Molecular Identification of Yeast Cells

For the non-aseptic trials performed in the Duran bottles, samples were analyzed in the beginning, the middle and in the end of the fermentation process, for yeast identification at strain level. One milliliter of each sample was aseptically transferred in 9 mL of sterile $\frac{1}{4}$ Ringer's solution and serially diluted in the same diluent. Each dilution with 0.1 mL was spread at yeast extract peptone dextrose (YPD) agar plates and inoculated at $T = 28$ °C for 48 h. A percentage of 20% of colonies was picked from the appropriate dilution and transferred at 10 mL of YPD broth according to the representative sampling scheme of Harrigan and McCance, which is still applied over time [31,32]. Total DNA was extracted from each culture [33]. Briefly, one milliliter of overnight culture was centrifuged (14,000 rpm) for 5 min at 4 °C then the pellet was resuspended in 0.5 mL buffer solution (1 M sorbitol, 0.1 M EDTA, pH 7.5) containing lyticase (2.5 U/mL) (lyticase from *Arthrobacter luteus*, Sigma–Aldrich, Germany) for yeast cell lysis. After centrifugation, the pellet was resuspended in 0.5 mL of buffer (50 mM Tris–HCl, 20 mM EDTA, pH 7.4) and incubated for 30 min at 65 °C with 50 μL of 10% SDS solution. Then, each sample was mixed with 0.2 mL potassium acetate (5M) (Merck) for 30 min and centrifuged (14,000 rpm) for 10 min at 4 °C. The supernatant was precipitated with 1 mL ice-cold isopropanol (Applichem) and then centrifuged (14,000 rpm) for 10 min at 4 °C. The final pellet was dried and resuspended in 50 μL sterile ddH20. The quantification and quality control of DNA extract was performed by means of a nanophotometer (Implen, Germany) at wavelengths of 260, 280 and 230 nm. Pure cultures of the inoculated yeast strains, *S. cerevisiae* LMBF Y-54 and *S. cerevisiae* Passion Fruit were used as controls strains for the molecular characterization.

A rep-PCR method was applied in order to unambiguously discriminate genotypes of different species and reach strain level. PCR-fingerprinting was performed with the primers $(GTG)_5$ [34]. The reaction involved initial denaturation at 94 °C for 4 min, followed by 35 cycles of the series of 94 °C for 15 s, 55 °C for 45 s, and 72 °C for 90 s, with a final cycle at 72 °C for 15 min. Amplification was carried out in a thermocycler (Applied Biosystems, Bedford, MA, USA). PCR products were separated by electrophoresis in a 1.3% agarose gel, 1 × TAE (40 mM Tris-acetate, 1 mM EDTA, pH 8.2) buffer at 80 V for 120 min, then were stained with ethidium bromide solution (1%) and finally digitalized under UV light (GelDoc system, Bio-Rad, Hercules, CA, USA). The similarity among digitalized profiles was calculated using the Pearson correlation and an average linkage (UPGMA) dendrogram was derived from the profiles. The 1 Kb (Invitrogen, Waltham, Massachusetts, USA) molecular weight marker was used to compare the sizes of the bands. Cluster analysis was performed using the Bionumerics software version 6.1 (Applied Maths, Sint-Martens-Latem, Belgium).

2.4. Sensory Assessment

Samples were evaluated by a group of 12 trained panelists with previous experience in wine sensory analysis [35–37]. The tests were conducted from 11:00 a.m. to 13:00 a.m. in individual booths. In more detail, each sample was served in a completely randomized presentation order and was evaluated in triplicate by each panelist. The judges were provided with 30 mL of each sample in standard wine glasses, marked with three-digit random numbers, at room temperature (T = 18–20 °C). The following olfactory attributes were evaluated: floral, fresh fruits, dry fruits, reduction, odor of oxidation, aroma intensity and overall aroma quality using a 5-point scale. Zero intensity of the attributes was marked on the left end of the scale whereas maximum intensity was marked on the right.

2.5. Data Analysis

Each experimental point of all the kinetics presented in the tables and figures is the mean value of two independent determinations, while the standard error (SE) for most experimental points was ≤17%. Data were plotted using Kaleidagraph 4.0 Version 2005 showing the mean values with the standard error mean.

Regarding statistical analysis of volatile contents and sensory results, analysis of variance (ANOVA) was performed using Statistica V.7 (Statsoft Inc., Tulsa, OK, USA) to determine whether the mean values differed between samples. Tukey's HSD was used as comparison tests when samples were significantly different after ANOVA ($p < 0.05$).

2.6. Nomenclature

X: Dry cell weight (DCW; total microbial mass) (g/L); Glc: Glucose (g/L); Fru: Fructose (g/L); TS: Total sugars (g/L); EtOH: ethanol (g/L); Glyc: Glycerol (g/L); IPS: intra-cellular polysaccharides (g/L); t: fermentation time (h); $Y_{EtOH/Glc}$: yield of ethanol production with respect to glucose consumed (g/g); $Y_{EtOH/TS}$: yield of ethanol production with respect to total sugars consumed (g/g); $Y_{IPS/X}$: intra-cellular polysaccharides in DCW (%, w/w); indices 0, max and cons show the initial and maximum and the consumed quantity of the elements in the experiments carried out; DOC: dissolved oxygen concentration (%, v/v).

3. Results and Discussion

3.1. Commercial- and Wild-Type Yeast Strains Perfrormemces in Synthetic Medium/Aerobic Experiments and Yeast Selection

Ethanol production in wine is based on the ability of yeast strains to catabolize six-carbon molecules present in must into ethanol, a two-carbon compound [11]. The bio processes where yeasts convert glucose to ethanol are known as the glycolytic pathway followed by ethanol fermentation [38]. Fermentation (*viz.* "anaerobic"-type transformation of glucose into ethanol) may occur despite the presence of O_2 in the culture medium in significant concentrations when the initial concentration of the employed carbohydrate (i.e., glucose and/or fructose and/or sucrose) is higher than a "critical" value [39]. In fact, for a remarkable number of yeast species (the so-called "conventional" yeasts), even with the significant presence of oxygen in the fermentation medium (i.e., DOC values $\geq 20\%\ v/v$ and in some cases $>50\%\ v/v$), if sugar concentration is higher than a critical (and in many instances not very high) concentration (e.g., *ca* 9 g/L or even lower), respiration is impossible; furthermore, despite the oxygen saturation conditions imposed, the microorganism shifts its metabolism completely towards the fermentative pathway and the subsequent biosynthesis and accumulation of ethanol into the medium. This phenomenon is known as the "Crabtree effect" (named after the English biochemist Herbert Grace Crabtree), Pasteur contrary effect, or as catabolic repression by glucose [12,40]. Specifically, for the mentioned type of yeasts (the "Crabtree"-positive ones), in somehow elevated sugar concentrations imposed in the culture medium, the mitochondria degenerate, the proportion of cellular sterols and fatty acids decrease and both the enzymes involved in the oxidative part of the metabolism (namely the Krebs cycle and the oxidative phosphorylation chain) and the constituents of respiratory chains are subjected to catabolite repression, leading to the elaboration of the ethanol fermentation despite oxygen-sufficient culture conditions [12,40,41].

All yeast strains tested in shake-flask trials converted glucose into ethanol (EtOH) and dry yeast biomass (X), despite aerobic conditions imposed into the medium (in all trials and irrespective of the strain, the initial glucose concentration and the culture time, the DOC was always $\geq 20\%\ v/v$, indicating sufficient aerobic conditions into the shake-flask environment [7,29,41], and the obtained results are depicted in Table 1(a) ($Glc_0 \approx 100$ g/L) and Table 1(b) ($Glc_0 \approx 200$ g/L). In all trials performed, non-negligible quantities of glucose were relatively rapidly consumed and converted into DCW and (mostly) ethanol; therefore, all selected yeast strains were referenced as positive to "Crabtree effect" in accordance with the literature [12], that by far considers the *S. cerevisiae* species as the most "classic" yeast strain in which this phenomenon occurs [12,40–44]. It is interesting to indicate that in the trials with $Glc_0 \approx 100$ g/L, with the exception of the wild-type non-commercial LMBF Y-54 strain that consumed the majority of glucose quantity that was found in the growth medium, all other LMBF Y- strains did not consume all available sugar quantity of the medium, whereas further incubation did not lead to increased glucose assimilation, but to degradation (oxidation) of ethanol, that in most cases was not accompanied by a DCW concentration increase (thus no diauxic growth occurred). Concerning X production of the wild-type non-commercial yeasts growing on glucose at $Glc_0 \approx 100$ g/L, by far the highest biomass producer was the strain LMBF Y-54 (X = 7.1 g/L; see Table 1). The X_{max} value of this strain was recorded at t = 121 h, being =10.5 g/L (kinetics non shown; this value was obtained after ethanol oxidation that occurred when glucose had been depleted from the medium). In fact, it is the so-called "ethanol make–accumulate–consume" phenomenon. This phenomenon (in fact a metabolic adaptation "strategy" developed by *Saccharomyces* strains under aerobic conditions) relies on the evolution of *Saccharomyces* cultures against their competitors, as ethanol is toxic to most other microbes. Therefore, it is considered in a (non-aseptic) sugar- and oxygen-rich environment that *Saccharomyces* strains eliminate their competitors by producing ethanol, but in a next fermentation step, they consume the previously generated ethanol, in order to create further DCW or maintain the already existing one (consumption of ethanol for energy of maintenance requirements) [12,41]. Alcohol dehydrogenase (Adh) catalyzes the acetaldehyde-to-ethanol conversion in both directions. Genes ADH1 (expressed constitutively) and ADH2 (expressed only when the internal sugar concentration drops) encode cytoplasmic Adh activity [12,40,41,43]. Concerning the commercial strains: Passion Fruit, Symphony and Cross X; all of these microorganisms converted glucose rapidly into ethanol with the highest quantities of glucose (*ca* 90% *w/w* of the available sugar) having been assimilated within the first 24–36 h after inoculation (Table 1(a)). In the trials with Glc_0 adjusted to around 100 g/L, the conversion yield $Y_{EtOH/Glc}$ presented variable values, with some wild-type strains (i.e., LMBF Y-25 and LMBF Y-35) presenting excellent values ($Y_{EtOH/Glc} \geq 0.45$ g/g, *viz.* $\geq 88\%$ of the maximum theoretical yield that is =0.51 g/g [12,13,41,43]). On the other hand, in similar types of experiments performed with other wild-type ("novel") or commercial-type *Saccharomyces cerevisiae* strains cultured in shake-flask trials on previously pasteurized natural grape musts under aerobic conditions, the conversion achieved yields $Y_{EtOH/TS}$ ranged between 0.28 and 0.40 g/g [24], that were values slightly or somehow lower compared to those obtained in the present study (Table 1). This indicates the potential of the employed strains in the current investigation regarding the conversion of glucose into ethanol under aerobic conditions. The Crabtree-effect (*viz.* the production of ethanol from glucose or other sugar fermentation under full aerobic conditions) has been studied in other conventional but non-*Saccharomyces* yeast strains, and the conversion yield $Y_{EtOH/TS}$ has been revealed to be 0.10 g/g for *Candida diddensiae*, 0.27 g/g for *Candida tropicalis* and 0.28 g/g for *Candida zemplinina* [44,45].

Table 1. Experimental results originated from kinetics of *S. cerevisiae* wild strains (LMBF Y-10, Y-25, Y-35 Y-54) and commercial strains (Symphony, Passion Fruit, Cross X) growing in shake-flask glucose-based synthetic media at initial glucose concentration $\approx$ 100 g/L (a) and $\approx$200 g/L (b) when the maximum concentration of ethanol ($EtOH_{max}$) was achieved. Representations of dry biomass (X, g/L), glucose consumed (Glc_{cons}, g/L), ethanol produced (EtOH, g/L) and ethanol conversion yield per sugar consumed ($Y_{EtOH/Glc}$, g/g). Each experimental point is the mean value of 2 independent measurements (SE for most experimental points $\leq$ 17%).

(a).

Strain	Time (h)	Glc_{cons} (g/L)	X (g/L)	EtOH (g/L)	Glyc (g/L)	$Y_{EtOH/Glc}$ (g/g)
S. cerevisiae LMBF Υ-35	72.0	66.9	2.5	30.1	1.4	0.45
S. cerevisiae LMBF Υ-25	51.0	57.0	2.9	26.7	2.7	0.47
S. cerevisiae LMBF Υ-54	52.0	112.2	7.1	41.0	1.0	0.37
S. cerevisiae LMBF Υ-10	48.0	78.0	3.2	30.4	0.9	0.39
S. cerevisiae Symphony	36.0	93.0	5.9	32.6	0.5	0.35
S. cerevisiae Cross X	34.0	89.1	4.0	40.0	1.9	0.45
S. cerevisiae Passion Fruit	24.0	92.9	5.1	43.1	1.6	0.46

(b).

Strain	Time (h)	Glc_{cons} (g/L)	X (g/L)	EtOH (g/L)	Glyc (g/L)	$Y_{EtOH/Glc}$ (g/g)
S. cerevisiae LMBF Υ-35	95.0	166.8	2.7	40.9	6.4	0.25
S. cerevisiae LMBF Υ-25	68.0	165.5	2.6	47.4	5.2	0.29
S. cerevisiae LMBF Υ-54	168.0	179.3	6.1	68.0	6.3	0.38
S. cerevisiae LMBF Υ-10	72.0	117.1	3.9	40.0	3.1	0.34
S. cerevisiae Symphony	48.0	214.6	8.1	60.9	1.4	0.28
S. cerevisiae Cross X	72.0	178.6	6.3	82.2	4.4	0.46
S. cerevisiae Passion Fruit	76.0	200.5	6.4	91.1	3.8	0.45

Adaptation to higher Glc_0 concentrations imposed in the medium (*viz. ca* 200 g/L) resulted in significant quantities of assimilated glucose irrespective of the strains (commercial or wild-type ones) implicated as microbial cell factories of the process (Table 1(b)). Apart from the strain LMBF Υ-10, in all other cases, consumed Glc quantities $\geq$ 160 g/L were recorded, suggesting the potential of the employed strains towards the so-called "very-high-gravity" alcoholic fermentation process [12,13,26]. It is also interesting to indicate that despite the significant consumption of glucose that was reported for many of the trials, the conversion yield $Y_{EtOH/Glc}$ was in a number of cases lower than the respective one reported for the experiments with $Glc_0 \approx$ 100 g/L (see Table 1). This has also been reported in other cases in which alcoholic fermentations had been performed in media in which all other culture parameters and components (i.e., initial nitrogen) remained constant and only glucose concentration increased [42,43], exactly as in the current investigation, demonstrating that potentially the conversion yield $Y_{EtOH/Glc}$ is negatively correlated with the increase in carbon excess in the medium (increment of Glc_0 concentration in the medium with the initial nitrogen remaining constant, evidently increases the initial C/N and, thus, the carbon excess in the medium [28,29,42]) On the other hand, given the fact that at the increasing Glc_0 concentrations, the initial concentration of glucose imposed on the medium is rather high ($\approx$200 g/L), besides potential application of nitrogen-limited conditions, other parameters such as oxidative stress and osmotic stress, that are both linked with high Glc_0 concentration media [46], may have negative impact upon the decrease in the $Y_{EtOH/Glc}$ values with a glucose concentration increment in the medium. In fact, as far as the ethanol fermentation under high Glc_0 concentrations by *Saccharomyces cerevisiae* strains is concerned, to alleviate the various environmental stresses imposed by the increased initial concentrations of sugar, trials in media rich in Mg^{2+} and organic nitrogen (i.e., peptone) seem of importance in order to achieve high ethanol concentrations and yields [47].

In all cases and in all fermentations performed, the intra-cellular polysaccharides were quantified, and $Y_{IPS/X}$ values ranging between 4.2–11.0% w/w were recorded for all strains, all fermentation periods and all culture conditions. The quantity of IPS per DCW (%, w/w) did not increase as the Glc_0 concentration in the medium increased, suggesting that under the present culture conditions and for the given strains tested, there was not any shift towards the synthesis of endopolysaccharides due to the increase in carbon excess in the medium [29,42]. This result does not comply with previously published data; in fact, in other studies in which yeasts showed both Crabtree-negative (i.e., *Rhodosporidium toruloides, Yarrowia lipolytica, Cryptococcus curvatus*) and Crabtree-positive (i.e., *Saccharomyces cerevisiae*) effect, the utilization of relatively high initial concentrations of sugar (i.e., ≥ 20 g/L) and the employment of carbon excess conditions (*viz.* the utilization of media in which the initial ratio of C/N in somehow high, i.e., ≥ 40–50 moles/moles) seem necessary in order to increase the quantity of IPS per unit of DCW [29,48–50], with $Y_{IPS/X}$ values in some cases being $\geq 55\%$ w/w. It is finally noted that similar types of experiments performed by a wide range of wild-type and commercial *Saccharomyces cerevisiae* strains cultured under aerobic conditions on YPD medium resulted in the synthesis of intra-cellular polysaccharides (mostly trehalose and glycogen) with $Y_{IPS/X}$ values of 9.0–18.0% w/w (values somehow similar with those reported in the current investigation) [25].

3.2. Wine-Making of Selected Strains/Microaerophilic/Anaerobic Experiments

As mentioned above, "novel" (*viz.* newly isolated and, therefore, non-commercial) *Saccharomyces cerevisiae* strains can present in some instances equal (or even better) biotechnological properties compared to the commercial strains (i.e., higher vitality, resistance to high concentrations of sugar, high potential of sugar uptake, high ethanol production, equal of increased content of intra-cellular polysaccharides), while these properties, linked also to the potential for the production of wines with not "homogenous" organoleptic characters, can present significant importance for the wine-making industries. One of the "novel" strains of the current study (*viz.* the strain LMBF Y54) presented very interesting technological characteristics regarding its sugar consumption and alcohol production properties. In fact, this wild-type strain was the best amongst the "novel" studied strains regarding its potential upon the previously mentioned characteristics (see Table 1). For this reason, this wild-type strain was chosen for further trials. Equally, one of the commercial strains that presented the best previously mentioned technological characteristics (*viz.* the strain Passion Fruit) was also selected (see Table 1). Both strains were further studied in wine-making conditions. These strains were cultivated in static conditions that were performed under microaerophilic (initially) and self-generated anaerobic (after the initial steps of the fermentation and the subsequent CO_2 production in the must and the flask) conditions (see also [14]) in 1.0 L Duran bottles in wine-making conditions using grape musts of the varieties Assyrtiko and Mavrotragano, and the achieved results are depicted in Table 2.

The wild yeast strain, *Saccharomyces cerevisiae* LMBF Y-54, achieved maximum sugar consumption in Mavrotragano must after *ca* 13 days of fermentation (Table 2). Kinetic analysis for both fermented musts demonstrated that glucose and fructose were assimilated with almost equivalent assimilation rates (see example in Assyrtiko fermentation in Figure 1a), that is a not frequently a phenomenon seen in similar types of fermentations [51].

Table 2. Experimental results originated from kinetics of *S. cerevisiae* LMBF Y-54 and Passion Fruit growing on Assyrtiko and Mavrotragano must at points when a maximum concentration of ethanol (EtOH$_{max}$) was achieved. Representations of residual glucose (Glc, g/L), residual fructose (Fru, g/L), total sugars consumed (TS$_{cons}$, g/L), ethanol produced (EtOH, g/L) and ethanol conversion yield per total sugars consumed ($Y_{EtOH/Glc}$, g/g). Each experimental point is the mean value of 2 independent measurements (SE for most experimental points $\leq$ 17%).

	Time (h)	Glc (g/L)	Fru (g/L)	TS$_{cons}$ (g/L)	EtOH (g/L)	Glyc (g/L)	$Y_{EtOH/TS}$ (g/g)
Assyrtiko must							
Passion Fruit; TS$_0$ $\approx$ 221.5 g/L	310	3.7	8.0	209.8	102.7	2.2	0.49
LMBF Y-54; TS$_0$ $\approx$ 221.5 g/L	310	3.9	3.7	213.9	106.3	3.5	0.50
Mavrotragano must							
Passion Fruit; TS$_0$ $\approx$ 213.5 g/L	286	1.8	5.8	205.9	97.8	5.2	0.47
LMBF Y-54; TS$_0$ $\approx$ 213.5 g/L	240	2.2	1.9	209.4	99.7	5.9	0.48

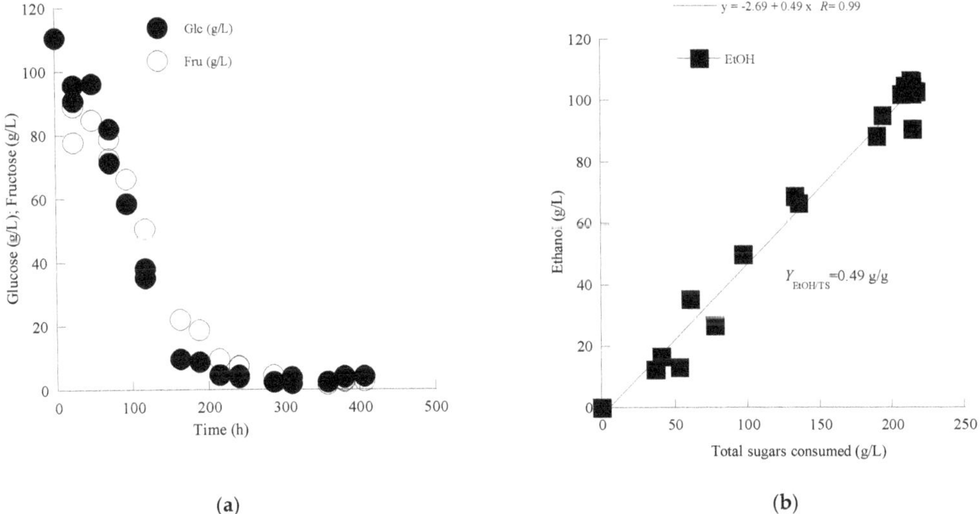

(a)

(b)

Figure 1. (a) Kinetics of glucose (Glu) and fructose (Fru) of the *S. cerevisiae* LMBF Y-54 strain in must from the grape variety Assyrtiko. Each experimental point is the mean value of 2 independent measurements (SE for most experimental points $\leq$ 17%). (b) Representation of ethanol produced vs. total sugars consumed of *S. cerevisiae* strain LMBF Y-54 in must from the grape variety Assyrtiko. for the whole set of data. Each experimental point is the mean value of 2 independent measurements (SE for most experimental points $\leq$ 17%).

Moreover, Mavrotragano fermentation was accompanied by the slightly lower EtOH$_{max}$ concentration achieved compared to Assyrtiko grape must (Table 2). This outcome could be attributed to the initial higher sugar content of Assyrtiko grape must. The global conversion yield of ethanol produced per unit of total sugar consumed ($Y_{EtOH/TS}$, in g/g), calculated by linear regression of the concentration of EtOH produced as a function of the quantity of TS consumed for the whole set of experimental data (see Figure 1b, for Assyrtiko fermentation) demonstrates very similar values with those reported in Table 2, that were calculated on the basis of the experimental point showing the highest ethanol concentration achieved and the respective remaining total sugar concentration value (*viz.* $Y_{EtOH/TS} = \frac{EtOH_{max}}{TS_0 - TS_t}$, where TS$_t$

is the remaining total sugar concentration in the must when the EtOH$_{max}$ concentration is achieved; see Table 2).

A well-established sugar consumption capability is a crucial step for wine fermentation as high sugar contents can cause sluggish or stuck fermentation rates. On the other hand, the final sugar content can cause undesirable sweetness in wines or even unwanted fermentation during storage. As a result, low final sugar content and high ethanol yield are mostly wanted in wine-making [7]. From all the above-mentioned analysis and taking into consideration the results achieved with the wild-type LMBF Y-54 strain (see Table 2), the high potential of this strain towards the production of wines with desired characteristics was demonstrated. Regarding glycerol production, a gradual increase was noted achieving a maximum value of 5.9–6.1 g/L by the end of fermentation for the case of Mavrotragano must, whereas for the Assyrtiko must, the respective concentrations at the end of growth were 3.5–3.8 g/L.

The commercial yeast strain, *Saccharomyces cerevisiae* Passion Fruit, presented similar experimental results with the wild-type LMBF Y-54 strain. In both types of grape musts, fermentations were carried out (see Table 2). On the other hand, in disagreement with the results for the LMBF Y-54 strain, the commercial strain demonstrated a higher assimilation rate of glucose compared to fructose (see example in Assyrtiko fermentation in Figure 2a), in accordance with the results reported for several wild-type or commercial yeast strains in wine-making conditions [51–53]. As previously mentioned, the global conversion yields $Y_{EtOH/TS}$ (see the yield for the case of Assyrtiko fermentation represented in Figure 2b) were almost the same as those that were calculated based on the experimental point showing the highest ethanol concentration achieved (see as previously: $Y_{EtOH/TS} = \frac{EtOH_{max}}{TS_0 - TS_t}$; Table 2). Finally, glycerol production showed an upward trend, with a maximum concentration of 5.2–5.5 g/L after 406 h of fermentation for the case of Mavrotragano must, whereas for the Assyrtiko must, the respective values at the end of growth were 2.2–2.6 g/L.

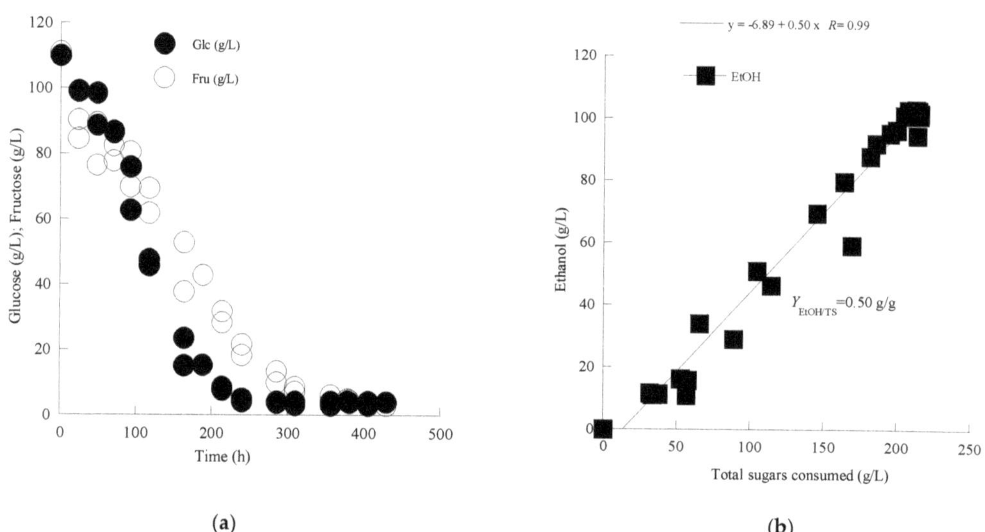

Figure 2. (**a**) Kinetics of glucose (Glu) and fructose (Fru) of the *S. cerevisiae* Passion Fruit strain in must from the grape variety Assyrtiko. Each experimental point is the mean value of 2 independent measurements (SE for most experimental points ≤ 17%). (**b**) Representation of ethanol produced vs. total sugars consumed of the *S. cerevisiae* Passion Fruit strain in must from the grape variety Assyrtiko. for the whole set of data. Each experimental point is the mean value of 2 independent measurements (SE for most experimental points ≤ 17%).

According to molecular typing with rep-PCR, all isolates showed the same finger-printing with the *Saccharomyces cerevisiae* inoculated strain in Assyrtiko must. The cluster analysis of the data set allowed the recognition of two main groups, branching at a similarity value of 90.0% in the case of the LMBF Y-54 strain and 87.9% in the case of the Passion Fruit strain (Figures 3 and 4). Likewise, PCR fingerprinting of *Saccharomyces cerevisiae* isolates with (GTG)5 in Mavrotragano must fermentation indeed yielded low genetic polymorphism for both inoculated strains. The obtained molecular patterns for the 30 isolates created two main groups, sharing a similarity value of 96.8% when LMBF Y-54 was inoculated and 95.8% when Passion Fruit was inoculated (Figures 5 and 6).

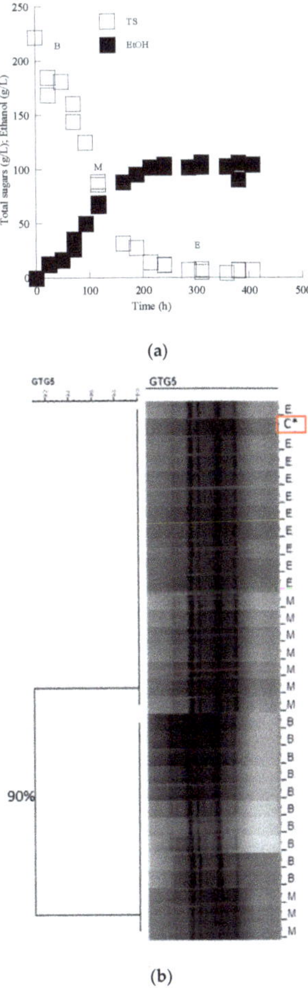

(a)

(b)

Figure 3. (**a**) Sugar and ethanol evolution of the *S. cerevisiae* LMBF Y-54 strain in must from the grape variety Assyrtiko. Each experimental point is the mean value of 2 independent measurements (SE for most experimental points ≤ 17%). (**b**) Dendrogram generated after cluster analysis of the digitized (GTG)5 PCR fingerprints of the 31 isolates from the beginning (B), middle (M) and end (E) of the fermentation kinetics in the grape must media. The control condition (C*) is the relative rep-PCR profile of the pure culture of the *S. cerevisiae* LMBF Y-54 strain. The dendrogram was constructed using the unweighted pair-group method using arithmetic averages with correlation levels expressed as percentage values of the Pearson correlation coefficient.

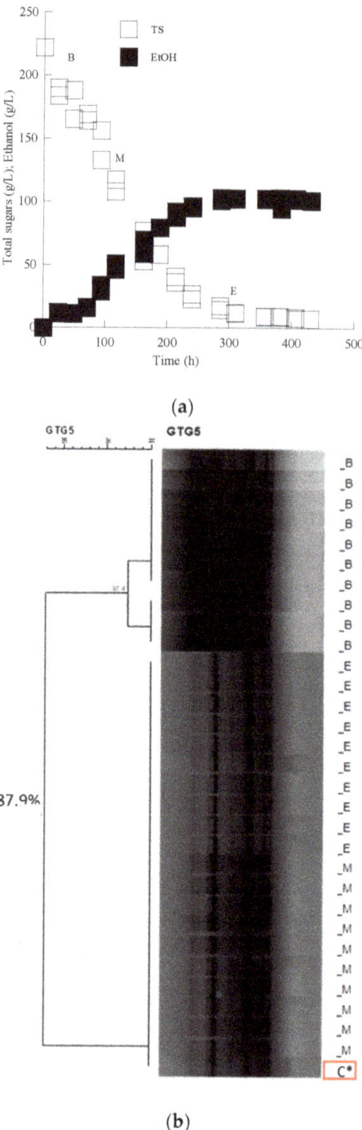

(a)

(b)

Figure 4. (**a**) Sugar and ethanol evolution of the *S. cerevisiae* Passion Fruit strain in must from the grape variety Assyrtiko. Each experimental point is the mean value of 2 independent measurements (SE for most experimental points ≤ 17%). (**b**) Dendrogram generated after cluster analysis of the digitized (GTG)5-PCR fingerprints of the 31 isolates from the beginning (B), middle (M) and end (E) of the fermentation kinetics in the grape must media. The control condition (C*) is the relative rep-PCR profile of the pure culture of the *S. cerevisiae* Passion Fruit strain. The dendrogram was constructed using the unweighted pair-group method using arithmetic averages with correlation levels expressed as percentage values of the Pearson correlation coefficient.

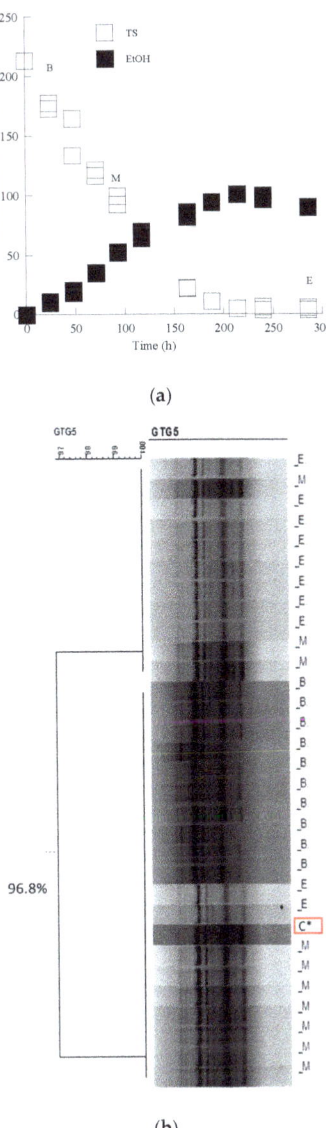

(a)

(b)

Figure 5. (**a**) Sugar and ethanol evolution of the *S. cerevisiae* LMBF Y-54 strain in must from the grape variety Mavrotragano. Each experimental point is the mean value of 2 independent measurements (SE for most experimental points ≤ 17%). (**b**) Dendrogram generated after cluster analysis of the digitized (GTG)5-PCR fingerprints of the 31 isolates from the beginning (B), middle (M) and end (E) of the fermentation kinetics in the grape must media. The control condition (C*) is the relative rep-PCR profile of the pure culture of the *S. cerevisiae* LMBF Y-54 strain. The dendrogram was constructed using the unweighted pair group method using arithmetic averages with correlation levels expressed as percentage values of the Pearson correlation coefficient.

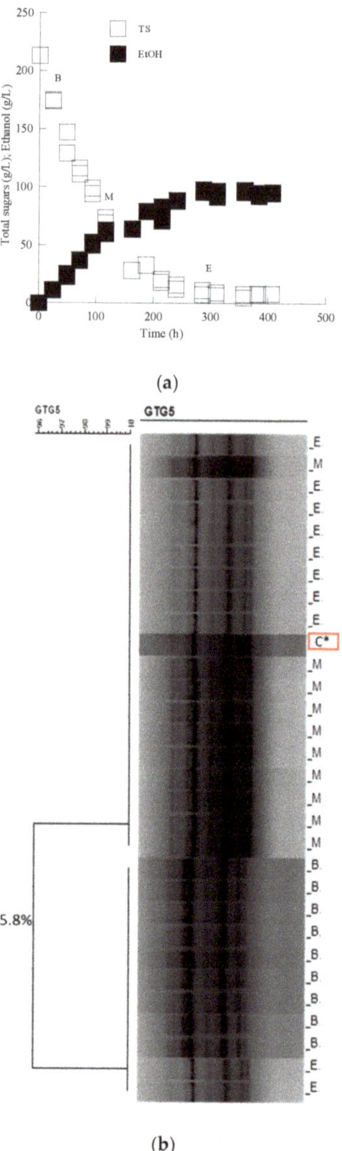

(a)

(b)

Figure 6. (**a**) Sugar and ethanol evolution of *S. cerevisiae* Passion Fruit in must from the grape variety Mavrotragano. Each experimental point is the mean value of 2 independent measurements (SE for most experimental points ≤ 17%). (**b**) Dendrogram generated after cluster analysis of the digitized (GTG)5-PCR fingerprints of the 31 isolates from the beginning (B), middle (M) and end (E) of the fermentation kinetics in the grape must media. The control condition (C*) is the relative rep-PCR profile of the pure culture of the *S. cerevisiae* Passion Fruit strain. The dendrogram was constructed using the unweighted pair-group method using arithmetic averages with correlation levels expressed as percentage values of the Pearson correlation coefficient.

The implicated strains (commercial and wild-type) presented significant ethanol production during growth of sugar-based media under aerobic and microaerophilic/anaerobic experiments. It appears that the trials under microaerophilic/anaerobic conditions favored the production of ethanol, with this type of result being strain dependent according to the results found in the literature [7,24]. For instance, Tronchoni et al. [24], in accordance with our results, demonstrated that the cultures of various *Saccharomyces cerevisiae* strains performed on previously pasteurized grape musts under aerobic conditions was always accompanied by the biosynthesis of ethanol with lower $Y_{EtOH/TS}$ values, compared to equivalent anaerobic experiments [24]. In fact, according to the authors, this was a strategy developed in order to produce wines under aerobic conditions, since this method was proposed to reduce the ethanol content of the produced wine. It is noted that low alcohol-titer wines ($\leq 10.5\%$ v/v) are novel products that due to multiple reasons (modern lifestyle, social reasons, economic motives, etc.) have been gradually gaining the interest of consumers and market in the last decade [24]. On the other hand, in other cases, cultures of wild-type *Saccharomyces cerevisiae* strains on grape musts or glucose-enriched media under aerobic conditions resulted in the production of ethanol in substantial quantities (in some cases in concentrations ≥ 110 g/L) with simultaneous very high $Y_{EtOH/TS}$ values (i.e., ≥ 0.40 g/g, in some instances these values were ≈ 0.49 g/g–very close to the maximum theoretical values) [26,42,43,52], providing evidence that trials of *Saccharomyces cerevisiae* under full aerobic conditions can also lead to the significant production of ethanol. In Table 3, the metrics for ethanol production and the initial sugar content of various *Saccharomyces cerevisiae* strains cultured on different carbon sources are presented and compared with results from the current study.

Table 3. Metrics for ethanol production of *Saccharomyces cerevisiae* different strains cultured on various carbon sources in comparison with results from the current study.

Yeast Strain	Carbon Source	Initial Sugar (g/L)	EtOH (g/L)	Reference
S. cerevisiae AXAZ-1	Microalgae biomass and raisin extract	257	111	[14]
S. cerevisiae AXAZ-1	Pomegranate residue hydrolysate	37	13	[17]
S. cerevisiae LMBF-Y 16	Grape must	250	112	[26]
S. cerevisiae LMBF-Y 18	Grape must	250	125	[26]
S. cerevisiae MAK-1	Grape musts	250	106–119	[52]
Bakers' yeast	Carob pod	200–350	62	[54]
S. cerevisiae	Carob Pod Extracts	200	95	[55]
S. cerevisiae NP01	Sweet sorghum juice	280–300	134	[56]
S. cerevisiae BY4741	Sweet sorghum juice	278.6	113	[57]
S. cerevisiae NP01	Sucrose	280	95	[58]
S. cerevisiae 27817	Glucose	50–200	5–91	[59]
S. cerevisiae 2.399	Glucose	32	13	[60]
S. cerevisiae 24860	Glucose	150	48	[61]
S. cerevisiae CMI237	Sugar	160	70	[62]
S. cerevisiae EC1118	Grape must	280	105	[63]
S. cerevisiae DBVPG 1014	Grape must	270	115	[64]
S. cerevisiae BP2-17	Grape must	225	89	[21]
S. cerevisiae BP2-33	Grape must	225	89	[21]
S. cerevisiae PP2-22	Grape must	225	86	[21]
S. cerevisiae Mpr2-42	Grape must	225	87	[21]
S. cerevisiae PR50	Grape must	220	72	[24]
S. cerevisiae PR543	Grape must	220	85	[24]

Table 3. *Cont.*

Yeast Strain	Carbon Source	Initial Sugar (g/L)	EtOH (g/L)	Reference
S. cerevisiae LMBF Y-54	Glucose	200	68.0	Current study
S. cerevisiae Cross X	Glucose	200	82.2	Current study
S. cerevisiae Passion Fruit	Glucose	200	91.1	Current study
S. cerevisiae LMBF Y-54	Grape must Assyrtiko	222	106.3	Current study
S. cerevisiae LMBF Y-54	Grape must Mavrotragano	214	99.7	Current study
S. cerevisiae Passion Fruit	Grape must Assyrtiko	222	102.7	Current study
S. cerevisiae Passion Fruit	Grape must Mavrotragano	214	97.8	Current study

3.3. Volatile Compounds–Wine Sensory Evaluation

At the end of grape must fermentations with the commercial Passion Fruit strain and the wild-type LMBF Y-54 strain cultured on Assyrtiko and Mavrotragano grape musts, the volatile compounds were analyzed with GC–MS (SPME) analysis. More than 100 volatiles have been identified, belonging to five major chemical group compounds from acids, aldehydes, ketones, esters and higher alcohols and as far as Assyrtiko is concerned the results are in agreement with other studies of the literature [7]. Mavrotragano is a rare indigenous *Vitis vinifera* Greek grape variety mostly unexploited, while the literature regarding its volatile composition is limited. The average (%) contribution of these groups of compounds for both Assyrtiko and Mavrotragano wines are presented in Figure 7. The ester group was comprised by the highest number of volatile compounds [65] and had the higher percentage of contribution to the volatile content of wines followed by higher alcohols. Esters are considered a significant group of volatile compounds as they pose a major impact on flavor and aroma of alcoholic beverages, while a plethora of esters is also produced during fermentation as a result of yeast metabolism [66,67]. Esters mainly contribute to the aromatic profile of wine via fruity and floral odors [65,67]. In the present study, no significant differences were found among the relative contents of esters between the commercial (Passion Fruit) and wild-type (LMBF Y-54) yeast strain in both white and red wines.

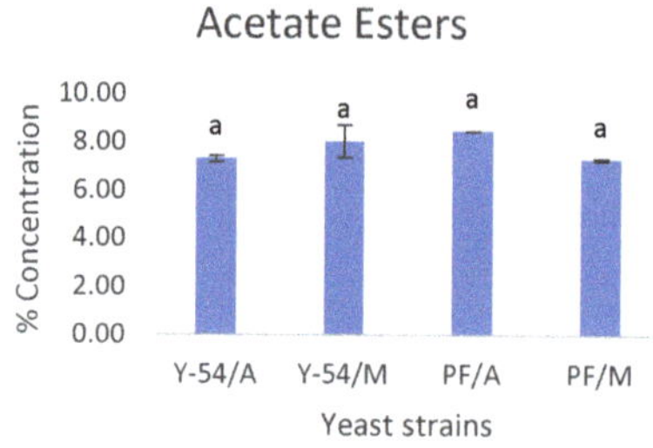

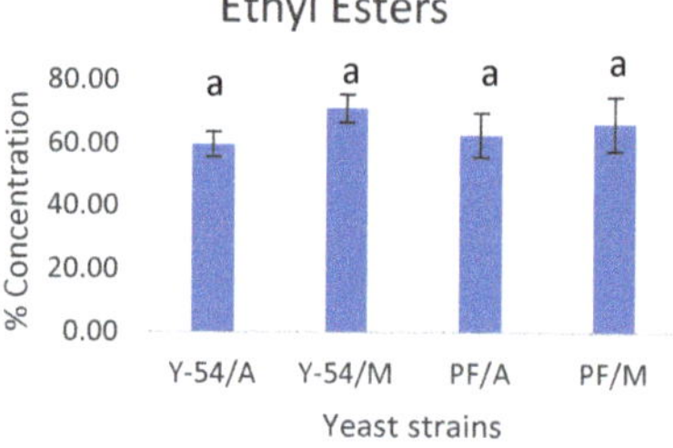

Figure 7. *Cont.*

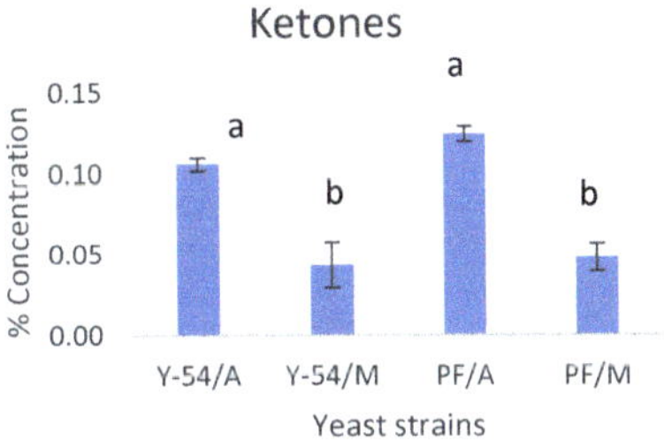

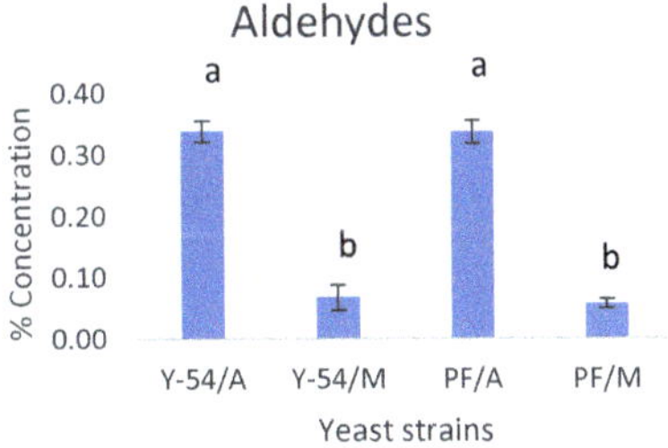

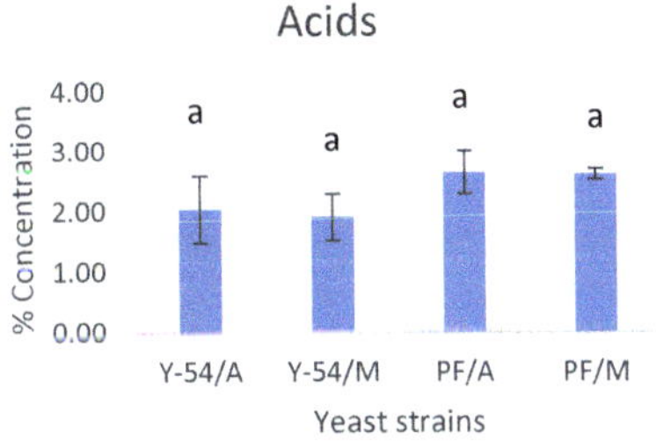

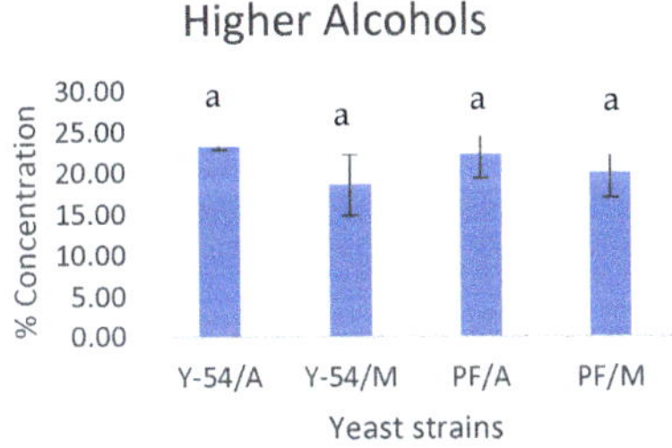

Figure 7. *Cont.*

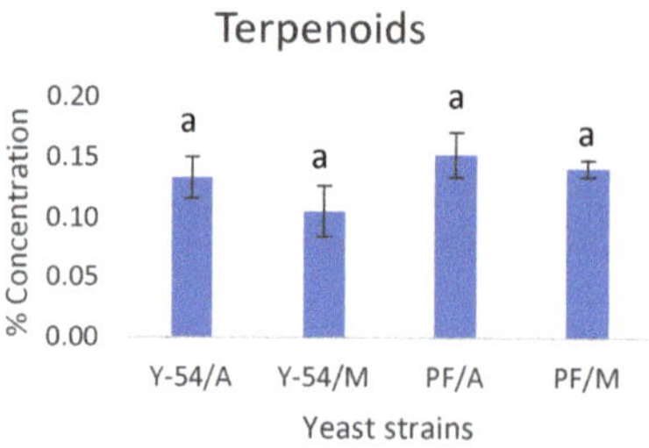

Figure 7. Percent concentrations of the most important groups of volatile compounds: esters (ethyl, acetic), ketones, aldehydes, fatty acids, higher alcohols and terpenoids. Y-54/A: Assyrtiko wine made with *S. cerevisiae* LMBF Y-54; Y-54/M: Mavrotragano wine made with *S. cerevisiae* LMBF Y-54; PF/A: Assyrtiko wine made with *S. cerevisiae* Passion Fruit; PF/M: Mavrotragano wine made with *S. cerevisiae* Passion Fruit. Columns with different letters are significantly different ($p < 0.05$).

Statistical differences were only observed among the white (Assyrtiko) and red wines (Mavrotragano) in terms of their ketone and aldehyde contents. In more detail, significantly lower contribution of the two groups of compounds mentioned above were observed in red wines irrespectively of the yeast strain used. This can be attributed to the higher phenolic content of red wines and consequently their higher antioxidant activity which exerts a protective role on oxidation phenomena, thus preventing the oxidation of carbonyl compounds to their corresponding acids [7]. Interestingly, the volatile composition of either red or white wines produced by the LMBF Y-54 and commercial strains did not differ significantly, verifying the high potential of the novel Y-54 strain to produce high-quality wines. The results from the GC–MS analysis are in line with the sensory data. Specifically, no statistically significant differences were observed between the results for the "control" wines (*viz.* the ones produced with the commercial Passion Fruit yeast strain implicated in the fermentation) and the wines made with the newly isolated and studied yeast strain (LMBF Y-54). Statistically significant differences were only observed concerning fresh and dried fruit odors; however, these differences were between red and white wines and not between the yeast stains studied in this experiment. Moreover, the aroma of all wines examined was of medium intensity with a fresh fruity and floral character and absence of defects, while the overall quality was rated as very high confirming the high potentiality of the novel yeast strain on the wine-making process (Figure 8).

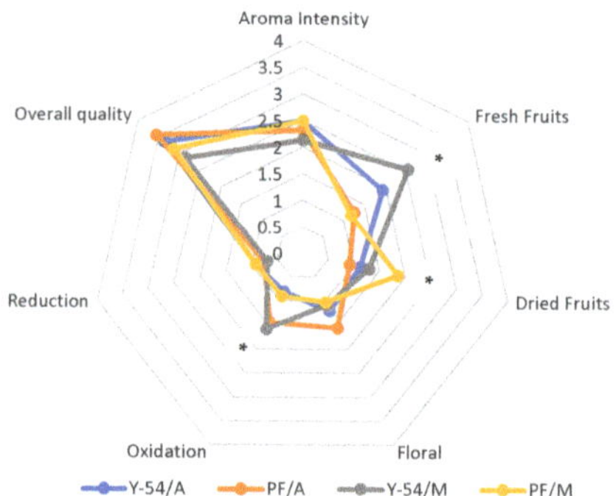

Figure 8. Sensory data of wine samples. Y-54/A: Assyrtiko wine made with *S. cerevisiae* LMBF Y-54; Y-54/M: Mavrotragano wine made with *S. cerevisiae* LMBF Y-54; PF/A: Assyrtiko wine made with *S. cerevisiae* Passion Fruit; PF/M: Mavrotragano wine made with *S. cerevisiae* Passion Fruit. *: Significant differences were found only between the values of red and white wine samples.

4. Conclusions

The exploitation of novel yeast strains in wine-making is considered as an important path in order to perform targeted must fermentation for unique and aromatic products. On the other hand, isolation of yeast strains capable of converting sugars into ethanol with high conversion yields (near to the maximum theoretical one that is =0.51 g of ethanol per g of consumed sugar) and high product concentrations (i.e., $\geq$100 g/L) presents significant importance for bioethanol-producing facilities. Targeted yeast metabolic compounds such as ethanol and glycerol are of paramount importance in wine-making as they provide the necessary alcohol content along with a thick mouthfeel in wine products.

The current study verifies the successful application of the novel and not previously systematically studied yeast strain *Saccharomyces cerevisiae* LMBF Y-54, for both the production of bioethanol from glucose-based media and red and white must fermentation. Interestingly, the cultivation under aerobic conditions resulted in lower production (in terms of both absolute (g/L) and relative (g of ethanol produced per g of sugar consumed)) compared to the trial performed under microaerophilic/anaerobic environment. This could be a first indication that this specific strain could be a good candidate for the production of wines with lower ethanol content, as suggested in other literature reports [24], producing wines of good quality while maintaining the "low-alcohol" phenotype. Likewise, in both tested musts from Santorini Island, the mentioned strain (LMBF Y-54) succeeded in dominating the indigenous yeast strains during all the wine-making fermentation processes. The novel strain compared to the commercial one showed optimum technological characteristics verifying its suitability for wine production and thus posing great potential for industrial applications. Nevertheless, further studies based on genomic approaches can provide supplementary information regarding the fermentative potential of the yeast strain and thus further valorize the obtained knowledge in future wine-making bioprocesses.

Author Contributions: Conceptualization: S.P.; Methodology, K.B., M.D., A.T. and S.K.; Validation: S.K., G.-J.E.N. and S.P.; Formal Analysis: M.D., A.T. and S.K.; Investigation: K.B. and M.D.; Resources: S.P. and G.-J.E.N.; Writing—Original Draft Preparation: M.D., A.T. and S.P.; Writing—Review & Editing: M.D., A.T., S.K. and S.P.; Supervision: S.K. and S.P.; Project Administration: G.-J.E.N., S.K. and S.P.; Funding acquisition: G.-J.E.N. and S.P. All authors have read and agreed to the published version of the manuscript.

Funding: The current investigation was financially supported by the project entitled "Exploitation of new natural microbial flora from Greek origin amenable for the production of high-quality wines" (Acronym: Oenovation, project code T1EΔK-04747) financed by the Ministry of National Education and Religious Affairs, Greece (project action: "Investigate–Create–Innovate 2014–2020, Intervention II").

Institutional Review Board Statement: Not applicable.

Informed Consent Statement: Not applicable.

Conflicts of Interest: The authors declare no conflict of interest.

References

1. Barnard, H.; Dooley, A.N.; Areshian, G.; Gasparyan, B.; Faull, K.F. Chemical evidence for wine production around 4000 BCE in the Late Chalcolithic Near Eastern highlands. *J. Archaeol. Sci.* **2011**, *38*, 977–984. [CrossRef]
2. de Lorgeril, M.; Salen, P. Wine ethanol, platelets, and Mediterranean diet. *Lancet* **1999**, *353*, 1067. [CrossRef]
3. Scolaro, B.; Alves Castro, I. Chapter 47—Red wine and atherosclerosis: Implications for the Mediterranean diet. In *The Mediterranean Diet*, 2nd ed.; Preedy, V.R., Watson, R.R., Eds.; Academic Press: Amsterdam, The Netherlands, 2020; pp. 337–344.
4. Iriti, M.; Varoni, E.M.; Vitalini, S. Chapter 18—Light, regular red wine consumption at main meals: A key cardioprotective element of traditional Mediterranean diet. In *The Mediterranean Diet*, 2nd ed.; Preedy, V.R., Watson, R.R., Eds.; Academic Press: Amsterdam, The Netherlands, 2020; pp. 179–189.
5. Urquiaga, I.; Strobel, P.; Perez, D.; Martinez, C.; Cuevas, A.; Castillo, O.; Marshall, G.; Rozowski, J.; Leighton, F. Mediterranean diet and red wine protect against oxidative damage in young volunteers. *Atherosclerosis* **2010**, *211*, 694–699. [CrossRef]
6. Comitini, F.; Capece, A.; Ciani, M.; Romano, P. New insights on the use of wine yeasts. *Curr. Opin. Food Sci.* **2017**, *13*, 44–49. [CrossRef]

7. Terpou, A.; Ganatsios, V.; Kanellaki, M.; Koutinas, A.A. Entrapped psychrotolerant yeast cells within pine sawdust for low temperature wine making: Impact on wine quality. *Microorganisms* **2020**, *8*, 764. [CrossRef] [PubMed]

8. Yıldırım, H.K. 2—Insights into the role of yeasts in alcoholic beverages. In *Microbial Biotechnology in Food and Health*; Ray, R.C., Ed.; Academic Press: Amsterdam, The Netherlands, 2021; pp. 21–52.

9. Capozzi, V.; Garofalo, C.; Chiriatti, M.A.; Grieco, F.; Spano, G. Microbial terroir and food innovation: The case of yeast biodiversity in wine. *Microbiol. Res.* **2015**, *181*, 75–83. [CrossRef] [PubMed]

10. Minebois, R.; Pérez-Torrado, R.; Querol, A. A time course metabolism comparison among *Saccharomyces cerevisiae*, *S. uvarum* and *S. kudriavzevii* species in wine fermentation. *Food Microbiol.* **2020**, *90*, 103484. [CrossRef] [PubMed]

11. Garrigues, S.; Salazar-Cerezo, S. Ethanol tolerance and production by yeasts. In *Encyclopedia of Mycology*; Zaragoza, Ó., Casadevall, A., Eds.; Elsevier: Oxford, UK, 2021; pp. 447–457.

12. Sarris, D.; Papanikolaou, S. Biotechnological production of ethanol: Biochemistry, processes and technologies. *Eng. Life Sci.* **2016**, *16*, 307–329. [CrossRef]

13. Roukas, T.; Kotzekidou, P. From food industry wastes to second generation bioethanol: A review. *Rev. Environ. Sci. Bio/Technol.* **2022**, *21*, 299–329. [CrossRef]

14. Tsolcha, O.N.; Patrinou, V.; Economou, C.N.; Dourou, M.; Aggelis, G.; Tekerlekopoulou, A.G. Utilization of biomass derived from cyanobacteria-based agro-industrial wastewater treatment and raisin residue extract for ethanol production. *Water* **2021**, *13*, 486. [CrossRef]

15. Thomas, K.C.; Hynes, S.H.; Jones, A.M.; Ingledew, W.M. Production of fuel alcohol from wheat by VHG technology. *Appl. Biochem. Biotechnol.* **1993**, *43*, 211–226. [CrossRef]

16. Mavrommati, M.; Daskalaki, A.; Papanikolaou, S.; Aggelis, G. Adaptive laboratory evolution principles and applications in industrial biotechnology. *Biotechnol. Adv.* **2022**, *54*, 107795. [CrossRef]

17. Dourou, M.; Economou, C.N.; Aggeli, L.; Janák, M.; Valdés, G.; Elezi, N.; Kakavas, D.; Papageorgiou, T.; Lianou, A.; Vayenas, D.V.; et al. Bioconversion of pomegranate residues into biofuels and bioactive lipids. *J. Clean. Prod.* **2021**, *323*, 129193. [CrossRef]

18. Camarasa, C.; Chiron, H.; Daboussi, F.; Della Valle, G.; Dumas, C.; Farines, V.; Floury, J.; Gagnaire, V.; Gorret, N.; Leonil, J.; et al. INRA's research in industrial biotechnology: For food, chemicals, materials and fuels. *Innov. Food Sci. Emerg. Technol.* **2018**, *46*, 140–152. [CrossRef]

19. Mohd Azhar, S.H.; Abdulla, R.; Jambo, S.A.; Marbawi, H.; Gansau, J.A.; Mohd Faik, A.A.; Rodrigues, K.F. Yeasts in sustainable bioethanol production: A review. *Biochem. Biophys. Rep.* **2017**, *10*, 52–61. [CrossRef]

20. Romano, P.; Fiore, C.; Paraggio, M.; Caruso, M.; Capece, A. Function of yeast species and strains in wine flavour. *Int. J. Food Microbiol.* **2003**, *86*, 169–180. [CrossRef]

21. Capece, A.; Pietrafesa, R.; Siesto, G.; Romaniello, R.; Condelli, N.; Romano, P. Selected indigenous *Saccharomyces cerevisiae* strains as profitable strategy to preserve typical traits of Primitivo wine. *Fermentation* **2019**, *5*, 87. [CrossRef]

22. Capece, A.; Romaniello, R.; Siesto, G.; Pietrafesa, R.; Massari, C.; Poeta, C.; Romano, P. Selection of indigenous *Saccharomyces cerevisiae* strains for Nero d'Avola wine and evaluation of selected starter implantation in pilot fermentation. *Int. J. Food Microbiol.* **2010**, *144*, 187–192. [CrossRef]

23. Feghali, N.; Bianco, A.; Zara, G.; Tabet, E.; Ghanem, C.; Budroni, M. Selection of Saccharomyces cerevisiae Starter Strain for Merwah Wine. *Fermentation* **2020**, *6*, 43. [CrossRef]

24. Tronchoni, J.; Gonzalez, R.; Guindal, A.M.; Calleja, E.; Morales, P. Exploring the suitability of *Saccharomyces cerevisiae* strains for winemaking under aerobic conditions. *Food Microbiol.* **2022**, *101*, 103893. [CrossRef]

25. Barrajón, N.; Arévalo-Villena, M.; Úbeda, J.; Briones, A. Enological properties in wild and commercial *Saccharomyces cerevisiae* yeasts: Relationship with competition during alcoholic fermentation. *World J. Microbiol. Biotechnol.* **2011**, *27*, 2703–2710. [CrossRef]

26. Terpou, A.; Dimopoulou, M.; Belka, A.; Kallithraka, S.; Nychas, G.-J.E.; Papanikolaou, S. Effect of Myclobutanil pesticide on the physiological behavior of two newly isolated *Saccharomyces cerevisiae* strains during very-high-rravity alcoholic fermentation. *Microorganisms* **2019**, *7*, 666. [CrossRef] [PubMed]

27. Székelyhidi, R.; Lakatos, E.; Sik, B.; Nagy, Á.; Varga, L.; Molnár, Z.; Kapcsándi, V. The beneficial effect of peppermint (*Mentha X Piperita* L.) and lemongrass (*Melissa officinalis* L.) dosage on total antioxidant and polyphenol content during alcoholic fermentation. *Food Chem. X* **2022**, *13*, 100226. [CrossRef] [PubMed]

28. Argyropoulos, D.; Psallida, C.; Sitareniou, P.; Flemetakis, E.; Diamantopoulou, P. Biochemical evaluation of *Agaricus* and *Pleurotus* strains in batch cultures for production optimization of valuable metabolites. *Microorganisms* **2022**, *10*, 964. [CrossRef]

29. Filippousi, R.; Tsouko, E.; Mordini, K.; Ladakis, D.; Koutinas, A.A.; Aggelis, G.; Papanikolaou, S. Sustainable arabitol production by a newly isolated *Debaryomyces prosopidis* strain cultivated on biodiesel-derived glycerol. *Carbon Resour. Convers.* **2022**, *5*, 92–99. [CrossRef]

30. Christofi, S.; Katsaros, G.; Mallouchos, A.; Cotea, V.; Kallithraka, S. Reducing SO_2 content in wine by combining High Pressure and glutathione addition. *OENO One* **2021**, *55*, 235–252. [CrossRef]

31. Doulgeraki, A.I.; Paramithiotis, S.; Kagkli, D.M.; Nychas, G.-J.E. Lactic acid bacteria population dynamics during minced beef storage under aerobic or modified atmosphere packaging conditions. *Food Microbiol.* **2010**, *27*, 1028–1034. [CrossRef]

32. Syrokou, M.K.; Themeli, C.; Paramithiotis, S.; Mataragas, M.; Bosnea, L.; Argyri, A.A.; Chorianopoulos, N.G.; Skandamis, P.N.; Drosinos, E.H. Microbial Ecology of Greek wheat Ssourdoughs, Iidentified by a culture-dependent and a culture-independent approach. *Foods* **2020**, *9*, 1603. [CrossRef]

33. Bonatsou, S.; Paramithiotis, S.; Panagou, E.Z. Evolution of yeast consortia during the fermentation of Kalamata Natural Black Olives upon two initial acidification treatments. *Front. Microbiol.* **2018**, *8*, 2673. [CrossRef]

34. da Silva-Filho, E.A.; Brito dos Santos, S.K.; Resende Ado, M.; de Morais, J.O.; de Morais, M.A., Jr.; Ardaillon Simões, D. Yeast population dynamics of industrial fuel-ethanol fermentation process assessed by PCR-fingerprinting. *Antonie Leeuwenhoek* **2005**, *88*, 13–23. [CrossRef]

35. Kallithraka, S.; Kotseridis, Y.; Kyraleou, M.; Proxenia, N.; Tsakiris, A.; Karapetrou, G. Analytical phenolic composition and sensory assessment of selected rare Greek cultivars after extended bottle ageing. *J. Sci. Food Agric.* **2015**, *95*, 1638–1647. [CrossRef]

36. Kyraleou, M.; Kallithraka, S.; Chira, K.; Tzanakouli, E.; Ligas, I.; Kotseridis, Y. Differentiation of wines treated with wood chips based on their phenolic content, volatile composition, and sensory parameters. *J. Food Sci.* **2015**, *80*, C2701–C2710. [CrossRef]

37. Kyraleou, M.; Kotseridis, Y.; Koundouras, S.; Chira, K.; Teissedre, P.-L.; Kallithraka, S. Effect of irrigation regime on perceived astringency and proanthocyanidin composition of skins and seeds of *Vitis vinifera* L. cv. Syrah grapes under semiarid conditions. *Food Chem.* **2016**, *203*, 292–300. [CrossRef]

38. van Dijken, J.P.; Scheffers, W.A. Redox balances in the metabolism of sugars by yeasts. *FEMS Microbiol. Lett.* **1986**, *32*, 199–224. [CrossRef]

39. de Kok, M.J.C.; Schaapherder, A.F.; Wüst, R.C.I.; Zuiderwijk, M.; Bakker, J.A.; Lindeman, J.H.N.; Le Dévédec, S.E. Circumventing the Crabtree effect in cell culture: A systematic review. *Mitochondrion* **2021**, *59*, 83–95. [CrossRef]

40. Piskur, J.; Rozpedowska, E.; Polakova, S.; Merico, A.; Compagno, C. How did *Saccharomyces* evolve to become a good brewer? *Trends Genet.* **2006**, *22*, 183–186. [CrossRef]

41. Hagman, A.; Säll, T.; Compagno, C.; Piskur, J. Yeast "make-accumulate-consume" life strategy evolved as a multi-step process that predates the whole genome duplication. *PLoS ONE* **2013**, *8*, e68734. [CrossRef]

42. Sarris, D.; Giannakis, M.; Philippoussis, A.; Komaitis, M.; Koutinas, A.A.; Papanikolaou, S. Conversions of olive mill wastewater-based media by *Saccharomyces cerevisiae* through sterile and non-sterile bioprocesses. *J. Chem. Technol. Biotechnol.* **2013**, *88*, 958–969. [CrossRef]

43. Sarris, D.; Matsakas, L.; Aggelis, G.; Koutinas, A.A.; Papanikolaou, S. Aerated vs. non-aerated conversions of molasses and olive mill wastewaters blends into bioethanol by *Saccharomyces cerevisiae* under non-aseptic conditions. *Ind. Crop. Prod.* **2014**, *56*, 83–93. [CrossRef]

44. Papanikolaou, S.; Rontou, M.; Belka, A.; Athenaki, M.; Gardeli, C.; Mallouchos, A.; Kalantzi, O.; Koutinas, A.A.; Kookos, I.K.; Zeng, A.P.; et al. Conversion of biodiesel-derived glycerol into biotechnological products of industrial significance by yeast and fungal strains. *Eng. Life Sci.* **2017**, *17*, 262–281. [CrossRef]

45. Pinto, L.; Malfeito-Ferreira, M.; Quintieri, L.; Silva, A.C.; Baruzzi, F. Growth and metabolite production of a grape sour rot yeast-bacterium consortium on different carbon sources. *Int. J. Food Microbiol.* **2019**, *296*, 65–74. [CrossRef] [PubMed]

46. Burphan, T.; Tatip, S.; Limcharoensuk, T.; Kangboonruang, K.; Boonchird, C.; Auesukaree, C. Enhancement of ethanol production in very high gravity fermentation by reducing fermentation-induced oxidative stress in *Saccharomyces cerevisiae*. *Sci. Rep.* **2018**, *8*, 13069. [CrossRef] [PubMed]

47. Wang, F.Q.; Gao, C.J.; Yang, C.Y.; Xu, P. Optimization of an ethanol production medium in very high gravity fermentation. *Biotechnol. Lett.* **2007**, *29*, 233–236. [CrossRef] [PubMed]

48. Evans, C.T.; Ratledge, C. Phosphofructokinase and the regulation of the flux of carbon from glucose to lipid in the oleaginous yeast *Rhodosporidium toruloides*. *Microbiology* **1984**, *130*, 3251–3264. [CrossRef]

49. Bhutada, G.; Kavšcek, M.; Ledesma-Amaro, R.; Thomas, S.; Rechberger, G.N.; Nicaud, J.M.; Natter, K. Sugar versus fat: Elimination of glycogen storage improves lipid accumulation in *Yarrowia lipolytica*. *FEMS Yeast Res.* **2017**, *17*, fox020. [CrossRef]

50. Diamantopoulou, P.; Filippousi, R.; Antoniou, D.; Varfi, E.; Xenopoulos, E.; Sarris, D.; Papanikolaou, S. Production of added-value microbial metabolites during growth of yeast strains on media composed of biodiesel-derived crude glycerol and glycerol/xylose blends. *FEMS Microbiol. Lett.* **2020**, *367*, fnaa063. [CrossRef]

51. Tronchoni, J.; Gamero, A.; Arroyo-López, F.N.; Barrio, E.; Querol, A. Differences in the glucose and fructose consumption profiles in diverse *Saccharomyces* wine species and their hybrids during grape juice fermentation. *Int. J. Food Microbiol.* **2009**, *134*, 237–243. [CrossRef]

52. Sarris, D.; Kotseridis, Y.; Linga, M.; Galiotou-Panayotou, M.; Papanikolaou, S. Enhanced ethanol production, volatile compound biosynthesis and fungicide removal during growth of a newly isolated *Saccharomyces cerevisiae* strain on enriched pasteurized grape musts. *Eng. Life Sci.* **2009**, *9*, 29–37. [CrossRef]

53. Christofi, S.; Papanikolaou, S.; Dimopoulou, M.; Terpou, A.; Cioroiu, I.B.; Cotea, V.; Kallithraka, S. Effect of yeast assimilable nitrogen content on fermentation kinetics, wine chemical composition and sensory character in the production of Assyrtiko wines. *Appl. Sci.* **2022**, *12*, 1405. [CrossRef]

54. Roukas, T. Continuous ethanol production from carob pod extract by immobilized *Saccharomyces cerevisiae* in a packed-bed reactor. *J. Chem. Technol. Biotechnol.* **1994**, *59*, 387–393. [CrossRef]

55. Sanchez-Segado, S.; Salar-García, M.J.; Ortiz-Martínez, V.M.; de los Ríos, A.P.; Hernández-Fernández, F.J.; Lozano-Blanco, L.J. Evaluation of ionic liquids as In Situ extraction agents during the alcoholic fermentation of carob pod extracts. *Fermentation* **2019**, *5*, 90. [CrossRef]

56. Phukoetphim, N.; Chan-u-tit, P.; Laopaiboon, P.; Laopaiboon, L. Improvement of bioethanol production from sweet sorghum juice under very high gravity fermentation: Effect of nitrogen, osmoprotectant, and aeration. *Energies* **2019**, *12*, 3620. [CrossRef]

57. Sasaki, K.; Tsuge, Y.; Sasaki, D.; Kawaguchi, H.; Sazuka, T.; Ogino, C.; Kondo, A. Repeated ethanol production from sweet sorghum juice concentrated by membrane separation. *Bioresour. Technol.* **2015**, *186*, 351–355. [CrossRef]

58. Chan-u-tit, P.; Laopaiboon, L.; Jaisil, P.; Laopaiboon, P. High level Eethanol production by nitrogen and osmoprotectant supplementation under very high gravity fermentation conditions. *Energies* **2013**, *6*, 884–899. [CrossRef]

59. Vallet, C.; Saïd, R.; Rabiller, C.; Martin, M.L. Natural Abundance Isotopic fractionation in the fermentation reaction: Influence of the nature of the yeast. *Bioorg. Chem.* **1996**, *24*, 319–330. [CrossRef]

60. Yu, Z.; Zhang, H. Ethanol fermentation of acid-hydrolyzed cellulosic pyrolysate with *Saccharomyces cerevisiae*. *Bioresour. Technol.* **2003**, *90*, 95–100. [CrossRef]

61. Najafpour, G.; Younesi, H.; Syahidah Ku Ismail, K. Ethanol fermentation in an immobilized cell reactor using *Saccharomyces cerevisiae*. *Bioresour. Technol.* **2004**, *92*, 251–260. [CrossRef]

62. Navarro, A.R.; del Sepúlveda, M.C.; Rubio, M.C. Bio-concentration of vinasse from the alcoholic fermentation of sugar cane molasses. *Waste Manag.* **2000**, *20*, 581–585. [CrossRef]

63. Martínez-Moreno, R.; Morales, P.; Gonzalez, R.; Mas, A.; Beltran, G. Biomass production and alcoholic fermentation performance of *Saccharomyces cerevisiae* as a function of nitrogen source. *FEMS Yeast Res.* **2012**, *12*, 477–485. [CrossRef]

64. Ciani, M.; Beco, L.; Comitini, F. Fermentation behaviour and metabolic interactions of multistarter wine yeast fermentations. *Int. J. Food Microbiol.* **2006**, *108*, 239–245. [CrossRef]

65. Kechagia, D.; Paraskevopoulos, Y.; Symeou, E.; Galiotou-Panayotou, M.; Kotseridis, Y. Influence of prefermentative treatments to the major volatile compounds of Assyrtiko wines. *J. Agric. Food Chem.* **2008**, *56*, 4555–4563. [CrossRef] [PubMed]

66. Fraile, P.; Garrido, J.; Ancín, C. Influence of a *Saccharomyces cerevisiae* selected strain in the volatile composition of rose wines. Evolution during fermentation. *J. Agric. Food Chem.* **2000**, *48*, 1789–1798. [CrossRef] [PubMed]

67. Ganatsios, V.; Terpou, A.; Gialleli, A.-I.; Kanellaki, M.; Bekatorou, A.; Koutinas, A.A. A ready-to-use freeze-dried juice and immobilized yeast mixture for low temperature sour cherry (*Prunus cerasus*) wine making. *Food Bioprod. Process.* **2019**, *117*, 373–379. [CrossRef]

 *fermentation*

MDPI

Article

Optimization of *cis*-9-Heptadecenoic Acid Production from the Oleaginous Yeast *Yarrowia lipolytica*

Wendy Al Sahyouni [1,2], Sally El Kantar [1,*], Anissa Khelfa [1], Young-Kyoung Park [3], Jean-Marc Nicaud [3], Nicolas Louka [2] and Mohamed Koubaa [1,*]

1   TIMR (Integrated Transformations of Renewable Matter), Centre de Recherche Royallieu, Université de Technologie de Compiègne, ESCOM—CS 60319, CEDEX, 60203 Compiègne, France; wendy.sahyouni@net.usj.edu.lb (W.A.S.); a.khelfa@escom.fr (A.K.)
2   Laboratoire CTA, UR TVA, Centre d'Analyses et de Recherche, Faculté des Sciences, Université Saint-Joseph, Beyrouth 1104 2020, Lebanon; nicolas.louka@usj.edu.lb
3   Micalis Institute, INRAE, AgroParisTech, Université Paris-Saclay, 78350 Jouy-en-Josas, France; yk16.park@gmail.com (Y.-K.P.); jean-marc.nicaud@inrae.fr (J.-M.N.)
*   Correspondence: sally.el-kantar@utc.fr (S.E.K.); m.koubaa@escom.fr (M.K.); Tel.: +33-344-238-841 (M.K.)

Abstract: Odd-chain fatty acids (OCFA) have been studied for their therapeutic and nutritional properties, as well as for their potential use in the chemical industry for the production of biofuel. Genetic modification strategies have demonstrated an improved production of OCFA by oleaginous microorganisms. In this study, the production of OCFA-enriched lipids by fermentation using a genetically engineered *Yarrowia lipolytica* strain was investigated. The major fatty acid produced by this strain was the *cis*-9-heptadecenoic acid (C17:1). Its biosynthesis was optimized using a design of experiment strategy involving a central composite design. The optimal responses maximizing the cell density (optical density at 600 nm) and the C17:1 content (%) in lipids were found using 52.4 g/L sucrose, 26.9 g/L glycerol, 10.4 g/L sodium acetate, 5 g/L sodium propionate, and 4 g/L yeast extract. Under these conditions, in a 5 L scale bioreactor, the respective contents of lipids and C17:1 in culture medium were 2.52 ± 0.05 and 0.82 ± 0.01 g/L after 96 h fermentation. The results obtained in this work pave the way toward the process upscale of C17:1 and encourage its industrial production.

Keywords: *Y. lipolytica*; fermentation; odd-chain fatty acids; *cis*-9-heptadecenoic acid; central composite design

Citation: Al Sahyouni, W.; El Kantar, S.; Khelfa, A.; Park, Y.-K.; Nicaud, J.-M.; Louka, N.; Koubaa, M. Optimization of *cis*-9-Heptadecenoic Acid Production from the Oleaginous Yeast *Yarrowia lipolytica*. *Fermentation* **2022**, *8*, 245. https://doi.org/10.3390/fermentation8060245

Academic Editor: Timothy J. Tse

Received: 2 May 2022
Accepted: 20 May 2022
Published: 25 May 2022

1. Introduction

Nowadays, microbial oils, also called single cell oils (SCOs), are considered promising renewable alternatives to fossil fuels and can be used for the production of biofuels or oleochemicals [1,2]. They present multiple advantages over plant oils or animal fats, such as non-competitiveness with the food crops, no influence of seasonal changes, and high scalability [3].

SCOs are produced by oleaginous microorganisms, such as yeast, bacteria, and microalgae [4,5]. The cultivation of microalgae is relatively costly and results in low biomass concentration [6]. In addition, the available data for the transformation strategies to engineer reliable models of algal strains capable of producing high lipid yields are still insufficient [7]. Moreover, the use of bacteria for oil production is also limited, since they accumulate lipids in their membrane, which makes the extraction of oil difficult [8]. Therefore, among the oleaginous microorganisms, yeasts are the most favorable in terms of oil production and accumulation.

Yarrowia lipolytica is a non-conventional oleaginous yeast that is grown in various types of substrates to produce value-added compounds, such as lipids [9], organic acids [10], alkanes [11], and polyalcohols [12]. In addition, *Y. lipolytica* has a well-studied genome and metabolism that allow the engineering of high lipid accumulation strains [13]. Therefore, for some engineered strains, lipid accumulation in *Y. lipolytica* can reach up to 80% dry weight basis [14].

Despite the consequent achievements in improving oil production by *Y. lipolytica*, its industrial exploitation is still facing the high cost of substrates and both upstream and downstream costly processing steps. To overcome the cost drawback, one of the solutions was to redirect the fatty acid synthesis toward the production of unusual and high-value-added fatty acids, such as odd-chain fatty acids (OCFA) [15,16]. The occurrence of OCFA in nature is very scarce despite their multiple applications (e.g., precursors for several compounds, such as biocides, plasticizers, and industrial chemicals) and the health-related benefits that make them be on the verge of commercialization [16–18]. In a recent work, *Y. lipolytica* was engineered to produce OCFA by using sodium propionate as precursor [16]. It was demonstrated that the deletion of the *PHD1* gene, encoding the 2-methylcitrate dehydratase in the methyl citrate cycle, prevented the degradation of propionyl-CoA, which promotes the production of OCFA. Producing OCFA was also successful in other microorganisms, which was well summarized in a recent review paper [19]. Among OCFA, the *cis*-9-heptadecenoic acid (C17:1) and its salts have been described as high-value-added compounds known for their anti-inflammatory effects and their efficacy in treating prophylaxis, psoriasis, allergies, and autoimmune diseases [17,20].

Due to the high potential of C17:1, the purpose of this study was to improve its production by fermentation using an engineered strain of *Y. lipolytica*. Experimental design methodology was used, meaning a central composite design for the study of five factors, including carbon and nitrogen substrates. Optimal conditions were tested and validated using a 5 L bioreactor.

2. Materials and Methods

2.1. Oleaginous Yeast Strain

The oleaginous yeast strain *Y. lipolytica* JMY7877 used in this study was constructed from OCFA-producing strain, JMY7780 [21]. The strain was further engineered for the utilization of sucrose by overexpressing two genes (*SUC2* and *HXK1*). This allows its ability to metabolize sucrose-rich industrial by-products, such as sugar beet and sugar cane molasses, which may contribute to reducing the production cost. In order to remove selective markers in JMY7780 strain, the Cre-lox system was used with JME547 plasmid (pUB-Cre-Hygro) [22]. After removing *URA3* and *LEU2* marker genes in JMY7780 strain, two gene expression cassettes, invertase *SUC2* from *Saccharomyces cerevisiae* (JME2347, JMP62-LEU2 ex-pTEF-ScSUC2-Tlip2 [23]), and hexokinase *HXK1* from *Y. lipolytica* (JME2103, JMP62-URA3 ex-pTEF-YlHXK1-Tlip2 [24]) were introduced to the genome of *Y. lipolytica*, resulting in the JMY7877 strain (Table 1. The integration of gene expression cassettes was verified by colony PCR with specific primer pairs (SUC2, pTEF-internal-Fw/SUC2-internal-Rev; HXK1, pTEF-internal-Fw/HXK1-internal-Rev) (Table 2).

Table 1. Plasmids and strains used in this study.

Strain	Description	Reference
E. coli		
JME547	pUB-Cre-Hygro	[22]
JME2103	JMP62-URA3 ex-pTEF-*YlHXK1*-Tlip2	[24]
JME2347	JMP62-LEU2 ex-pTEF-*ScSUC2*-Tlip2	[23]

Table 1. *Cont.*

Strain	Description	Reference
Y. lipolytica		
JMY7780	Δ*phd1* Δ*mfe1* Δ*tgl4* pTEF-*DGA2* pTEF-*GPD1* hp4d-*LDP1-URA3* ex pTEF-*Repct*-LEU2 ex	[21]
JMY7829	Δ*phd1* Δ*mfe1* Δ*tgl4* pTEF-*DGA2* pTEF-*GPD1* hp4d-*LDP1* pTEF-*Repct*	This study
JMY7877	JMY7829 pTEF-*ScSUC2-LEU2* ex pTEF-*YlHXK1-URA3* ex	This study

Table 2. Sequence of primers used in this study.

Name	Sequence
pTEF-internal-Fw	TCTGGAATCTACGCTTGTTCA
SUC2-internal-Rev	GCAGATTCTAGCTTCCAGGAC
HXK1-internal-Rev	CTCATCTTCTCGAAGGTCTGCTG

2.2. Culture Conditions

2.2.1. Growth Media Composition

Growth media was composed of sucrose and glycerol as main carbon substrates, yeast extract as organic nitrogen substrate, ammonium chloride as inorganic nitrogen substrate, sodium acetate as precursor of acetyl-CoA, sodium propionate as precursor of propionyl-CoA for OCFA synthesis, and saline solution 1x (for 1000x concentration, the composition was the following for 1 L added, in the order: H_3PO_4 85% liquid: 107 g, water: 650 mL, KCl: 20 g, NaCl: 20 g, $MgSO_4.7H_2O$: 27 g, $ZnSO_4.7H_2O$: 7.7 g, $MnSO_4.H_2O$: 0.47 g, $CoCl_2.6H_2O$: 0.3 g, $CuSO_4.5H_2O$: 0.6 g, $Na_2MoO_4.2H_2O$): 0.094 g, H_3BO: 0.3 g, water up to 1 L).

Na_2HPO_4 (0.05 M) and KH_2PO_4 (0.05 M) were added to the media to adjust the pH to 6 when the cultures were performed in Erlenmeyer flasks, whereas for the cultures performed in the bioreactor, the pH was adjusted automatically using NaOH (1 M) and H_2SO_4 (1 M).

2.2.2. Yeast Culture in Erlenmeyer Flasks (Design of Experiment)

A central composite experimental design was used to optimize the concentrations of five variables (substrates concentrations): sucrose (X_1, (20–100) g/L), pure glycerol (X_2, (10–50) g/L), sodium acetate (X_3, (10–30) g/L), sodium propionate (X_4, (3–7) g/L), and yeast extract (X_5, (1–5) g/L). The ranges of concentrations were set according to a previous screening design (not shown in this work). The responses studied for this experimental design were the maximization of the cell density (optical density (OD) at 600 nm) (Y_1) and the C17:1 content (%) in lipids (Y_2). The central composite design (2^{5-1} + star) resulted in 26 experimental points (16 points of the fractional factorial design and 10 star points), 4 center points (repetitions for the statistical analysis), and 6 test points for the validation of the models. Considering 5 variables and 2 responses, the experimental data were fitted to obtain a second-degree regression model with the form of Equation (1):

$$Y = b_0 + \sum_{i=1}^{5} b_i X_i + \sum_{i=1}^{4} \sum_{j=i+1}^{5} b_{ij} X_i X_j + \sum_{i=1}^{5} b_{ii} X_i^2 \quad \text{with } j > i \tag{1}$$

where Y is the predicted response, X_i refers to the variables from 1 to 5, b_0 is the constant coefficient, b_i refers to the linear term coefficients, b_{ij} presents the interaction term coefficients, and b_{ii} presents the quadratic effect term coefficients. The analyses of the experimental design, the optimal path, and the desirability functions to reach the optimal composition of the fermentation medium were performed using NEMRODW software (v.2017). The general scheme of experiments is presented in Figure 1.

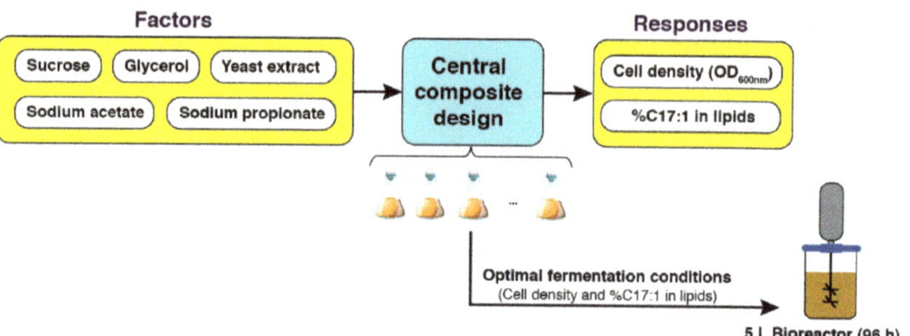

Figure 1. Experimental set-up for the optimization of culture conditions to maximize cell density (OD$_{600nm}$) and the C17:1 content (%) in lipids.

The fermentation experiments of the experimental design were performed in 500 mL Erlenmeyer flasks, each containing 100 mL of culture medium. In addition to the five variables studied, NH$_4$Cl (0.5 g/L) and saline solution 1 × (0.1 mL/100 mL) were added to each flask. These concentrations were kept constant for all the experiments according to a screening design performed prior to this study and demonstrating no impact of these constituents, when varying their concentrations, on growth and C17:1 content (%).

Inoculation of the culture media was performed by adding 2 mL of a stock cell culture (prepared from one colony in a modified YPD medium (sucrose 20 g/L, yeast extract 10 g/L, peptone 20 g/L, KH$_2$PO$_4$ 0.05 M, and Na$_2$HPO$_4$ 0.05 M) to each flask (initial OD$_{600nm}$ = 0.08 ± 0.01), 24 h incubation at 170 rpm and 28 °C, and storage at −80 °C in 25% glycerol). The Erlenmeyer flasks of the experimental design (performed 12 by 12) were incubated in a shaker at 28 °C, 170 rpm agitation, and for 96 h. Samples were taken periodically from each flask to measure cell density at 600 nm using a cell density meter (Ultrospec 10, Biochrom). C17:1 content analysis, recovered from cells, was performed by centrifuging 1 mL of each sample for 10 min at 13,500 rpm (Micro Star 12, VWR, Rosny-Sous-Bois, France). All the pellets were washed 3 times with 1 mL ultrapure water, followed by centrifugation under the same conditions. After freeze drying (Alpha 1–4 LD Plus, Martin Christ, Osterode am Harz, Germany) for 3 days, all the pellets were processed for lipid extraction and fatty acid composition analysis.

2.2.3. Yeast Culture in Bioreactor

The optimal conditions giving the maximal cell density and C17:1 content found using the experimental design were first verified using 500 mL Erlenmeyer flasks (100 mL culture volume) over 96 h, at 28 °C temperature and 170 rpm agitation, and then up-scaled to a 5 L bioreactor (Biostat B Plus, Sartorius Stedim Biotech, Göttingen, Germany). The twin bioreactors each containing 4.75 L of the culture medium were first sterilized at 121 °C for 20 min using an HMC HV 110L autoclave (HMC Europe GmbH, Tüßling, Germany). After cooling down to 28 °C, saturating the medium with oxygen, and calibrating the pH to 6, a volume of 250 mL mid-exponential-phase pre-cultures (each cultivated for 24 h in a 1 L Erlenmeyer flask, as described in Section 2.2.1) was added to inoculate the medium. The temperature was maintained at 28 °C using a double jacket and circulating cooling water, whereas the agitation speed and the air flow rate were set to 400 rpm and 2.5 L/min, respectively (the values were set in order to avoid foam formation). During fermentation, samples were taken from the bioreactor and used to determine cell density (OD$_{600nm}$) and C17:1 content (%) in lipids. A non-inoculated medium was used for the blank of the spectrophotometer and for the dilution of the samples to measure the OD$_{600nm}$. In addition, 1 mL of each sample was centrifuged for 10 min at 13,500 rpm. The supernatant was stored at −20 °C and used later to determine the substrates consumption kinetics, whereas the pellet was washed with 1 mL ultrapure water, freeze dried, and used to quantify the biomass dry weight and the C17:1 content in lipids.

2.3. Determination of Substrate Consumption during Fermentation

The concentrations of glycerol, sucrose, glucose, and fructose (generated from sucrose hydrolysis by the action of the extracellular invertase produced by the yeast), as well as sodium acetate and sodium propionate, were determined by HPLC using an Agilent 1260 Infinity II system equipped with an Aminex HPX-87H column coupled to a refractive index (RI) detector. The mobile phases used were ultrapure water for the analysis of sugars and glycerol, and sulfuric acid (5 mM) for the analysis of sodium acetate and sodium propionate in isocratic mode with a flow rate of 0.6 mL/min. The temperature of the column was fixed at 35 °C. For all analyses, the samples were filtered before injection using 0.45 μm filters, diluted in ultrapure water (for sugars and glycerol quantification) and in sulfuric acid (5 mM) (for sodium acetate and sodium propionate quantification), and injected with a volume of 10 μL. The compounds' assignation and quantification were performed using pre-established standard curves of glycerol, sucrose, glucose, fructose, sodium acetate, and sodium propionate.

2.4. Lipid Extraction and Fatty Acids Analysis

2.4.1. Lipid Extraction from *Y. lipolytica* Biomass

Yeast biomass samples (taken during fermentation) obtained after freeze drying were weighted and used to extract the accumulated lipids. For each sample, an amount of ≈30 mg dry biomass was introduced into a 2 mL screw tube in the presence of two stainless-steel beads (4.9 mm diameter). Lipid extraction was performed by adding 1 mL of cyclohexane/isopropanol mixture (2:1, *v:v*), followed by cell disruption for 5 min at 30 Hz using a bead miller (Retsch MM400, Haan, Germany). After centrifugation for 15 min at 13,500 rpm (MicroStar12 centrifuge, VWR, Rosny-Sous-Bois, France), the supernatant was transferred to a 10 mL glass tube, and the extraction was repeated twice under the same conditions. The supernatants collected (3 mL) were evaporated under nitrogen flow at 60 °C, and the extracted lipid was subjected to transmethylation.

In addition, lipid content (dry weight basis) accumulated in yeasts after 96 h fermentation conducted in the 5 L bioreactors was determined using Soxhlet's method, as previously reported [25]. Briefly, 5 g of lyophilized biomass was grinded in the bead miller (Retsch MM400, Haan, Germany) for 3 × 5 min at 30 Hz and then washed with 250 mL *n*-hexane for 24 h. After evaporation of *n*-hexane at 60 °C and reduced pressure using a rotary evaporation system, oil content (%) was determined, according to Equation (2).

$$Oil\ content\ (\%) = \frac{mass\ of\ recovered\ oil}{mass\ of\ the\ sample} \times 100 \tag{2}$$

2.4.2. Preparation and Quantification of Fatty Acid Methyl Esters (FAMEs)

For each tube containing the extracted lipids, 300 μL toluene (Carlo Erba, Val-de-Reuil, France) and 1 mL methanolic HCl (3 M) (Sigma Aldrich, Saint-Quentin-Fallavier, France) were added. After flushing with nitrogen to avoid oxidation, the tubes were tightly capped and placed in an oven at 80 °C for 2 h for the transmethylation process. The reaction was then stopped by adding 500 μL sodium bisulfite (Acros Organics, Fisher Scientific France, Illkirch, France) (5% *w/v* in water), and the samples were vortexed for 10 s. To extract FAMEs, 1.7 mL cyclohexane (Carlo Erba, Val-de-Reuil, France) was added, followed by rigorous vortexing for 1 min and centrifugation for 2 min at 2000 rpm. From the upper phase containing FAMEs, 100 μL was taken, evaporated under nitrogen flow at room temperature, and resuspended in 1 mL cyclohexane.

FAMEs were analyzed using gas chromatography equipment (Agilent Technologies 7890A, Les Ulis, France) coupled with a flame ionization detector (GC-FID). The GC oven equipped with a capillary column 50% cyanopropyl-50% methyl polysiloxane

(60 m × 320 μm × 0.25 μm) was first set at 150 °C and held for 2 min, and then increased to 220 °C at a ramp of 1.5 °C/min and held for 30 min. Helium was used as carrier gas at a flow rate of 30 mL/min, and injection volume was 2 μL. Standard solutions of *cis*-9-heptadecenoic acid (Cayman Chemical Company, Ann Arbor, MI, USA), *cis*-10-nonadecenoic acid (Merck, Fontenay Sous Bois, France), methyl nonadecanoate (Merck, Fontenay Sous Bois, France), and Supelco 37 FAME mix (Merck, Fontenay Sous Bois, France) containing, among others, *cis*-10-heptadecenoic acid, were used to determine the retention time of FAMEs. The standards ordered in the acid form were transmethylated, similarly to the extracted lipids.

2.5. Statistical Analyses

One-way analysis of variance (ANOVA) was used to determine the significant differences using StatPlus V6 software for Macintosh systems.

3. Results and Discussion

3.1. Impact of the Fermentation Conditions on Yeast Growth and C17:1 Biosynthesis

3.1.1. Models' Validation and Fitting

The central composite design was used for the optimization of the culture conditions of *Y. lipolytica*. Two responses were maximized: cell density (OD_{600nm}) and % of C17:1 in lipids. Table 3 presents the experimental conditions, as well as the measured responses of the experimental design.

Table 3. Experimental plan and measured responses. Coded variables are provided in parentheses.

Experiment	Sucrose (g/L)	Glycerol (g/L)	Sodium Acetate (g/L)	Sodium Propionate (g/L)	Yeast Extract (g/L)	Cell Density (OD_{600nm})	C17:1 Content (%)
1	40 (−1)	20 (−1)	15 (−1)	4 (−1)	4 (+ 1)	19.2	31.7
2	80 (+1)	20 (−1)	15 (−1)	4 (−1)	2 (−1)	14.0	31.2
3	40 (−1)	40 (+1)	15 (−1)	4 (−1)	2 (−1)	14.4	29.3
4	80 (+1)	40 (+1)	15 (−1)	4 (−1)	4 (+1)	17.8	26.4
5	40 (−1)	20 (−1)	25 (+1)	4 (−1)	2 (−1)	16.8	27.4
6	80 (+1)	20 (−1)	25 (+1)	4 (−1)	4 (+1)	19.4	19.7
7	40 (−1)	40 (+1)	25 (+1)	4 (−1)	4 (+1)	19.2	21.3
8	80 (+1)	40 (+1)	25 (+1)	4 (−1)	2 (−1)	14.8	21.1
9	40 (−1)	20 (−1)	15 (−1)	6 (+1)	2 (−1)	13.0	34.7
10	80 (+1)	20 (−1)	15 (−1)	6 (+1)	4 (+1)	17.0	30.7
11	40 (−1)	40 (+1)	15 (−1)	6 (+1)	4 (+1)	17.0	30.4
12	80 (+1)	40 (+1)	15 (−1)	6 (+1)	2 (−1)	12.2	34.2
13	40 (−1)	20 (−1)	25 (+1)	6 (+1)	4 (+1)	20.0	26.5
14	80 (+1)	20 (−1)	25 (+1)	6 (+1)	2 (−1)	15.0	29.3
15	40 (−1)	40 (+1)	25 (+1)	6 (+1)	2 (−1)	15.6	26.4
16	80 (+1)	40 (+1)	25 (+1)	6 (+1)	4 (+1)	16.0	25.1
17	20 (−2)	30 (0)	20 (0)	5 (0)	3 (0)	17.6	29.5
18	100 (+2)	30 (0)	20 (0)	5 (0)	3 (0)	15.8	28.3
19	60 (0)	10 (−2)	20 (0)	5 (0)	3 (0)	17.8	29.4
20	60 (0)	50 (+2)	20 (0)	5 (0)	3 (0)	16.0	25.8
21	60 (0)	30 (0)	10 (−2)	5 (0)	3 (0)	13.8	36.6
22	60 (0)	30 (0)	30 (+2)	5 (0)	3 (0)	17.6	23.1
23	60 (0)	30 (0)	20 (0)	3 (−2)	3 (0)	18.2	21.5
24	60 (0)	30 (0)	20 (0)	7 (+2)	3 (0)	16.2	32.9
25	60 (0)	30 (0)	20 (0)	5 (0)	1 (−2)	12.0	38.2
26	60 (0)	30 (0)	20 (0)	5 (0)	5 (+2)	16.4	28.7
27	60 (0)	30 (0)	20 (0)	5 (0)	3 (0)	15.0	32.1

Table 3. *Cont.*

Experiment	Sucrose (g/L)	Glycerol (g/L)	Sodium Acetate (g/L)	Sodium Propionate (g/L)	Yeast Extract (g/L)	Cell Density (OD_{600nm})	C17:1 Content (%)
28	60 (0)	30 (0)	20 (0)	5 (0)	3 (0)	14.6	32.2
29	60 (0)	30 (0)	20 (0)	5 (0)	3 (0)	14.8	31.2
30	60 (0)	30 (0)	20 (0)	5 (0)	3 (0)	13.2	32.0
31	42.7 (−0.866)	25 (−0.5)	18.2 (−0.3535)	4.7 (−0.2739)	2.8 (−0.2236)	14.0	32.3
32	77.3 (+0.866)	25 (−0.5)	18.2 (−0.3535)	4.7 (−0.2739)	2.8 (−0.2236)	13.2	32.4
33	60 (0)	40 (+1)	18.2 (−0.3535)	4.7 (−0.2739)	2.8 (−0.2236)	13.6	31.6
34	60 (0)	30 (0)	25.3 (+1.0607)	4.7 (−0.2739)	2.8 (−0.2236)	15.0	28.4
35	60 (0)	30 (0)	20 (0)	6.1 (+1.0955)	2.8 (−0.2236)	12.8	32.4
36	60 (0)	30 (0)	20 (0)	5 (0)	4.1 (+1.118)	15.8	30.1

The experimental plan was validated using the test points corresponding to the experiments 31–36. The analysis of the residues showed non-significant differences between the measured responses and the predicted ones (Table 4). The measured responses of the test points were then included in the models provided using NEMRODW software as follows (* means significant coefficient):

$$Y_{\text{Cell density (OD600 nm)}} = 13.95 - 0.53\,X_1 - 0.44\,X_2 + 0.84\,X_3 - 0.6\,X_4 + 1.64\,X_5 - 0.11\,X_1 \times X_2 - 0.23\,X_1 \times X_3 - 0.11\,X_1 \times X_4 - 0.09\,X_1 \times X_5 - 0.25\,X_2 \times X_3 - 0.07\,X_2 \times X_4 - 0.24\,X_2 \times X_5 + 0.15\,X_3 \times X_4 - 0.32\,X_3 \times X_5 - 0.08\,X_4 \times X_5 + 0.62^*\,(X_1)^2 + 0.68\,(X_2)^2 + 0.39\,(X_3)^2 + 0.74\,(X_4)^2 + 0.02\,(X_5)^2 \tag{3}$$

$$Y_{\%C17:1} = 32.19 - 0.46\,X_1 - 0.96\,X_2 - 3.27\,X_3 + 2.11\,X_4 - 1.73\,X_5 + 0.52\,X_1 \times X_2 - 0.2\,X_1 \times X_3 + 0.77\,X_1 \times X_4 - 0.39\,X_1 \times X_5 - 0.11\,X_2 \times X_3 + 0.41\,X_2 \times X_4 + 0.37\,X_2 \times X_5 + 0.39\,X_3 \times X_4 \quad 0.1\,X_3 \times X_5 - 0.11\,X_4 \times X_5 - 0.95\,(X_1)^2 - 1.25\,(X_2)^2 - 0.68\,(X_3)^2 - 1.40\,(X_4)^2 + 0.17\,(X_5)^2 \tag{4}$$

Table 4. Validation of the models using the test points.

	Cell Density (OD_{600nm})		% C17:1	
Experiment	Y_{measured}	$Y_{\text{predicted}}$	Y_{measured}	$Y_{\text{predicted}}$
31	14	15.02	32.29	33.27
32	13.2	14.44	32.41	31.76
33	13.6	14.31	31.63	30.49
34	15	15.51	28.39	27.45
35	12.8	14.27	32.43	33.30
36	15.8	16.14	30.07	30.49

According to the Pareto charts presented in Figure 2a,b, all the constituents had significant linear effects for both response parameters OD_{600nm} and C17:1. Figure 2a shows negative linear effects and positive quadratic effects for sucrose, glycerol, and sodium propionate, while sodium acetate and yeast extract had positive linear effects without any quadratic effects. Figure 2b shows negative linear effects and negative quadratic effects for sucrose and glycerol, a positive linear effect and a negative quadratic effect for sodium propionate, while sodium acetate and yeast extract had negative linear effects without any quadratic effects. The inserts in Figure 2b illustrate the variations of OD_{600nm} and C17:1, respectively, according to each constituent between the low level (−1) and the high level (+1); the other constituents are fixed at the central level (0). Furthermore, Figure 2 does not show any significant interaction effects between the constituents.

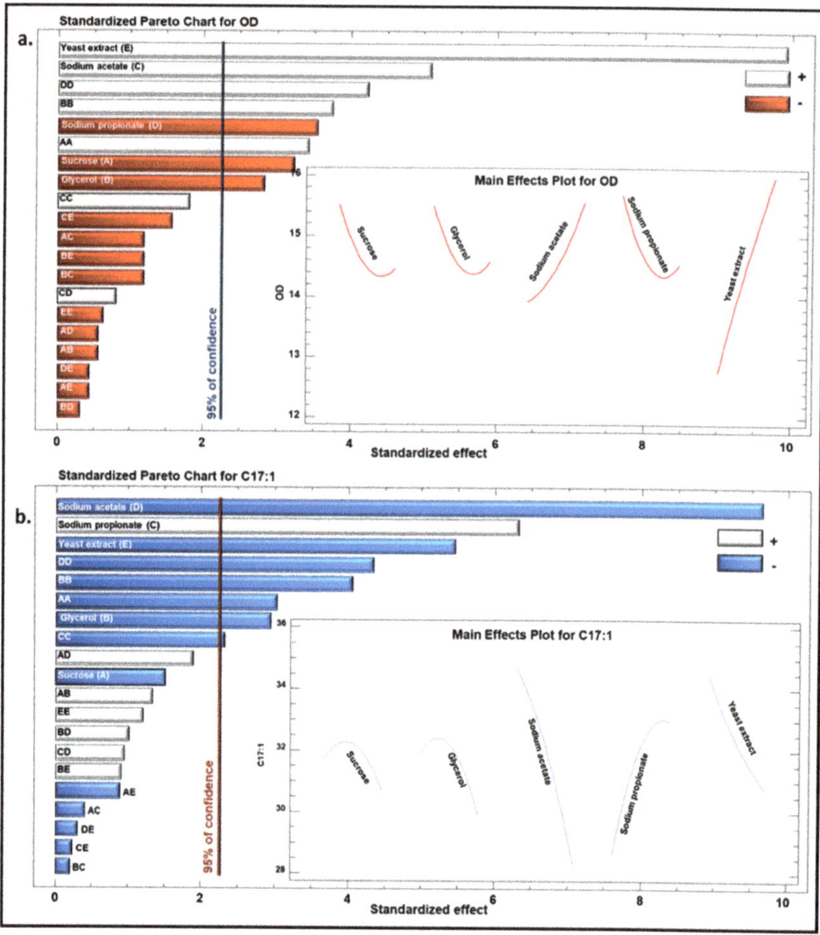

Figure 2. Standardized Pareto chart for OD_{600nm} (**a**) and for % of C17:1 (**b**). Vertical lines indicate statistically significant parameters with more than 95% confidence. (+) indicates positive effect, (−) indicates negative effect. Inserts show the variation of response parameters, OD_{600nm} (**a**) and % of C17:1 (**b**), as a function of each ingredient.

In addition, the two inserts of Figure 2a,b demonstrate that sucrose and glycerol had nearly the same effect on the two response parameters OD_{600nm} and % of C17:1. In this sense, the increase in these two constituents led to a decrease in the growth of microorganisms (OD_{600nm}) and in the production of C17:1. The other three constituents had an opposite effect on the two response parameters: the increase in sodium acetate and yeast extract increased the OD_{600nm} and lowered their oil content, while the increase in sodium propionate decreased the OD_{600nm} and increased oil production.

Results of ANOVA (Tables 5 and 6) show that there are significant differences ($p < 0.05$) between the regression and the residue, whereas no significant differences ($p > 0.05$) were observed between the lack of fit and the pure error. These results reveal the validation of the models reported in Equations (3) and (4). In addition, the R^2 values were equal to 0.942 and 0.969 for the responses of cell density (OD_{600nm}), and C17:1 content, respectively, indicating a good fit of the models.

Table 5. The analysis of variance (ANOVA) for cell density. * means significant differences ($p < 0.05$).

Source of Variation	Sum of Squares	Degrees of Freedom	Mean Squares	F-Ratio	*p*-Value
Regression	149.27	20	7.46	12.22	<0.01 *
Residue	9.16	15	0.61		
Lack of fit	7.16	12	0.59	0.89	61.9
Pure error	2.0	3	0.66		
Total	158.43	35			

Table 6. The analysis of variance (ANOVA) for C17:1 content (%). * means significant differences ($p < 0.05$).

Source of Variation	Sum of Squares	Degrees of Freedom	Mean Squares	F-Ratio	*p*-Value
Regression	637.21	20	31.86	23.34	<0.01 *
Residue	20.47	15	1.36		
Lack of fit	19.87	12	1.65	8.37	5.3
Pure error	0.59	3	0.19		
Total	657.68	35			

3.1.2. Optimal Culture Conditions Analyses

As mentioned before, the objectives of the current study were the maximization of both cell density (OD$_{600nm}$) and C17:1 content (%) in lipids. The optimal paths provided by NEMRODW software for each response are given in Figure 3.

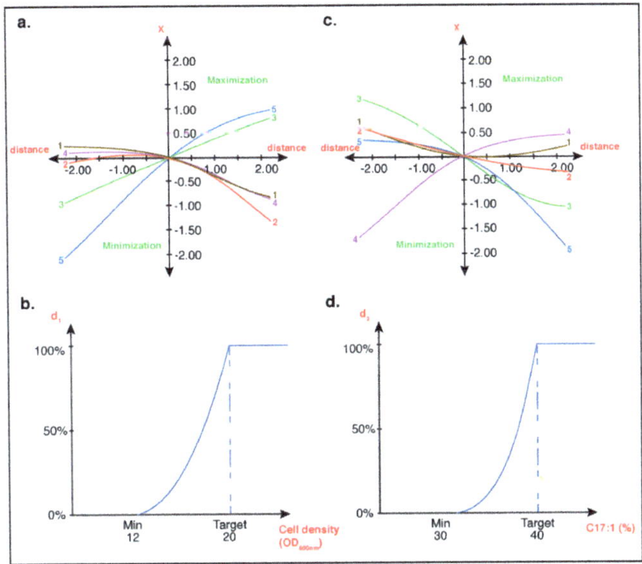

Figure 3. (a) Optimal path for the response to cell density (OD$_{600nm}$), (b) Desirability function for the optimization of the response to cell density (OD$_{600nm}$), (c) Optimal path for the response to C17:1 content (%), (d) Desirability function for the optimization of the response to C17:1 content (%).

Figure 3 shows that the maximization of cell density requires an increase in the concentrations of sodium acetate (factor 3) and yeast extract (factor 5), and a decrease in the concentrations of sucrose (factor 1), glycerol (factor 2), and sodium propionate (factor 4). On the other hand, the maximization of C17:1 content requires an increase in the concentrations of sucrose (factor 1) and sodium propionate (factor 4), and a decrease in the concentrations of glycerol (factor 2), sodium acetate (factor 3), and yeast extract (factor 4). It should be noted that yeast extract is a source of nitrogen that favors cell growth and limits the accumulation of lipids. Several review papers have shown that TAG lipogenesis begins when a non-carbon nutrient, such as phosphorus, oxygen, sulfur, and especially nitrogen, becomes growth limiting [26–31].

During nitrogen starvation, the activation of lipogenesis begins with a decrease by $\approx 95\%$ in the concentration of adenosine monophosphate (AMP) due to an increase in the activity of AMP deaminase. This decrease in AMP is accompanied by the deactivation of the AMP-regulated mitochondrial isocitrate dehydrogenase ICDH (NAD^+), which creates an excess of mitochondrial citrate. This metabolite will then be transported from the mitochondria to the cytosol and will serve as a substrate for the production of cytosolic acetyl-CoA by ATP-citrate lyase (ACL) [32,33]. Similarly, increasing both the concentrations of sucrose and glycerol contributes to decreasing cell growth. This is probably due to an increase in osmotic pressure, which can affect oxygen uptake by the cells and decrease the culture growth [34–38]. This fact was previously reported by many authors. For instance, a decrease in the growth rate of *Saccharomyces cerevisiae* was observed when the concentration of glucose was increased from 200 g/L to 300 g/L [39]. Similarly, the impact of sugar concentration on both biomass accumulation and lipid content in *Trichosporon fermentans* was studied [40]. It was reported that the highest values were obtained using 15% total sugar concentration. Beyond it, a decrease in biomass accumulation and lipid content was observed. In addition, when *Y. lipolytica* Po1g was cultivated in defatted rice bran hydrolysate with sugar concentrations varying from 20 g/L to 40 g/L, the maximum biomass accumulation and lipid content were obtained at 30 g/L. Further rise in the total sugar concentration inhibited the growth of yeast and resulted in lower lipid content [41]. Similar trend was reported for glycerol concentration. For example, the growth of oleaginous yeast *Candida viswanathii* Y-E4 was limited when glycerol concentration exceeded 15% in the fermentation medium [42].

On the other hand, sodium acetate and sodium propionate are the respective precursors of acetyl-CoA and propionyl-CoA. Increasing the concentration of sodium acetate will contribute to a decrease in OCFA; however, it is required to constitute the fatty acid backbone and for the synthesis of lipids in the cell membranes, whereas increasing the concentration of sodium propionate will contribute to increasing the content in OCFA, but it will reduce cell growth due to its toxicity at high concentrations. For example, it has been shown in a previous study that propionate had an inhibitory effect on the growth of *Y. lipolytica* JMY2900 when its concentration exceeded 10 g/L [16]. In another work, the strain *Y. lipolytica* CICC31596 was more sensitive to the presence of propionate and had an inhibitory effect at 5 g/L [43]. Due to the abovementioned observations, a compromise must be made to determine the optimal concentrations of the different compounds constituting the culture medium. For this purpose, the desirability function was used in NEMRODW software to find the optimal conditions allowing maximizing both cell density (OD_{600nm}) and the content in C17:1 (Figure 2). The optimal substrates concentrations generated are the following: (sucrose) = 52.4 g/L, (glycerol) = 26.9 g/L, (sodium acetate) = 10.4 g/L, (sodium propionate) = 5 g/L, and (yeast extract) = 4 g/L. They correspond in coded variables to $X_1 = -0.386$, $X_2 = -0.305$, $X_3 = -1.92$, $X_4 = 0.044$, and $X_5 = 1.012$, respectively. Under these conditions, the predicted responses are: cell density (OD_{600nm}) = 16.3 $\pm$ 1.54 and C17:1 content (%) = 34.7% $\pm$ 2.31.

3.2. Fermentation under the Optimal Culture Conditions

The validation of the optimal culture conditions was performed by conducting fermentations using the optimal substrates' concentrations, first in 500 mL Erlenmeyer flasks (100 mL culture medium) and then in continuous stirred-tank bioreactors (5 L culture medium). After 96 h of fermentation, the measured cell density (OD_{600nm}) was 17.86 $\pm$ 0.34 in Erlenmeyer flasks and 17.52 $\pm$ 0.28 in the bioreactor. These values concur with the predicted ones found using NEMRODW (16.3 ± 1.54), showing no significant differences. The concentrations of glycerol, sucrose, glucose, fructose, sodium acetate, and sodium propionate were determined during fermentation in 5 L bioreactors up to 96 h (Figure 4).

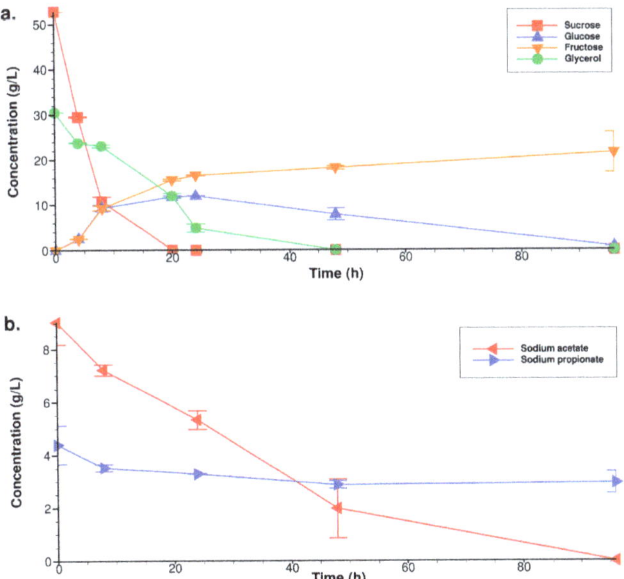

Figure 4. Growth kinetics and substrates consumption during the fermentations conducted in a 5 L bioreactor under the optimal conditions of (sucrose) = 52.4 g/L, (glycerol) = 26.9 g/L, (sodium acetate) = 10.4 g/L, (sodium propionate) = 5 g/L, and (yeast extract) = 4 g/L. (**a**) Consumption kinetics of sucrose, glucose, fructose, and glycerol. (**b**) Consumption kinetics of sodium acetate and sodium propionate.

Results show that sucrose was actively hydrolyzed by the secreted invertase in the medium to generate glucose and fructose. The strain used, JMY7877, showed more affinity to glucose rather than fructose, as it was completely consumed after 96 h fermentation, whereas the concentration of fructose remained relatively high, at ≈20 g/L. The same behavior of glucose consumption was observed for glycerol and sodium acetate, which were completely consumed after 48 h and 96 h of fermentation, respectively. However, the consumption of sodium propionate was lower, reaching ≈3 g/L in the medium after 96 h fermentation. Increasing the membrane transport of fructose and sodium propionate by genetic engineering could be a potential way to improve cell growth and lipid accumulation in yeast.

Lipid content after 96 h of fermentation in 5 L bioreactors, analyzed using the Soxhlet method, was 2.52 ± 0.05 g/L of the culture medium. Fatty acid composition was also analyzed using the lyophilized biomass (Table 7). The results obtained showed non-significant differences between the cultures performed in Erlenmeyer flasks and in 5 L bioreactors. The respective C17:1 content in lipids was 31.62% ± 1.15 and 31.62% ± 0.94, which was non-significantly different ($p < 0.05$) from the predicted value found using NEMRODW software (34.7% ± 2.31). The C17:1 obtained was higher than that reported by Zhang et al. (the highest value reported was 26.42% ± 0.03) [44] and lower than that reported by Kolouchová et al. (37% but low lipid content of 0.23 g/L) [45]. OCFA content in both cultures represents ≈ 60% of the total fatty acids in lipids, which corresponds to ≈ 1.5 g/L of culture medium. In a recent work, Park et al. (2021) found 1.87 g OCFA/L of culture medium; however, the fermentation time was longer (192 h) compared to one used in this study (96 h), and the main carbon source was glucose.

Table 7. Fatty acid composition of lipids extracted from *Y. lipolytica* cultures after 96 h fermentation in Erlenmeyer flasks and in 5 L bioreactors. Culture conditions were those provided by the design of experiment approach: (sucrose) = 52.4 g/L, (glycerol) = 26.9 g/L, (sodium acetate) = 10.4 g/L, (sodium propionate) = 5 g/L, and (yeast extract) = 4 g/L.

FAME	% in Erlenmeyer	% in 5 L Bioreactors
C15:0	5.03 ± 0.19	5.37 ± 0.34
C16:0	7.78 ± 0.15	9.37 ± 1.59
C16:1	1.00 ± 0.06	1.29 ± 0.30
C17:0	19.71 ± 0.42	18.00 ± 1.71
C17:1	31.62 ± 1.15	31.62 ± 0.94
C18:0	10.02 ± 0.24	9.85 ± 0.27
C18:1	17.30 ± 0.27	17.87 ± 0.56
C18:2	2.59 ± 0.18	2.61 ± 0.17
C19:0	2.72 ± 0.06	2.64 ± 0.10
C19:1	1.41 ± 0.07	1.39 ± 0.13

4. Conclusions

In this study, the production of *cis*-9-heptadecenoic acid (C17:1) by *Y. lipolytica* was optimized using a central composite design. When performing the experiments in 5 L bioreactors, and after 96 h fermentation using the optimal composition, the respective concentrations of lipids and OCFA were 2.52 g/L and 1.5 g/L of the culture medium. The content of C17:1 in lipids was 0.82 g/L of the culture medium. Further process improvements could be attained by either performing the experiments in fed-batch mode or using longer fermentation time (accumulating higher lipid content in the cells). Replacing pure sucrose and glycerol with low-cost by-products, obtained from the sugar industry and from oil biorefinery, respectively, will further reduce the production cost. In addition, genetic engineering modifications, such as improving the membrane transport of fructose and propionate, may lead to improved lipid and OCFA contents in yeast cells.

Author Contributions: Conceptualization, M.K. and J.-M.N.; methodology, W.A.S., S.E.K. and Y.-K.P.; software, N.L.; validation, M.K., N.L. and S.E.K.; investigation, M.K. and S.E.K.; writing—original draft preparation, M.K., S.E.K. and W.A.S.; writing—review and editing, M.K., N.L., J.-M.N., Y.-K.P., S.E.K., A.K. and W.A.S.; supervision, M.K.; funding acquisition, M.K. All authors have read and agreed to the published version of the manuscript.

Funding: This work was performed within the project YaLiOl supported by the ANR grant "ANR-20-CE43-0007" of the French National Research Agency (ANR) in France.

Institutional Review Board Statement: Not applicable.

Informed Consent Statement: Not applicable.

Conflicts of Interest: The authors declare no conflict of interest.

References

1. Abghari, A.; Chen, S. *Yarrowia lipolytica* as an Oleaginous Cell Factory Platform for Production of Fatty Acid-Based Biofuel and Bioproducts. *Front. Energy Res.* **2014**, *2*, 21. [CrossRef]
2. Koubaa, M.; Imatoukene, N.; Drévillon, L.; Vorobiev, E. Current Insights in Yeast Cell Disruption Technologies for Oil Recovery: A Review. *Chem. Eng. Process.-Process Intensif.* **2020**, *150*, 107868. [CrossRef]
3. Li, Q.; Du, W.; Liu, D. Perspectives of Microbial Oils for Biodiesel Production. *Appl. Microbiol. Biotechnol.* **2008**, *80*, 749–756. [CrossRef] [PubMed]
4. Cohen, Z.; Ratledge, C. *Single Cell Oils: Microbial and Algal Oils*; AOCS Press: Champaign, IL, USA, 2010; ISBN 978-1-893997-73-8.
5. Thevenieau, F.; Nicaud, J.-M. Microorganisms as Sources of Oils. *OCL* **2013**, *20*, D603. [CrossRef]
6. Li, Y.; Horsman, M.; Wu, N.; Lan, C.Q.; Dubois-Calero, N. Biofuels from Microalgae. *Biotechnol. Prog.* **2008**, *24*, 815–820. [CrossRef] [PubMed]
7. Xue, J.; Balamurugan, S.; Li, T.; Cai, J.X.; Chen, T.T.; Wang, X.; Yang, W.D.; Li, H.Y. Biotechnological Approaches to Enhance Biofuel Producing Potential of Microalgae. *Fuel* **2021**, *302*, 121169. [CrossRef]

8. Meng, X.; Yang, J.; Xu, X.; Zhang, L.; Nie, Q.; Xian, M. Biodiesel Production from Oleaginous Microorganisms. *Renew. Energy* **2009**, *34*, 1–5. [CrossRef]

9. Amalia, L.; Zhang, Y.-H.; Ju, Y.-H.; Tsai, S.-L. Enhanced Lipid Production in *Yarrowia lipolytica* Po1g by over-Expressing Lro1 Gene under Two Different Promoters. *Appl. Biochem. Biotechnol.* **2020**, *191*, 104–111. [CrossRef]

10. Papanikolaou, S.; Muniglia, L.; Chevalot, I.; Aggelis, G.; Marc, I. *Yarrowia lipolytica* as a Potential Producer of Citric Acid from Raw Glycerol. *J. Appl. Microbiol.* **2002**, *92*, 737–744. [CrossRef]

11. Blazeck, J.; Liu, L.; Knight, R.; Alper, H.S. Heterologous Production of Pentane in the Oleaginous Yeast *Yarrowia lipolytica*. *J. Biotechnol.* **2013**, *165*, 184–194. [CrossRef]

12. Rakicka, M.; Rywińska, A.; Cybulski, K.; Rymowicz, W. Enhanced Production of Erythritol and Mannitol by *Yarrowia lipolytica* in Media Containing Surfactants. *Braz. J. Microbiol.* **2016**, *47*, 417–423. [CrossRef] [PubMed]

13. Ledesma-Amaro, R.; Dulermo, T.; Nicaud, J.M. Engineering *Yarrowia lipolytica* to Produce Biodiesel from Raw Starch. *Biotechnol. Biofuels* **2015**, *8*, 148. [CrossRef] [PubMed]

14. Nicaud, J.-M. Yarrowia lipolytica. *Yeast* **2012**, *29*, 409–418. [CrossRef]

15. Ledesma-Amaro, R.; Nicaud, J.-M. *Yarrowia lipolytica* as a Biotechnological Chassis to Produce Usual and Unusual Fatty Acids. *Prog. Lipid Res.* **2016**, *61*, 40–50. [CrossRef]

16. Park, Y.-K.; Dulermo, T.; Ledesma-Amaro, R.; Nicaud, J.-M. Optimization of Odd Chain Fatty Acid Production by *Yarrowia lipolytica*. *Biotechnol. Biofuels* **2018**, *11*, 158. [CrossRef] [PubMed]

17. Degwert, J.; Jacob, J.; Steckel, F. Use of Cis-9-Heptadecenoic Acid for Treating Psoriasis and Allergies. International Patent Application No. WO1994021247A1, 13 January 1998.

18. Clausen, C.A.; Coleman, R.D.; Yang, V.W. Fatty Acid-Based Formulations for Wood Protection against Mold and Sapstain. *For. Prod. J.* **2010**, *60*, 301–304. [CrossRef]

19. Zhang, L.-S.; Liang, S.; Zong, M.-H.; Yang, J.-G.; Lou, W.-Y. Microbial Synthesis of Functional Odd-Chain Fatty Acids: A Review. *World J. Microbiol. Biotechnol.* **2020**, *36*, 35. [CrossRef]

20. Jenkins, B.; West, J.A.; Koulman, A. A Review of Odd-Chain Fatty Acid Metabolism and the Role of Pentadecanoic Acid (C15:0) and Heptadecanoic Acid (C17:0) in Health and Disease. *Molecules* **2015**, *20*, 2425–2444. [CrossRef]

21. Park, Y.-K.; Bordes, F.; Letisse, F.; Nicaud, J.-M. Engineering Precursor Pools for Increasing Production of Odd-Chain Fatty Acids in *Yarrowia lipolytica*. *Metab. Eng. Commun.* **2021**, *12*, e00158. [CrossRef]

22. Fickers, P.; Le Dall, M.T.; Gaillardin, C.; Thonart, P.; Nicaud, J.M. New Disruption Cassettes for Rapid Gene Disruption and Marker Rescue in the Yeast *Yarrowia lipolytica*. *J. Microbiol. Methods* **2003**, *55*, 727–737. [CrossRef]

23. Lazar, Z.; Rossignol, T.; Verbeke, J.; Crutz-Le Coq, A.-M.; Nicaud, J.-M.; Robak, M. Optimized Invertase Expression and Secretion Cassette for Improving *Yarrowia lipolytica* Growth on Sucrose for Industrial Applications. *J. Ind. Microbiol. Biotechnol.* **2013**, *40*, 1273–1283. [CrossRef] [PubMed]

24. Lazar, Z.; Dulermo, T.; Neuvéglise, C.; Crutz-Le Coq, A.-M.; Nicaud, J. M. Hexokinase—A Limiting Factor in Lipid Production from Fructose in *Yarrowia lipolytica*. *Metab. Eng.* **2014**, *26*, 89–99. [CrossRef] [PubMed]

25. AFNOR. *Céréales et Produits Céréaliers: Détermination de la Teneur en Matières Grasses Totales*; Paris La Défense: Pairs, France, 1986.

26. Beopoulos, A.; Cescut, J.; Haddouche, R.; Uribelarrea, J.-L.; Molina-Jouve, C.; Nicaud, J.-M. *Yarrowia lipolytica* as a Model for Bio-Oil Production. *Prog. Lipid Res.* **2009**, *48*, 375–387. [CrossRef]

27. Rios, L.F.; Klein, B.C.; Luz, L.F.; Maciel Filho, R.; Wolf Maciel, M.R. Nitrogen Starvation for Lipid Accumulation in the Microalga Species *Desmodesmus* sp. *Appl. Biochem. Biotechnol.* **2015**, *175*, 469–476. [CrossRef] [PubMed]

28. Calvey, C.H.; Su, Y.-K.; Willis, L.B.; McGee, M.; Jeffries, T.W. Nitrogen Limitation, Oxygen Limitation, and Lipid Accumulation in *Lipomyces starkeyi*. *Bioresour. Technol.* **2016**, *200*, 780–788. [CrossRef]

29. González-García, Y.; Rábago-Panduro, L.M.; French, T.; Camacho-Córdova, D.I.; Gutiérrez-González, P.; Córdova, J. High Lipids Accumulation in *Rhodosporidium toruloides* by Applying Single and Multiple Nutrients Limitation in a Simple Chemically Defined Medium. *Ann. Microbiol.* **2017**, *67*, 519–527. [CrossRef]

30. Fakankun, I.; Mirzaei, M.; Levin, D.B. Impact of Culture Conditions on Neutral Lipid Production by Oleaginous Yeast. *Methods Mol. Biol.* **2019**, *1995*, 311–325.

31. Wang, X.; Fosse, H.K.; Li, K.; Chauton, M.S.; Vadstein, O.; Reitan, K.I. Influence of Nitrogen Limitation on Lipid Accumulation and EPA and DHA Content in Four Marine Microalgae for Possible Use in Aqauafeed. *Front. Mar. Sci.* **2019**, *6*, 95. [CrossRef]

32. Boulton, C.A.; Ratledge, C. 1981 Correlation of Lipid Accumulation in Yeasts with Possession of ATP: Citrate Lyase. *Microbiology* **1981**, *127*, 169–176. [CrossRef]

33. Dulermo, T.; Lazar, Z.; Dulermo, R.; Rakicka, M.; Haddouche, R.; Nicaud, J.-M. Analysis of ATP-Citrate Lyase and Malic Enzyme Mutants of *Yarrowia lipolytica* Points out the Importance of Mannitol Metabolism in Fatty Acid Synthesis. *Biochim. Biophys. Acta (BBA)-Mol. Cell Biol. Lipids* **2015**, *1851*, 1107–1117. [CrossRef]

34. Meesters, P.A.E.P.; Huijberts, G.N.M.; Eggink, G. High-Cell-Density Cultivation of the Lipid Accumulating Yeast *Cryptococcus curvatus* Using Glycerol as a Carbon Source. *Appl. Microbiol. Biotechnol.* **1996**, *45*, 575–579. [CrossRef]

35. Li, Y.; Zhao, Z.; Bai, F. High-Density Cultivation of Oleaginous Yeast *Rhodosporidium toruloides* Y4 in Fed-Batch Culture. *Enzym. Microb. Technol.* **2007**, *41*, 312–317. [CrossRef]

36. Papanikolaou, S.; Fakas, S.; Fick, M.; Chevalot, I.; Galiotou-Panayotou, M.; Komaitis, M.; Marc, I.; Aggelis, G. Biotechnological Valorisation of Raw Glycerol Discharged after Bio-Diesel (Fatty Acid Methyl Esters) Manufacturing Process: Production of 1,3-Propanediol, Citric Acid and Single Cell Oil. *Biomass Bioenergy* **2008**, *32*, 60–71. [CrossRef]
37. Liang, Y.; Cui, Y.; Trushenski, J.; Blackburn, J.W. Converting Crude Glycerol Derived from Yellow Grease to Lipids through Yeast Fermentation. *Bioresour. Technol.* **2010**, *101*, 7581–7586. [CrossRef]
38. Back, A.; Rossignol, T.; Krier, F.; Nicaud, J.-M.; Dhulster, P. High-Throughput Fermentation Screening for the Yeast *Yarrowia lipolytica* with Real-Time Monitoring of Biomass and Lipid Production. *Microb. Cell Factories* **2016**, *15*, 147. [CrossRef]
39. D'Amato, D.; Corbo, M.R.; Nobile, M.A.D.; Sinigaglia, M. Effects of Temperature, Ammonium and Glucose Concentrations on Yeast Growth in a Model Wine System. *Int. J. Food Sci. Technol.* **2006**, *41*, 1152–1157. [CrossRef]
40. Zhu, L.Y.; Zong, M.H.; Wu, H. Efficient Lipid Production with *Trichosporon fermentans* and Its Use for Biodiesel Preparation. *Bioresour. Technol.* **2008**, *99*, 7881–7885. [CrossRef]
41. Tsigie, Y.A.; Wang, C.-Y.; Kasim, N.S.; Diem, Q.-D.; Huynh, L.-H.; Ho, Q.-P.; Truong, C.-T.; Ju, Y.-H. Oil Production from *Yarrowia lipolytica* Po1g Using Rice Bran Hydrolysate. *J. Biomed. Biotechnol.* **2012**, *2012*, e378384. [CrossRef]
42. Guerfali, M.; Ayadi, I.; Sassi, H.-E.; Belhassen, A.; Gargouri, A.; Belghith, H. Biodiesel-Derived Crude Glycerol as Alternative Feedstock for Single Cell Oil Production by the Oleaginous Yeast *Candida viswanathii* Y-E4. *Ind. Crops Prod.* **2020**, *145*, 112103. [CrossRef]
43. Gao, R.; Li, Z.; Zhou, X.; Cheng, S.; Zheng, L. Oleaginous Yeast *Yarrowia lipolytica* Culture with Synthetic and Food Waste-Derived Volatile Fatty Acids for Lipid Production. *Biotechnol. Biofuels* **2017**, *10*, 247. [CrossRef]
44. Zhang, L.-S.; Xu, P.; Chu, M.-Y.; Zong, M.-H.; Yang, J.-G.; Lou, W.-Y. Using 1-Propanol to Significantly Enhance the Production of Valuable Odd-Chain Fatty Acids by *Rhodococcus opacus* PD630. *World J. Microbiol. Biotechnol.* **2019**, *35*, 164. [CrossRef] [PubMed]
45. Kolouchová, I.; Schreiberová, O.; Sigler, K.; Masák, J.; Řezanka, T. Biotransformation of Volatile Fatty Acids by Oleaginous and Non-Oleaginous Yeast Species. *FEMS Yeast Res.* **2015**, *15*, fov076. [CrossRef] [PubMed]

MDPI AG
Grosspeteranlage 5
4052 Basel
Switzerland
Tel.: +41 61 683 77 34

Fermentation Editorial Office
E-mail: fermentation@mdpi.com
www.mdpi.com/journal/fermentation